MEASURING TREES AND FORE!

Second Edition

MEASURING TREES AND FORESTS

Second Edition

Michael S. Philip

Honorary Reader,
Department of Forestry,
University of Aberdeen,
UK

CAB INTERNATIONAL

CAB INTERNATIONAL
Wallingford
Oxon OX10 8DE
UK

Tel: Wallingford (0491) 832111
Telex: 847964 (COMAGG G)
Telecom Gold/Dialcom: 84: CAU001
Fax: (0491) 833508

A catalogue entry for this book is available from the British Library

ISBN 0 85198 883 0

Typeset by Solidus (Bristol) Limited, UK
Printed and bound in the UK at the University Press, Cambridge

Contents

Sections for Advanced Students

This book is aimed at students of forestry at all levels, but there are certain complex sections which are aimed specifically at advanced students. These are arranged on the page as here so they can be easily identified.

ACKNOWLEDGEMENTS

As was stated in the first edition, I have great pleasure in acknowledging all the help and encouragement provided by: Aberdeen University in giving me leave of absence to work in Tanzania; Professor John Matthews, Head of the Department of Forestry in the University of Aberdeen, for his cooperation in making arrangements for my temporary replacement, and for carrying on certain specialized aspects of my teaching and research commitments; the Norwegian Agency for International Aid (NORAID) for sponsoring the publication of this book, and especially to professor Dr Gustav Klem for his assistance and encouragement; past students, foresters and associates in both the Universities of Aberdeen and Dar es Salaam in assisting, enforcing and stimulating my efforts to learn; Andreas Fitje of the Agricultural University of Norway; Howard Wright, of the Commonwealth University of Aberdeen, for scrutinizing the draft typescript and making critical and constructive comments; and, not least, the staff of Aberdeen University Press.

This revised text owes much to the first edition which, unfortunately, was not easily available to many who asked for it. I am, however, grateful for the support of the University of Dar es Salaam and Sokoine Agricultural University in Tanzania to the earlier publication. Changes in this new edition are mainly directed at removing the bias towards students in Africa, in the addition of recent references and some reflection of the current interest and recent work in biomass and fodder measurement, sampling with unequal probabilities and growth modelling – especially in mixed tropical forest. The additional material has been chosen to introduce students and provide some references to these subjects rather than to comprise comprehensive cover. I regret the lack of many examples from Malaysia; this is due to the fact that I have little familiarity with its flora.

I am particularly indebted to Howard Wright at the Oxford Forestry Institute, to Dr Alan Petty at the Forestry Department, University of Aberdeen, and to Dr Gary Murchison of Lakehead University, Ontario, and to other reviewers for their pressure to return to work and complete the new version; without their support it would not have been finished. I am also indebted to CAB International for their expertise in setting up the complex print, and in publishing and distributing the book.

INTRODUCTION

Mensuration means measurement of length, mass and time. Forest mensuration incorporates principles and practices perfected by land surveyors, map makers or cartographers, aerial photograph interpreters and photogrammetrists, mathematicians, statisticians, engineers and foresters. This text borrows from all these and the reason for writing yet another textbook on forest mensuration is to select the blend of these disciplines most suited to the needs of today's forestry training centres, especially those in developing countries. The references quoted, including recent forestry journals, should be available to most students with access to a library. The text does not include any detail of land surveying, remote sensing or engineering techniques. Modern general texts on remote sensing and airphoto interpretation are in the bibliography under Howard (1991) and Avery & Berlin (1992).

Forest mensuration is the tool that provides facts about the forest crops, or individual trees, or parcels of felled timber to sellers, buyers, planners, managers or researchers. They ask the questions: the forest mensurationist provides, or tries to provide, answers. The seller must know how much timber he has to sell and the buyer how much he is purchasing; the planner and manager need to know how much timber is ready for sale now and in the future under different management strategies; the researcher needs to know the state of the tree, crop or forest before and after he has applied some treatment, so that he can assess the effect of his intervention.

Trees may be described quantitatively by many measurements, or parameters, the commonest of which are:

- age
- diameter – over or under bark
- cross-sectional area – calculated from diameter
- length or height
- form or shape – trees are not cylindrical
- taper or the rate of change of diameter with length
- volume over or under bark; volume may be calculated to varying top diameters as often the top of the tree is cut off and left unused in the forest. As a result only part of the tree may have any value. Special rules

(scaling rules) may be used to predict the outturn or volume of usable wood
- crown width – a parameter that can be measured both in the field and on aerial photographs
- wood density or biomass, subdivided into stemwood, branchwood, twigs and foliage for animal bedding and fodder.

Other qualitative characters are also used, e.g. species, log quality, straightness, etc. As yet, most measurements refer to the aerial parts of the tree although recent advances in harvesting practices include roots, and studies in whole tree mensuration have been pioneered by Young (1967, 1968).

Tree crops and forests are more varied; examples of measurements common to many are:

- area – surveyed or estimated from maps or aerial photographs
- crop structure – in terms of species, age and diameter frequencies, etc.
- total basal area per hectare
- total volume per hectare
- total biomass, dry weight per hectare
- total energy resource per hectare.

For even-aged, uniform plantations of a single species the following are also frequently used:

- average volume per tree
- average basal area per tree
- diameter of tree of average basal area
- average height per tree.

In addition to describing the current state of a tree or crop, the forest mensurationist is concerned with measuring and predicting the growth of trees and crops, that is the change in the parameters with time.

1 MEASUREMENTS

1.1 THE ROLE OF MATHEMATICS AND STATISTICS

1.1.1 Definition of Terms

Forest mensuration involves simple mathematics; serious students must master this science and make numbers serve them to describe a tree or crop, or present a picture of relationships – such as the change in a tree's volume with time. Students must seek awareness of the expected order of size of the numbers that they are handling; for example, they should know that whereas a standing volume of 300 m^3 per hectare is likely, 3000 m^3 per hectare is certainly wrong. They must be able to add, subtract, divide and multiply easily and without error, or without operator error when using an electronic calculator.

Mathematical calculations are simplified by:

- careful layout
- the use of specially designed forms
- unambiguous figures
- careful alignment of decimal points, thousands, millions, etc.
- noting the units concerned at all relevant stages in a calculation.

Thought must always be given to the meaning of the numbers being used.

The following terms will recur in this text and require definition:

- parameter
- measurement
- estimate
- errors
- accuracy
- precision
- bias
- expected value
- population
- sample.

Not all authorities concur with a single definition of each term and confusion may result from reading different texts employing different meanings.

A **parameter** is a characteristic of a population that usually can be expressed in a numerical form.

Examples of parameters
- tree height or volume
- forest area
- vehicle weight

A **measurement** is a reading or value for a particular parameter, usually obtained with an instrument.

Examples of measurements
- the length of a piece of wood
- the height of a tree
- the diameter of a cylinder

These are parameters that may be measured using tapes, calipers, etc.

An **estimate** is a value for a parameter obtained indirectly, often with the help of measurements. An estimate is based either on the assumption that the part or parts measured represent the whole, or that the measurements combined with some assumption of a model describing their relationship to the parameter, provide together a definition of that parameter.

Examples of estimates
(a) The average diameter of a stand of trees may be estimated by measuring the diameter of 1% of the trees in that stand
(b) The volume of a tree may be estimated by assuming the shape of a cone and measuring the diameter at the base and the height

Errors may be:

- *human and erratic* errors or mistakes that can be avoided, e.g. misreading a tape, or using an instrument with a moving measuring arm that sticks
- *systematic* errors that affect a measurement in a regular and essentially predictable way, e.g. when using an old tape that has stretched
- *random* errors that are irregular and vary in size and sign; random errors arise from many sources, e.g. from personal judgement in reading the scale of an instrument or from variation in the parameter being measured (see Section 1.1.3).

Accuracy is measured by the difference between a 'true' value and the measurement or estimate of that 'true' value. The 'true' value can be determined only by very careful measurement with instruments of established accuracy. The accuracy of an estimate can only be ascertained if the true value is known.

In Example (a) above
- if all the diameters can be measured; or

In Example (b) above

- if some alternative method of obtaining the true volume is available, e.g. by immersion in a xylometer or tank

Usually in forestry, instruments are checked for accuracy against known standards.

Example

- 20 m tapes are checked against marks set in concrete

The work of field and laboratory observers is checked by repeating the observations using independent parties. Providing that a measurement is made with properly calibrated instruments maintained and handled correctly, then it will usually be more accurate than an estimation.

Precision is the degree to which measurements approach or estimate the average and is a function of the calibration of instruments, the variation that exists in a population and the fact that an estimation is based on an assumption. The measure of the random errors that affect precision is the statistical parameter *standard error.* The smaller the standard error of an estimate the more precise is that estimate. The precision of an estimate based on sampling can be increased by increasing the number of sampling units.

A **biased measurement** or estimation is inaccurate; bias refers to a systematic error that affects all the measurements in the same way.

Examples of biased measurements

- If lengths are measured with a tape that has been broken and joined so that it is 4 cm short, then all lengths measured to beyond the break will be biased and be recorded as 4 cm greater than their true length
- If all measurements are rounded down to the complete unit, e.g. 7.6 cm, 7.9 cm and 7.1 cm are all recorded as 7 cm, then the total and average of the measurements will be biased.

Bias is measured by the difference between a 'true' value and the measurement or estimate of that 'true' value. If a bias is consistent, then the measurement or estimate may be corrected for the bias and its effect removed. If the bias is small compared to the precision, the bias is not the main reason for the inaccuracy and may be unimportant.

The term **expected value** has a special meaning when used in the statistical sense. It is the result to be expected from a large number of trials. Hence the expected value of a sample is the population average (μ).

$$\text{i.e. } \mu = E(x) = \sum^{N} (x_i \cdot p_i)$$

and total value of whole population:

$$\text{TV} = E\left(\frac{x_i}{p_i}\right) = \frac{1}{N}\sum^{N}\left(\frac{x_i}{p_i}\right)$$

Where:
μ = population average
$E(x)$, $E(x_i/p_i)$ = expected values
x_i = the value of the ith unit in the population
N = number of units in the population
p_i = the probability of occurrence of the value of x_i

EXAMPLE 1 The calculation of a population average and total population of tree diameters (cm)

$N = 10$

$x_i = 15, 22, 17, 20, 19, 18, 16, 21, 17, 22$

$p_i = \frac{1}{N} = \frac{1}{10} = 0.1$ for each tree

$\mu = E(x) = \sum^{N}(x_i p_i)$

$(15)(0.1)+(22)(0.1)+(17)(0.1)+(20)(0.1)+(19)(0.1)+$

$= 1.5 \quad +2.2 \quad +1.7 \quad +2.0 \quad +1.9 \quad +$

$(18)(0.1)+(16)(0.1)+(21)(0.1)+(17)(0.1)+(22)(0.1)$

$= 1.8 \quad +1.6 \quad +2.1 \quad +1.7 \quad +2.2$

The true population average or expected value

$\mu = 18.7$ cm

Total value of whole population (TV)

$$E(TV) = E\left(\frac{x_i}{p_i}\right) = \frac{1}{N}\sum^{N}\left(\frac{x_i}{p_i}\right)$$

$= [(15)(10)+(22)(10)+(17)(10)+(20)(10)+(19)(10)+$

$(18)(10) +(16)(10)+(21)(10)+(17)(10)+(22)(10)]/10$

$= (1870/10)$

$= 187$ cm

More usually the population mean and total value are calculated by:

$$\mu = \frac{1}{N}\sum^{N} x_i = \frac{1}{10}(187) = 18.7 \text{ cm}$$

$$TV = N\mu \quad = \sum^{N} x_i \quad = 187 \text{ cm}$$

In the statistical sense, a **population** is an assembly of individual units that is usually formed in order to measure the units quantitatively.

Examples of populations

- all trees greater than 20 cm diameter at breast height in a particular forest
- all trees of *Pterocarpus angolensis* in a defined area of miombo woodland
- the stock of logs in a sawmill on a particular day
- all the people living in the boundaries of a township
- all the seeds within a sack

Populations may be small or large, restricted to a small area or very extensive, homogeneous or heterogeneous.

In the statistical sense, a **sample** is a representative part of a whole population. A small sample of a few individuals will adequately represent a homogeneous population, whereas a more numerous sample will be required to represent a heterogeneous one.

1.1.2 Accuracy in Calculation

Most numbers used in mensuration are either integers or approximations of pure numbers.

Example
If ten trees are measured as a sample, then the number 10 is a discrete number or integer. In contrast if we estimate the volume of a tree as 1.2 m^3 then this is an approximation implying that the limits within which the estimate lies are 1.15 and 1.25 m^3.

The accuracy of the approximation is indicated either by the denominator in a fraction or the number of significant digits in a decimalized number.

Examples of limits
$23\frac{1}{2}$ implies that the value lies between the limits of $23\frac{1}{4}$ and $23\frac{3}{4}$
$23\frac{1}{4}$ implies that the value lies between the limits of $23\frac{1}{8}$ and $23\frac{3}{8}$
23.5 implies that the value lies between the limits of 23.45 and 23.55
23.50 implies that the value lies between the limits of 23.495 and 23.505

A *significant digit* is any digit denoting the true size of the unit; significant digits are the digits reading from left to right in a number beginning at the first non-zero digit up to the last digit on the right which may be a zero. A transformation of the units in which a number is expressed does not affect the number of significant digits.

Example of transformation of units
15.60 cm^2 = 0.001560 m^2
Both figures have four significant digits.

Consequently numbers of decimal places are unimportant, but the number of significant digits indicates the precision of the approximation.

For cross-sectional areas of less than 1 m^2, diameters recorded to the nearest centimetre allow only one significant digit in the approximation of the corresponding basal area.

Examples of the limits of cross-sectional areas

- The limits of a diameter recorded as 10 cm are 9.5 and 10.5 cm whose corresponding basal areas are 70.9 and 86.6 cm^2 or 0.00709 and 0.00866 m^2. The correct approximation for a diameter of 10 cm must be 0.01 m^2
- For a diameter recorded as 80 cm the limits are 79.5 and 80.5 cm, and the

corresponding basal areas are 0.4964 and 0.5089 m^2. The correct approximation for a diameter of 80 cm must be 0.5 m^2

- For a diameter of 160 cm whose limits are 159.5 and 160.5 cm with corresponding basal areas of 1.998 and 2.023 m^2 the correct approximation is 2.0 m^2

Tables of areas of circles frequently quote four significant figures. They must not be used thoughtlessly nor applied to individual trees without amendment. Diameters must be read to the nearest 0.1 mm if the basal area is to contain four significant figures.

Conclusion

Calculations in forest mensuration must indicate whether the results are 'expected values' when large numbers of such values are involved or are estimates relating to a limited number of trees; in the latter case the number of significant digits must be realistic.

Normal practice is to round field records to the nearest smallest unit whose measurement or later use is feasible. Provided that care is taken so that marginal records are rounded up and down alternately, no bias results. Rounding does increase variance, but the amount is small and unimportant.

EXAMPLE 2 Rounding data

	True cm	Record to nearest 1 cm	
	7.35	7	
	8.82	9	
	9.50	10	rounded up to the even number 10
	7.89	8	
	6.50	6	rounded down to the even number 6
	9.65	10	
	8.53	9	
	7.14	7	
	8.41	8	
	6.18	6	
Total	79.97	80	
Average	8.00	8 cm	

Sometimes records are made to the completed unit. If unadjusted, the total and average will be biased. The expected value of this bias (in a large number of cases) will be half the completed unit.

EXAMPLE 3 Bias in data recorded to a completed unit

	True cm	Recorded to the completed cm	Expected bias
	7.35	7	−0.5
	8.82	8	−0.5
	9.50	9	−0.5
	7.89	7	−0.5
	6.50	6	−0.5
	9.65	9	−0.5
	8.53	8	−0.5
	7.14	7	−0.5
	8.41	8	−0.5
	6.18	6	−0.5
Total	79.97	75	−5.0
Average	8.00	7.5	−0.5
Adjusted average		7.5−(−0.5) = 8.0 cm	

FOR THE ADVANCED STUDENT

1.1.2.1 Rules for Assessing the Number of Significant Digits in Calculations

In pure mathematics the guides are:

Addition and subtraction

Numbers must be aligned by the decimal point, and addition or subtraction can take place at and to the left of the first position from the right where significant digits coincide. The number of significant digits in the result can never be greater than the number in the largest component, but may be less. The correct number of significant digits in a result may be established by repeating the calculations, first using the upper and then the lower limit of the least precise element. The maximum number of significant figures justified is that which provides an identical result from both calculations.

EXAMPLE 4(a) Calculation of numbers of significant digits

```
  799.53
+  84.789
+   1.0
 ______
=885.319 which must be corrected to 885
```

Example 4(a) continued

because 799.53 lies between 799.525 and 799.535
84.789 lies between 84.7885 and 84.7895
1.0 lies between 0.95 and 1.05

Total lies between 885.2635 and 885.3745

or between 885.26 and 885.37 with 5 sig. digits
or 885.3 and 885.4 with 4 sig. digits
or 885 and 885 with 3 sig. digits

Similarly

1,078,523.0
− 79,412.78

= 999,110.22 which must be corrected to 999,110

Multiplication and division

The number of significant digits in the product or quotient equals or is one digit less than the number in the factor with the least significant digits. In practice involving a series of multiplications and divisions, one extra digit may be retained at each stage and the final answer corrected. When integers and constants such as π or e are involved, their number of significant digits should be made to agree with the number in the previous operation so that precision is maintained.

EXAMPLE 4(b) Calculations

1,023.56 × 2.7 = 2,763.612 which must be corrected to 2800

799.12 × 10 where 10 is an integer is equivalent to 799.12 × 10.00 = 7991.20 which must be corrected to 7991.2

$(\pi/4)(15^2)(1/10^4)m^2$ is equivalent to (3.14/4.00)(15)(15)(1/10,000) = 0.0176625 which must be corrected to not more than two significant figures because of the factor 15. In fact only one significant digit is justified in this example.

In order to be certain of the correct number of significant figures in a result of a series of multiplications, calculations using the upper and lower limits of the components must be done.

However, in most forestry practices we are dealing with populations of trees and the expected values from very large number of individuals and rarely with individuals.

Expected values may contain the full number of decimal places generated in the calculation. Providing that more than 30 trees are involved, then one extra figure to that indicated by the guide from pure mathematics may be included in the total and average with a probability of 0.95 that it is significant. Care must be taken not to distort the accuracy of estimates of volumes when heights are recorded to the nearest ½ or ¼ m. As a rule of thumb in a complex situation, when height is recorded to the nearest ½ m, the approximation to volume will have one significant digit less than the basal area on which the volume approximation is based.

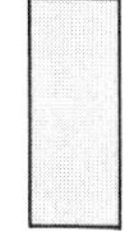

1.1.3 Errors of Measurement and the Measurement of Variation

Errors may arise because of:

- human errors, e.g. recording 30 in place of 20, or use of incorrect techniques or careless use of instruments
- instrument errors, e.g. lengths measured with a tape that has been broken and re-joined short
- errors due to the nature of the object being measured, e.g. errors in measuring the circumference of a tree such as *Newtonia buchananii*, which has large plank buttresses

EXAMPLE 5 Calculation of volume

Average basal area from 30 trees = 0.025 m^2 (limits 0.0245–0.0255)

Average height from 30 trees recorded to the nearest ½ m = 20 m (limits 19¾–20¼)

Form factor = 0.33 (limits 0.325–0.335)

Volume = (basal area)(height)(form factor) = (0.025)(20)(0.33)
= 0.165 m^3

at lower limit

= (0.0245)(19¾)(0.325)

= 0.1572594 m^3

at higher limit

= (0.0255)(20¼)(0.335)

= 0.1729856 m^3

Therefore volume = 0.2 m^3 with only one significant figure, but the expected value from a large number of such examples is 0.165 m^3

N.B. The value 0.2 m^3 will not introduce bias provided all such data are treated similarly and rounding errors are symmetrically distributed

Errors of the first two categories should be eliminated or at least minimized by adequate staff training in instrument maintenance and mensuration procedures. Errors in the third category can only be minimized by applying the most suitable techniques; this is the job of the mensurationist.

Errors may also be categorized as:

- *Random errors* – either occurring symmetrically on either side of the true value and of random magnitude, small errors being more frequent than large errors: such errors do not affect the expected value of the mean; or occurring only on one side of the true value, but of random magnitude, small errors being more frequent than large errors: such errors will give a bias to an estimate of the mean.
- *Systematic errors* – occurring on one side of the true value and of regular magnitude

Examples of errors

- *Random errors – symmetrical.* The difference between successive measurements of the horizontal distance between two points. The average of a large number of such measurements is unbiased
- *Random errors – one side only.* The errors in the calculation of the cross-sectional area of a tree based on assuming circularity, i.e. an irregular bias
- *Systematic errors.* Errors arising in measures of length using the wrong point of origin on the scale of the tape. This gives rise to a regular bias

Analogous to the random differences between measurements of the same object are the random differences between the measurements of different individuals in the same group or population. Usually a few individuals are very large, a few very small and most are near the average for the group. Very many biological populations have what the statisticians have termed a *Normal* distribution of size around the average (synonym 'mean'). A Normal distribution often results when the causes of the variation in size are numerous and independent. For example, the diameters of trees in an even-aged plantation vary because each tree has a different genotype, is growing on a different patch of soil, was a different size when planted, was planted by different people, has competed with different neighbours, etc. As a result diameters vary and the distribution of the frequencies of different diameters in the population may be predicted because random differences like random errors are expected to have a Normal distribution.

Variation is characterized by the statistical measure *standard deviation,* which is the square root of the *variance.* Variance is also named the *mean squared deviation.* Foresters usually estimate the variance and the standard deviation from small samples in order to reduce the cost of measurement. Each individual is called a sampling unit, e.g. each tree whose diameter is measured is a sampling unit.

In a Normal distribution approximately 67% of the observations are expected to lie within the range of $(\mu - \sigma)$ to $(\mu + \sigma)$ and 95% within the range of $(\mu - 2\sigma)$ to $(\mu + 2\sigma)$,

Where:
μ = population mean
σ = standard deviation

Formula

$$s^2 = \sum^{n} \frac{(x_i - \bar{x})^2}{n - 1}$$

Where:
s^2 is the estimate of variance based on n sampling units.
The units may be numbered 1,2,3, . . ., i, . . ., n;
x_i is the value of the ith sampling unit;
$\bar{x}$ is the estimated mean of the population based on the n sampling units

N.B. The origin of the synonym for variance – the mean squared deviation – is obvious from this formula.

The three parameters, that is the type of distribution, the mean and the variance, characterize and describe uniquely a population. This is their purpose. (See Section 5.1.1.)

Further reading

ON ERRORS

Loetsch, F. *et al.* (1973) *Forest Inventory*, vol. II. Section 42.

ON THE NORMAL POPULATION

Snedecor, W.G. and Cochran, G.W. (1967). *Statistical Methods.* Chapter 2. See also Chapter 5.

EXAMPLE 6 Calculation of variance and standard deviation from a small sample

Data: diameters of trees, cm

	Diameter x_i	$(x_i-\bar{x})$	$(x_i-\bar{x})^2$	x_i^2
	24	−1	1	576
	25	0	0	625
	24	−1	1	576
	27	2	4	729
	30	5	25	900
	26	1	1	676
	21	−4	16	441
	23	−2	4	529
	21	−4	16	441
	24	−1	1	576
	30	5	25	900
	25	0	0	625
	24	−1	1	576
	27	2	4	729
	23	−2	4	529
	26	1	1	676
Total $\sum^{n}$	400	0	104	10,104

n = 16 sampling units

$$\bar{x} = \left(\sum^{n} x_i\right)/n$$

$= 25$ cm

Example 6 continued

$s^2 = 104/15$

$= 6.93 \qquad s = 2.6$ cm

A more usual formula that is quicker and avoids rounding errors in $(x_i - \bar{x})$ is

$$s^2 = \frac{\sum^n x_i^2 - \frac{\left(\sum^n x_i\right)^2}{n}}{n-1} = \frac{10{,}104 - \frac{160{,}000}{16}}{15}$$

$$= 6.93$$

1.2 The Choice of Where and How to Measure Forest Produce

Trees may be measured for many purposes, in many ways, and at many points in the chain of production from the nursery to the time and place of final utilization.

Examples of feasible measurement points and methods

- On standing trees – measurements taken either in the field or on aerial photographs
- On felled trees, in the full tree-length, volume estimated over or under bark
- On felled trees cross-cut and divided into different product groups, e.g. saw timber, small round wood for pulp, foliage for fodder, etc. Different products may be measured in different ways
- Stacked at roadside, stacked volume being measured
- Piled on lorry and weighed

The choice of the method of measurement depends on several factors, e.g.

- the purpose of the measurement
- the form of the produce, i.e. log, chip, etc.
- the quantity and total value of the produce
- the relative precision and cost of different methods of measurement

Generally the more valuable the products being measured the greater is the justifiable expenditure in order to maximize the net worth to the owner.

Trees or parts of trees may be chipped in the forest and loaded into bulk carriers; weight is then the cheapest and most feasible method of measuring removals from the forest provided that a weigh-bridge is available. Pulp wood removals are also most frequently weighed.

Often trees and, especially, small round wood are cut into pieces of equal length and stacked at the roadside to await loading. Then measurements based on the cross-sectional areas of the wood pieces in the stack are feasible, fast and cheap. The limits to use vary greatly under different circumstances; for example, material that is left uncollected in the forest in some regions may have a high value in other parts of the world.

Once trees have been felled then accurate diameter measurements at many points along the bole are both feasible and relatively cheap, and provide accurate estimates of bole volume. Occasionally the measurement of volume by water displacement on immersion in a xylometer is feasible but is usually restricted to very special investigations.

In some parts of the world one of the most useful products of the forest is fodder for animals to tide them over seasons where more nutritive vegetation is not available; for example, many species of oak *Quercus* spp. are valuable fodder trees in the Himalayas. Assessment of fodder is not only in terms of green weight but also in palatability and nutritive value.

In comparison with felled forest produce, the measurement of standing trees and the estimation of weight or volume at a given level of precision is much more costly. Such measurements are necessary for standing sales, in forest inventories, in silvicultural research and in growth studies, for example.

Different methods for measurement and field procedures may be used to derive estimates of volume and weight. Estimates obtained by one method are not usually comparable with those derived from a different method; this must be remembered. Therefore the description of the method used is an essential part of an estimate of parameters of trees and forest produce; comparisons are only valid and useful when field procedures are the same.

This text cannot detail all the different conventions and procedures in forest and tree measurement adopted in different countries. However, it will stress those aspects of the principles underlying the techniques that affect their cost, accuracy and precision. Then as the technology of harvesting and processing develops and new measurement techniques are devised to supplement or replace current ones, their suitability may be judged using these or similar principles.

Further reading

Edwards, P.N. (1983) *Timber Measurement: A Field Guide.* Forestry Commission Booklet no. 49.

Hamilton, G.J. (1975) *Forest Mensuration Handbook,* Procedure 5. Forestry Commission Booklet no. 39.

Husch, B. *et al.* (1982) *Forest Mensuration* (3rd edn). Chapters 1–3.

1.2.1 Calorific Values and Density of Wood

The most widespread and important use of wood is as a domestic fuel. The suitability of different species varies with many factors such as tree dimension, absence of thorns, ease of splitting when green or dry, rate of burning, etc. This is in spite of the fact that the calorific value of all woody species is relatively constant at 4.5 kilocalories per gram, dry weight. In contrast, the calorific value in terms of unit volume varies with the density and moisture content of the wood.

Basic density of wood is defined as the oven dry weight per unit of green volume. Green density is defined as the weight of newly felled wood per unit of green volume. Basic wood density varies with:

- species, genotype and site
- age
- position in tree
- rate of growth and, therefore, site
- presence or absence of knots, reaction wood and other variation in the structure of the wood, etc.

Green density in addition varies with:

- season of the year

The density of the wood of felled trees and round wood lies between the corresponding basic and green densities, and varies additionally with:

- the time since felling
- treatment since felling, i.e. exposed to the sun, storage under sprays, etc.

Although high basic densities impart high calorific values per unit of volume, which is a desirable property from the point of view of one who stores and uses wood, the mensurationist is more concerned with the green density of felled trees.

Further reading

ON METHODS OF ESTIMATING WOOD DENSITY

Besley, L. (1967) The importance of variation and the measurement of density and moisture. In: *Wood Measurement Conference Proceedings,* pp. 112–42. Faculty of Forestry, University of Toronto, Tech. Rep. no. 7.

Kollman, F.F.P. and Côté, W.A. Jr (1968) *Principles of Wood Science and Technology.* George Allen & Unwin, London.

1.2.2 Weight or Volume?

The history of the development of wood measurement techniques shows clearly that changes reflect different demands from the wood processing industries. When logs were adzed into squares, the volume of round logs was estimated in units that approximated to the size of the square that could be manufactured from them; the outmoded hoppus system is an example of this type of measure.

When sawing became common the forest and saw milling industries produced a variety of measures for scaling logs and estimating their outturn of sawn wood. An example of this type of measure is the board foot and various American rules such as the Scribner and Doyle's Log Rules. Now, when most forest produce is broken down into small pieces or chips or even to the individual wood fibres and reconstituted as particle or fibre board or pulp, or used as a basis for fermentation, these measures or rules are irrelevant and biased.

Obviously the unit of measurement must reflect the use of the measurements (and therefore the use of the wood).

Other than for research purposes, trees are measured

- to provide a value for buying or selling
- to provide information for the forest manager on stock levels and changes

The two most widely used parameters are *volume* and *weight*. They are only mutually deducible one from the other if the wood density is known. Their respective advantages and disadvantages may be summarized thus:

Volume

Advantages
- Most growth studies and inventories have used this parameter: this facilitates growing stock accounting and the prediction of outturn
- The standing and felled crops may be measured in the same units using standard conversion tables
- Crop measurement can be done in the forest under the direct control of the forest supervisor

Disadvantages
- Volume cannot easily be measured directly
- Estimates of volume may be of varying precision; comparisons are only feasible if the procedures used in estimation are standardized
- Volume and value may be poorly related

Oven-dry weight

Advantages
- Measures what the fibre or pulp manufacturer or fermenter wants to buy and reflects his values
- Draws attention to the manufacturers' basis of valuation

Disadvantages
- Can only be estimated indirectly – either through sampling or by prediction – from green weight or volume
- Costs of sampling are high as dry weight or basic density varies with:
  — species and genotype
  — age of tree
  — rate of growth
  — site
  — position in tree
  — bark percentage
  — presence of defects – reaction wood, knots, etc.
- The combination of removals measured by dry weight with standing crop data measured by volume makes recording and predicting growing stock changes imprecise and difficult

Green weight

Advantages
- Fast with a low variable cost per load
- Reliable – the process of recording the weight can be automated and designed to give a printed record
- The measurement is direct – using a weigh-bridge

- The results of re-measuring the same produce are expected to be consistent
- Basing sale prices on delivered weight encourages producers to deliver fresh wood. This may be advantageous to the processor (for example less energy is required to chip green compared to air dried wood)

Disadvantages

- The weigh-bridge is expensive (high capital cost)
- The location of the weigh-bridge is often far from the forest. Some wood may be removed from the forest and never weighed. It is also difficult to relate the weights of removals with numbers of trees felled or volume prepared
- The weight of the water in the wood as well as basic density are additional sources of variation that makes comparison of prices for different lots difficult
- The combination of removals measured by weight with standing crop data measured by volume makes recording and predicting growing stock changes imprecise
- Where roots are harvested with the bole, weights may be distorted by the inclusion of soil, stones, etc.

Conclusion

Where weighing facilities are available, the most convenient unit for the forest manager is green weight. The manager must institute a reliable sampling scheme to ensure correct conversion to volume so that changes in the growing stock can be analysed. However, in those areas where weighing is not feasible or suited to the conditions of sale, volume must be the parameter recorded by the forest manager.

1.2.3 Some Other Forest Products

The forest produces not only wood, it also produces bark, foliage, resins and gums, fibres, medicinal plants, etc. Whenever the production of these has to be measured or predicted, then mensurational techniques have to be developed and adapted to suit the particular circumstances. In most instances the principles already in use in forest mensuration will serve to guide a forester to improvise suitable techniques to meet his requirements.

2 MEASURING SINGLE TREES

2.1 INTRODUCTION

Although other parts of the tree and other products of the forest are important, wood remains a very important, if not the most important, product. Even when wood is sold by weight, the forest manager will use volume data, for example volumes of crops derived from inventories or predictions of volume growth from yield tables, and most measurements made on standing crops will be expressed in volume. Indeed, frequently conversion to weight will be based on volume. So, even with changing processing techniques, tree volume is still a most important tree parameter, as are the measures of diameter, height and form (or shape) from which volume estimates are derived. These same primary parameters, especially diameter, are also used in estimating fodder production, bark volume, etc. Therefore it is not surprising that almost the whole of the existing literature is concerned with these three parameters.

The volumes of trees or logs are based on the formula:

$$v = ghf \text{ m}^3$$

This formula states the three parameters that are required to calculate the volume (v) of a tree (or log)

g = cross-sectional area, m^2
h = height (or length), m
f = a coefficient employed to reduce the volume of a cylinder ($v = gh$) to that of the tree or log.

This remains a theoretical approach as f cannot be measured but has to be calculated as the ratio v:gh after v has been measured directly, or estimated by means of mathematical functions or from volume tables.

2.2 CROSS-SECTIONAL AREA (g)

Just as volume (v) is derived from g, h and f, so the cross-sectional area itself is a derived parameter. The shape of the outline of the cross-section of a tree or log may vary enormously; it may be star-shaped, e.g. *Newtonia buchananii,* involuted e.g. *Balanites aegyptiaca* or approximately circular, e.g. *Pterocarpus angolensis* or *Fagus sylvatica* (Figure 1).

FIGURE 1 Examples of the outlines of trees

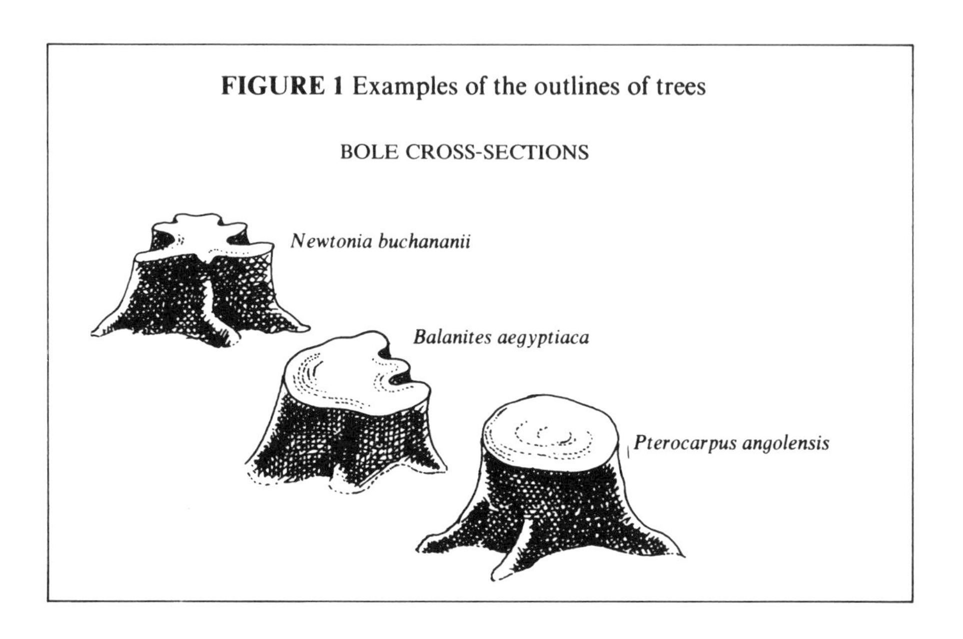

2.2.1 Principles in Cross-section Area Estimation

In most instances the difficulty inherent in measuring the cross-sectional area of an irregular bole is overcome by assuming that it is equivalent to that of an ellipse or circle whose dimensions are derived from measurements made on the tree. Very often a circle is assumed and the cross-sectional area estimated by

$$g = \frac{\pi}{4} d^2 \text{ where } d\text{, the diameter, is recorded in metres}$$

or

$$g = \frac{c^2}{4\pi} \text{ where } c\text{, the circumference, is recorded in metres}$$

The diameter used may be a single measurement or derived from the average of two measurements on the tree. The two measurements may be:

either the maximum and minimum

or the maximum and the diameter at right angles to that maximum diameter

or any two diameters at right angles, the first usually chosen systematically with reference to some known direction such as north, towards the plot centre, or the slope of the ground

$$\text{then } d = \frac{(d_1 + d_2)}{2}$$

The third option is the most commonly used and usually the most satisfactory.

2.2.2 Errors in Cross-section Area Estimation

For a single tree with an irregular cross-section an estimate of cross-sectional area derived from a single diameter measurement may be very inaccurate; the true area may be over- or underestimated. Hence a single diameter measurement is not recommended for individual tree measurement. (However, as we shall see in Chapter 3 – Measuring Tree Crops – when estimating the total cross-sectional area of a large number of trees from a single diameter measurement on each, the overestimates of some are balanced by the underestimates of others; in such cases single measurements are satisfactory.) Normally any two diameters at right angles provide adequately accurate and unbiased estimates for a single tree.

An estimate of the cross-sectional area of a tree based on a measurement of girth is usually satisfactory but is known to be slightly biased; the sectional area of trees that are not truly circular will be overestimated. Normally this overestimation is small.

Further reading

Chaturvedi, M.D. (1926) *Measurements of the Cubical Contents of Forest Crops.* Oxford Forestry Memoir No. 4.

2.2.3 Standardization of Measurement Procedures

Because it is usually impracticable to measure exactly the cross-sectional area of a tree or log and estimates are subject to errors, comparability can only be achieved by standardizing measurement procedures.

Usually one or two diameters or the girth is measured. Where some irregularity of the bole profile affects the point of measurement (p.o.m.), measurements are taken at equal distances on either side and beyond the affected section, and the results averaged.

When employing the formula $v = ghf$ on standing trees a diameter of girth is needed at ground level. Such a measurement is impracticable as the outline and cross-sectional area at ground level are affected by root swell. Therefore measurements representing the cross-sectional area at ground level are taken higher up above root swell and where the outline is usually more regular. A convenient working height of 1.3 m above ground level is generally accepted as this standard height and is referred to as *breast height*. The cross-sectional area at this point is called basal area, designated by the internationally recognized symbol, g. The basal area may be measured over or under bark.

In sample plots and inventories especially, further instructions are needed to ensure the consistency and comparability of field measurements. Examples of such instructions are:

- The 1.3 m to breast height will be measured from soil level at the highest point at the base of the tree, but litter and tree debris at ground level must be removed before the p.o.m. is determined
- On trees forking below 1.3 m from ground level, or where the distortion caused by the fork affects the measurement at 1.3 m, each stem will be measured and recorded separately
- Trees forking above 1.3 m will have only one measurement at breast height
- Before measuring, all loose bark scales, lichen, ferns, etc. will be scraped off with a knife

In tropical high forest buttresses and flutes are common. Special instructions to cope with such trees must be issued to field teams to ensure comparability in the records. Normally basal area is determined above buttress, over flutes, etc. Where they extend far up the bole of the tree some arbitrary height (such as 3 m above ground level, agl) has to be chosen from the point of view of the feasibility of obtaining the measurement in the field. In these circumstances girth is often easier to measure than diameter; field crews quickly acquire considerable dexterity in passing tapes around boles above buttresses using forked sticks.

2.2.4 Measuring Diameter or Girth

Commonly diameter and girth is measured to estimate cross-sectional area at:

- the ends of logs
- the mid-length of logs
- breast height or above buttress on standing trees

Diameter is usually measured with calipers; more recently optical instruments to measure diameter at points out of reach of normal calipers have been developed. Calipers normally record diameters (in cm or mm) but may be calibrated so that cross-sectional area (in m^2) can be read directly. Girth or circumference is usually measured with a tape calibrated in centimetres, or in units of π centimetres so that the reading is the equivalent diameter, or directly as cross-sectional area (in m^2).

Both diameter and girth may be measured over or under bark – in the latter case either by measuring bark thickness or by removing the bark at the point of measurement. Measuring under bark girth with a tape presents some practical difficulties on trees with copious exudate, so measuring bark thickness is more practicable.

In plantations the decision between measuring diameter and girth is unimportant as either method is practicable and capable of yielding satisfactory results. In tropical high forest where there are large trees girth is usually easier to measure.

2.2.4.1 *Calipers*

Calipers must:

- Be light in weight but sturdy and stable under all weather conditions. Most calipers are made of steel or aluminium alloy. Wood is not an ideal material as it lacks stability under conditions of varying moisture vapour deficits
- Have the fixed and moveable arms lying in the same plane and at right angles to the bar at the moment when pressure is applied to the two arms, squeezing them on to the bole of the tree, and the diameter recorded. Maladjustment of the arm is the commonest fault with calipers and results in a systematic error in the diameter (see Figure 2b)

FIGURE 2 Diameter measurement with calipers

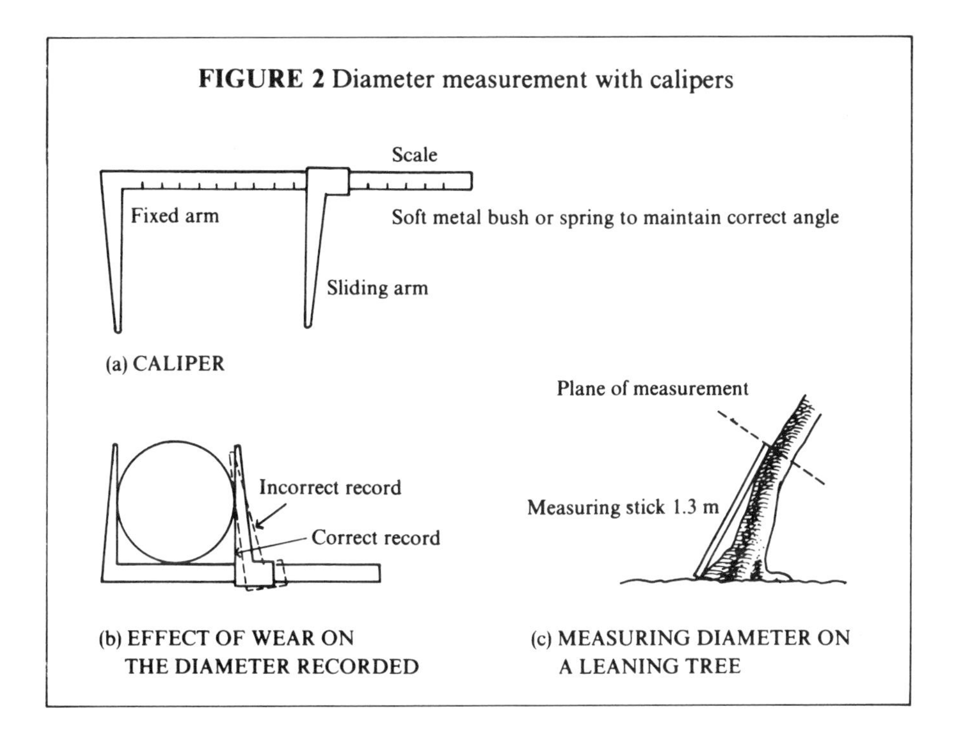

Operators must be trained to ensure that:

- the calipers are properly maintained and accurate
- the point of measurement is located correctly
- the plane of the calipers is at right angles to the longitudinal axis of the tree (see Figure 2c)
- the correct pressure is applied at the moment of measurement
- the bar of the caliper is pressed against the bole in order to minimize the effects of maladjustment of the moveable arm

Recently sophisticated calipers using electronics to provide automatic readings and data recordings have been developed – especially in Sweden (Jonsson, 1981; Eugene, 1989).

Calipers are an impractical instrument for use on very large trees; then a

Biltmore stick or adaptation of the same principle can be used to provide rather imprecise measurements. A Biltmore stick is a straight rule specially graduated to read diameters. The scale is not linear. The stick is held horizontally touching the bole of the tree and at a fixed distance from the observer (usually a convenient arm's length). The stick is held so that the end with the origin of the scale (usually the right hand end) is aligned with the tree profile and the observer's eye. Without moving the head or scale, the reading of the opposite edge of the tree on the scale is taken; this is the tree's diameter.

FIGURE 3 **The Biltmore stick**

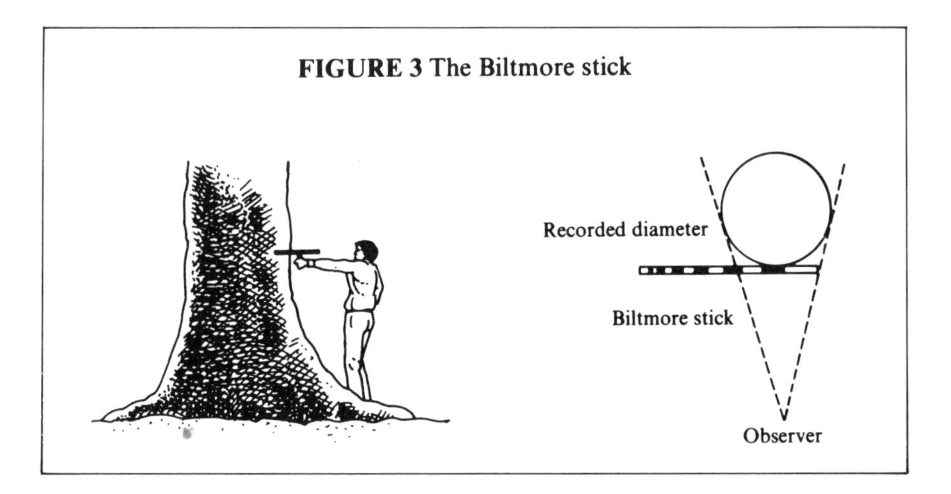

A discussion of the graduations is given in Appendix 1. The precision of the measurements depends on the size of the tree and the ability of the observer to position his eye at the correct distance from the scale.

2.2.4.2 Girth Tapes

Tapes must be:

- resistant to change in length under severe conditions of abrasion, temperature and humidity. Most modern tapes are of stout plastic, but their disadvantage is that they are easily torn. Very accurate research measurements may be done with steel tapes, but they are easily kinked and they then break, and the calibrations quickly become obscured or worn off.

Operators must be trained to ensure that:

- the tapes are not bent or twisted and are kept clean
- the point of measurement is correctly located
- the plane of the tape is at right angles to the longitudinal axis of the tree
- the correct tension is applied to the ends of the tape
- the dimensions are read correctly using the correct point of origin of the measurements. There is no standard position on tapes for the zero mark, so particular care has to be taken

- the tape is positioned against the bole of the tree and all climbers, loose bark scales, moss etc. are removed around the whole circumference of the tree

Theoretically the tape is biased and gives an overestimate, but field tests suggest that in practice calipers are also biased at least to the same degree because it is easy to press the caliper arms into soft bark. As with calipers, automatically recording tapes have been developed in order to avoid booking errors and to facilitate data handling without copying or re-entry into a data base.

Comparison of the characteristics of calipers and tapes

Calipers	*Tapes*
Rather cumbersome to carry	Compact and easy to carry
Theoretically unbiased	Theoretically biased
Consistency between repeated measurement attained only with difficulty	Repeated measurements obtainable with a high degree of consistency
Expensive	Cheap
Durable but need care and maintenance	Not very durable
Fast and with care relatively free from operator error	Not quite so fast and easy to handle, and more care needed to ensure tape not twisted and in contact with bole
Easier to use on felled logs	Often difficult to pass under heavy logs

2.2.4.3 *Optical Methods of Measuring Diameter*

Most surveying levels have stadia marks engraved on the lens so that they are superimposed on the image when viewed through the telescope. These stadia normally form a fixed angle at the eye piece of 1/100th of a radian. Then the horizontal distance from the instrument to a levelling rod is 100 times the vertical distance subtended by this fixed angle on the levelling rod. This vertical distance can be read directly on the levelling rod by subtracting the reading at the lower stadia from the reading at the upper one. Measuring length by this means is called tacheometry.

The same principle can be used to demarcate circular sample plots. A suitable target is located at the plot centre. The observer positions himself on the plot boundary by observing the target with a fixed angle gauge and adjusting his position so that the target appears to fill exactly the gauge (i.e. appears as the same width and therefore subtends the same angle).

- The advantage of the method is that as it uses optical principles no measuring tape has to be passed backwards and forwards from the centre to the perimeter as the boundary is demarcated
- The disadvantages are:
  - an angle gauge and matching target are needed for each plot size
  - a correction must be applied on sloping ground

— there is a danger of personal bias
— the error with a hand held instrument is likely to be greater than when using a tape

EXAMPLE 7 Horizontal distance measure using a fixed angle

Reading at upper stadia	6.73 m
Reading at lower	5.29 m
Difference	1.44 m
Distance	(1.44)(100) = 144 m

EXAMPLE 8 Plot demarcation with target and fixed angle gauge

The radius (L) of a plot of area (a):

$$L = \sqrt{\frac{a10^4}{\pi}}$$

But the fixed angle of the gauge,

$$\theta = \frac{d}{L} \text{ radians}$$

Therefore $d = \theta L$

Where:

L = radius of plot, m
a = area of plot, ha
d = diameter of target, m

Example: if $a = 0.01$ ha and $\theta = 0.02$ radians

then $L = \sqrt{\frac{(0.01)\,10^4}{\pi}}$

$= \sqrt{\frac{100}{\pi}}$

$= 5.64$ m

and $d = (0.02)(5.64)$

$= 0.1128$ m or 11.28 cm

The same principle can be used to measure the diameter of a tree on level ground. In this case the diameter of the tree to be measured is unknown. The measurer positions himself so that the tree appears to fill the angle gauge (that is subtends the same angle as the gauge). Then the diameter of the tree

$$d = \theta L \quad \text{N.B. } d \text{ is in the same units as } L$$

Where:
θ = the fixed angle in radians

L = the horizontal distance to the point of measurement

N.B. These relationships are approximate and are only practicable when θ is very small (less than 0.04 radians). Moreover this method is only used when it is impossible to use a caliper or tape, because of the low precision and risk of error.

A considerably more important application is in measuring tree diameters at different points up the bole of a standing tree for volume estimation (see Section 2.7.2.2.2). This avoids either felling or climbing the tree. The Spiegel relaskop is an instrument that has been designed especially for this task. Amongst other scales it incorporates a hypsometer giving direct readings of heights from either 20 m, 25 m or 30 m distances and fixed angle scales of 0.020, 0.028 and 0.040 radians. These scales have built in compensation so that when the instrument is tilted to read a diameter high up the bole and, therefore the viewing distance lengthens, the angle is corrected by reducing the width of the scale. A detailed description of the instrument and its use is given in Section 3.1.4, after height measurement has been discussed.

Other optical instruments that have been developed are the Wheeler pentaprism (Wheeler, 1962), and the Barr and Stroud dendrometer. Descriptions of these are given in *Forest Inventory* by Loetsch *et al.* (1973).

2.2.5 The Importance of Cross-section Area Estimation

The importance of diameter and girth in tree measurement is obvious. Also

- Diameter and girth are the easiest parameters of tree size to measure in the field, and a high level of consistency in these measurements can be obtained
- The sum of the basal areas of all trees in a crop is a useful measure of stocking
- In uniform plantations of a single species volume is closely related to basal area
- As volume is derived from the square of the diameter or girth, the proportional increase in volume is approximately equal to *twice* the proportional increase in diameter, plus the proportional increase in height, plus the proportional change in the form (see Section 2.8)
- In irregular crops of mixed species or age a table showing the frequency for the different diameter classes is a very useful means of characterizing the structure of the crop

EXAMPLE 9 Illustrating the contribution for growth in diameter to growth in volume

In 1975 the dimensions of a tree were:

$d = 0.560$ m
$h = 20$ m
$g = 0.25$ m^2
$f = 0.5$
$v = (0.25)(20)(0.5) = 2.5$ m^3

FIGURE 4 Tacheometry

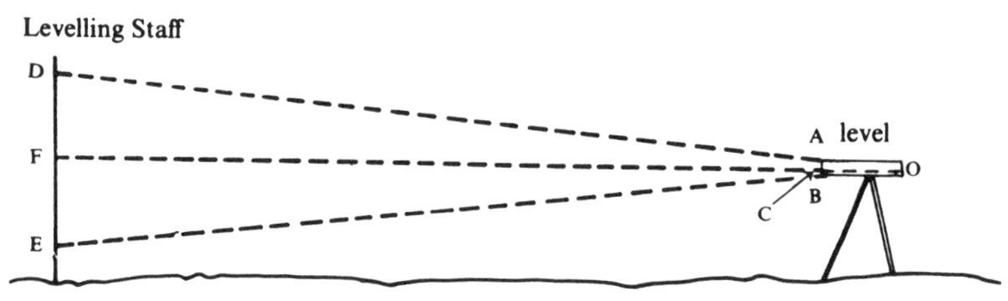

$$\triangle AOB \equiv \triangle DOE; \frac{AB}{OC} = \frac{DE}{OF} = \frac{1}{100} \Rightarrow OF = 100(DE)$$

(a) FIXED ANGLE TACHEOMETRY

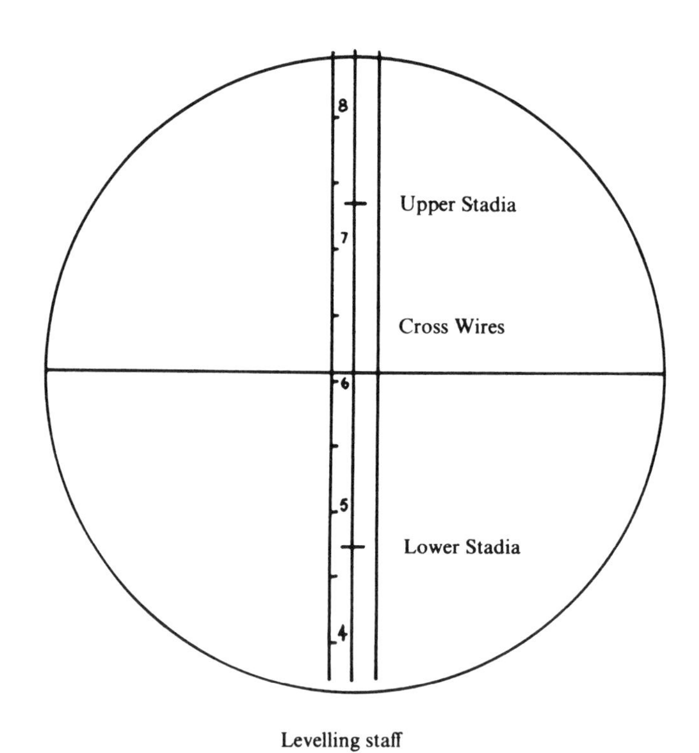

Levelling staff

(b) VIEW THROUGH THE LEVEL

Example 9 continued

Between 1975 and 1980

d increases to 0.620 m
$\Delta d = 0.060$ m or 10% approximately
g increases to 0.30 m^2
$\Delta g = 0.05$ m^2 or 20%
h increases to 22 m
$\Delta h = 2$ m or 10%
f remains unchanged

In 1980 $v = (0.30)(22)(0.5) = 3.3$ m^3
i.e. $\Delta v = 0.8$ m or approximately 30%

Alternatively the increase in volume may be calculated from the relative increase in each parameter, namely d, h and f:

or $\Delta v \triangleq 2\Delta d + \Delta h + \Delta f$

$\Delta v \triangleq 2(10\%) + 10\% + 0 = 30\%$

or $\Delta v \triangleq \Delta g + \Delta h + \Delta f$

$\Delta v \triangleq 20\% + 10\% + 0 = 30\%$

Where

Δ = increase or change
$\triangleq$ = approximately equal to

N.B. This example uses expected values

2.3 HEIGHT (h)

Height is a measured not a derived parameter of the tree. Its accurate measurement is important as it is one of the three variables in the estimation of tree volume:

$$v = ghf$$

Height is measured from ground level and may be measured to many different points up the tree, e.g.

- Total height – that is to the highest growing point
- Height to the point where the stem is 7 cm diameter over bark
- Height to the top of the buttresses
- Height to the point where the diameter over bark is half that of the diameter over bark at breast height

Additional commonly used definitions of points for height measurements and lengths of tree boles are given in Figure 5. When measuring tree heights, usually the vertical height above the ground and not the length along a leaning bole is measured. This may result in a slight bias and underestimate of standing volumes. A correction based on the measurement of the angle of lean is feasible but is not usually employed in extensive surveys of forest crops.

2.3.1 The Importance of Height Measurements

Height is an important variable as at a given age it reflects the fertility of the site. Sites with tall trees of a given age and species are more fertile and productive than sites with shorter trees of the same age. This generalization applies to plantations, miombo woodland or savanna and natural high forest.

FIGURE 5 Height measurements

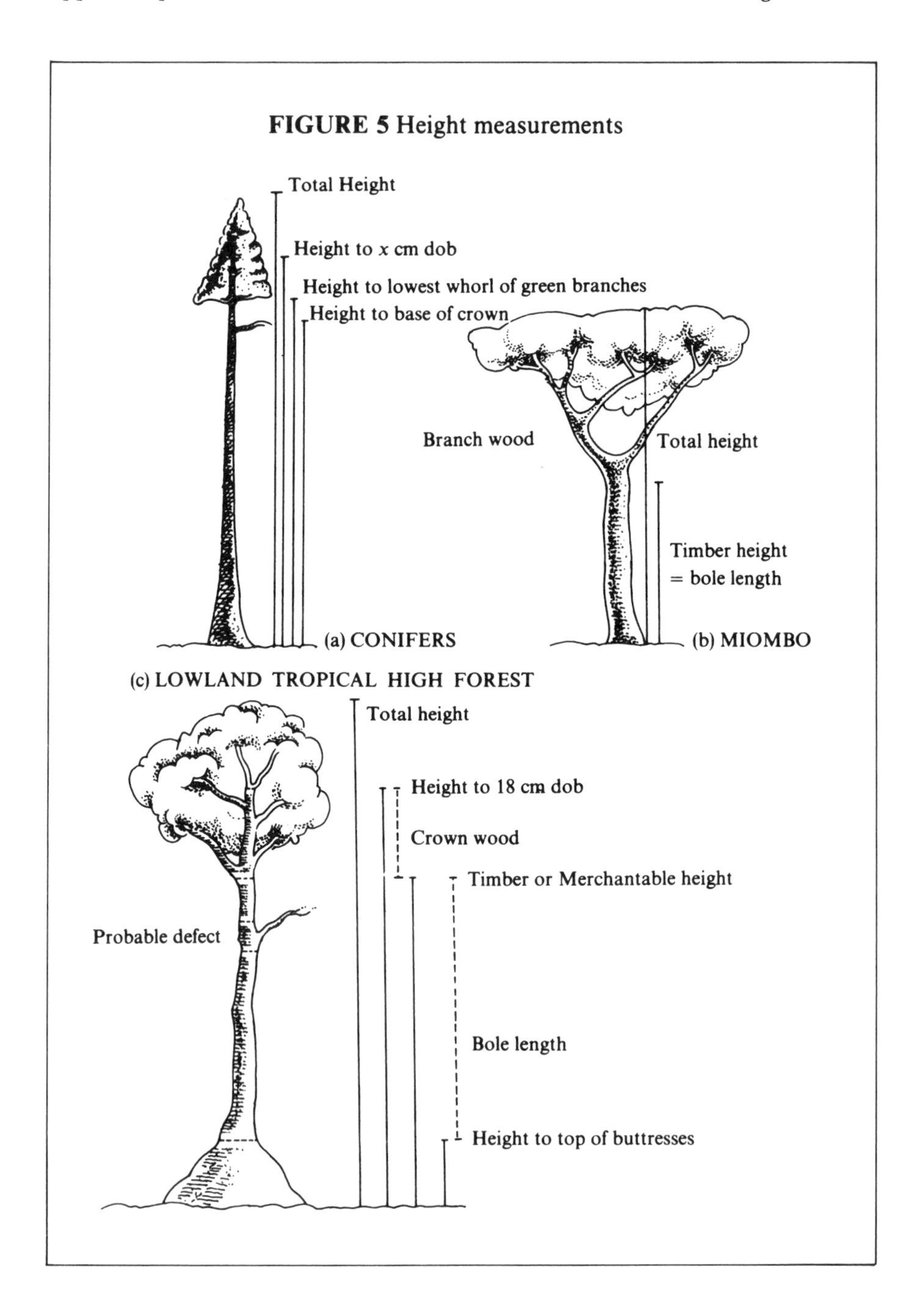

In plantations average height is assumed to be independent of treatment and stocking. This is an oversimplification of a complex situation, but is a useful and often-used assumption, though its validity in the montane softwood plantations in East Africa has been queried (Veldhoen, 1968, 1970; Karani, 1978); similar doubts have been expressed for many sites and forest types. Trees are rarely 'drawn up' by intense competition – but often, because of very narrow crowns, lightly tapered boles and small diameters in dense plantations, they give this impression. When grown in dense stands, trees of a species that tends to fork and have a short bole – such as teak, *Tectona grandis*, or beech, *Fagus sylvatica*, and many other hardwoods – may have longer boles than if grown under more open conditions; however, usually, they do not differ markedly in total height at the same age.

2.3.2 Methods of Measuring Tree Height

(excluding the measurements of tree heights on aerial photographs. For further reading on the measurement of tree heights on aerial photographs see Spurr, 1960; Husch *et al.*, 1982; Howard, 1991)

In order to measure the total height of a tree the top *must* be visible from a point from which it is feasible to see most of the bole. It is usually very difficult to measure total height in dense stands of trees with rounded crowns, i.e. in dense tropical forest and, often, in miombo woodland (see Figure 5). Measurement may also be very difficult in young conifer stands before thinning has opened gaps in the canopy. Instruments that measure tree height are called hypsometers. Clinometers that measure slope may also be adapted.

Methods may be classified as:

1. Direct methods – involving climbing or using height measuring rods.
2. Indirect methods:
   (a) using geometric principles of similar triangles.
   (b) using trigonometric principles and measuring a distance and the angles to the top and base of the tree.

2.3.2.1 *Direct Methods of Measuring Tree Height*

Relatively small trees may be climbed and their heights measured directly using a tape. This method is infrequently used because it is costly.

Trees growing less than about 7 m tall, especially those growing in dense stands, may be measured using a pole graduated in decimetres. The main difficulty is in aligning the end of the pole and the tip of the tree. If the pole can be held so that it lies along the stem of the tree, little or no error need result. However, if because of branches it is impossible to locate the pole near the bole, considerable error may result (through parallax – Figure 6). This may be minimized if a sprung arm is attached at the top of the pole. (The arm must be equipped with some closing device so that it is not ripped away when the pole is withdrawn through the crown.) Normally the pole is shorter than the height

FIGURE 6 Tree measuring rods

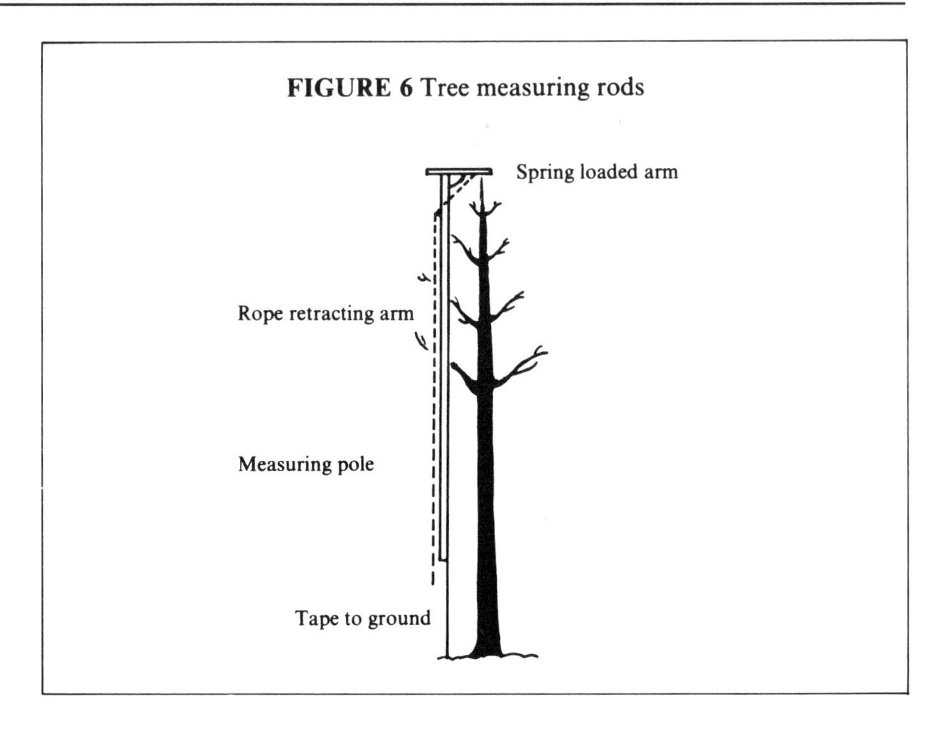

of the tree being measured but equipped with an extendable tape to ground level (Strickland & Binns, 1975).

Direct methods are normally the only feasible way to measure the total height of trees with rounded crowns.

2.3.2.2 *Indirect Methods of Measuring Tree Height*

2.3.2.2.1 Methods employing geometric principles

The simplest method of measuring the total height of a tree whose tip and base are visible from a distance equal to the height of the tree, is to use the isosceles triangle principle. In an isosceles triangle two sides are of equal length. Therefore the measurer grasps a stick and holds it vertical so that the length above his fist is equal to the length from his eye to his fist. Then keeping the stick vertical (approximately at right angles to his arm) and his arm horizontal and straight, the observer positions himself so that without moving his head, the top of the stick is aligned with the top of the tree and, at the same time, the top of his fist is aligned with the bottom of the tree. Then the height of the tree above the level of the observer's fist is equal to the horizontal distance measured from the observation point to the tree (see Figure 7). The accuracy of this method is not great, but estimates within 10% of the correct value are quite feasible.

Christen's Hypsometer also employs geometric principles. A stick of known length (e.g. 5 m) is placed upright against the tree to be measured. The observer uses a hypsometer graduated for the length of the target stick and

FIGURE 7 The isosceles triangle method of measuring tree height

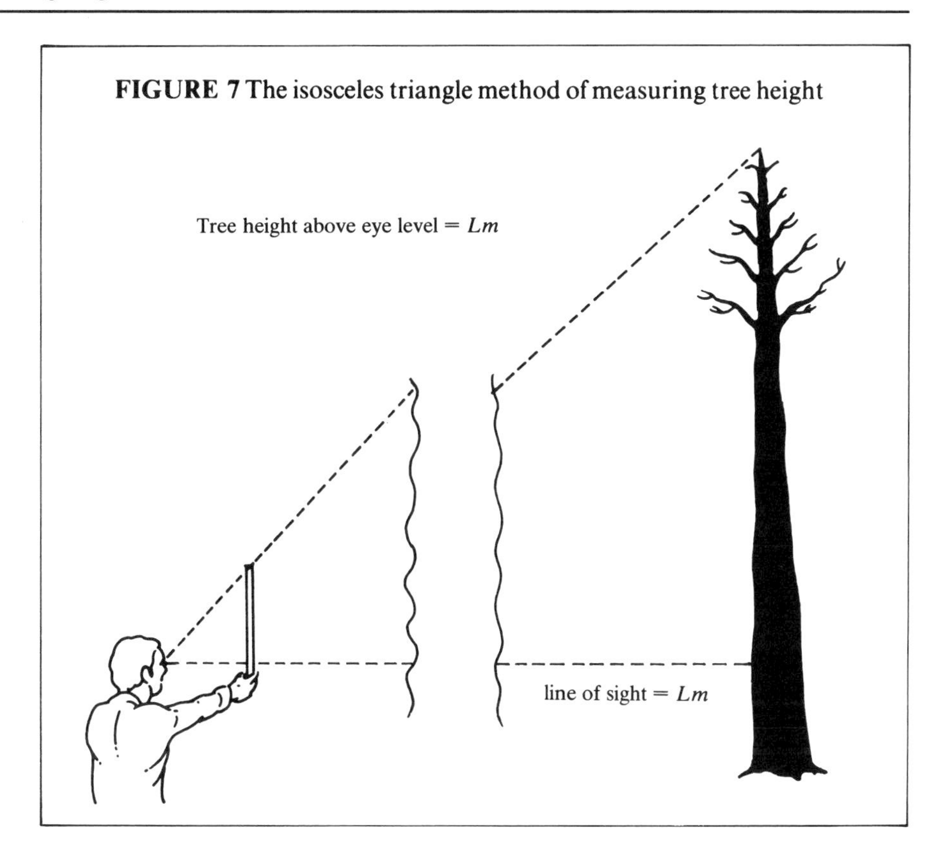

stands a suitable distance away so that the top of the tree subtends an angle of between 35° and 45° from the horizontal (see Section 2.3.3). He then holds the hypsometer so that it hangs vertically between his eye and the tree. The hypsometer is then moved nearer his eye or nearer the tree until the tree is made to fill exactly the viewing gauge of the hypsometer (see Figure 8a). Then the height of the tree is read on the hypsometer graduations against a point in line with the top of the sighting stick.

Similar principles are used in the simple instruments illustrated in Figure 8d, e and f. With that shown in Figure 8d, an assistant marks the point on the tree bole corresponding to X on the gauge. The tree height is ten times the height of this point above ground level. The instrument shown in Figure 8e is useful when estimating bole lengths to be pruned; for example if two-thirds of total height is to be pruned, then Y on the gauge is marked two-thirds of the distance up from A towards B. If the tree appears as in the illustration then the live crown below Y must be pruned. Even a simple ruler may be used as in Figure 8f with a target stick of say 5 m. The ruler is held so the base of the ruler is aligned with the base of the tree, the 5 unit mark on the ruler is aligned with the top of the 5 m stick, and the ruler is vertical. Then the scale reading on the ruler in line with the top of the tree provides an estimate of height in metres.

When using these simple instruments great care must be taken to hold the scale vertical and motionless. The observer should support his arm with a stick or against a tree.

In Figure 8b the hypsometer DEF is vertical and parallel to the tree bole ABC. Then

$\Delta GDF \equiv \Delta GAC$

$\Delta GDE \equiv \Delta GAB$ Δ triangle

$\Delta GEF \equiv \Delta GBC$ $\equiv$ is similar to

$\therefore AC/BC = DF/EF$ or $AC = BC \cdot DF/EF$

N.B. When the tree height is the same as the length of the target stick (Figure 8c), then AC/BC = 1 = DF/EF. Therefore the minimum reading on the hypsometer is identical with the length of the appropriate target stick. When the tree height is twice that of the target stick, then AC/BC = 2 = DF/EF or EF = DF/2, i.e. the mid-point on the hypsometer scale must be graduated to read twice the length of the target stick. Continuing this reasoning for the case where the tree height is four times the length of the stick, indicates that the scale readings on the hypsometer are not linear and that accuracy decreases with tree height and increases with the length of the viewing gauge. If a long viewing gauge is used then the instrument may have to be held more than arm's length away from the observer's eye in order to position the tree to fill the gauge; then it has to be held suspended from a pole (Buchner *et al.*, 1977).

2.3.2.2.2 Methods employing trigonometric principles

The observer stands at a point (see Figure 9, at D) from which he can see clearly both the top and the base of the tree. The instrument is sighted at the base of the tree – at C – and the angle θ_1 read and recorded; the instrument is then rotated and sighted at the top of the tree – at A – and again the angle from the horizontal – θ_2 – is read and recorded. (On many instruments readings for angles below the horizontal are prefaced with a −ve sign.) The readings of these two angles and the horizontal distance from the observer to the tree – DB in Figure 9 – are the raw data from which tree height is calculated.

In Figure 9 the line of sight DB is horizontal. Assuming the tree is vertical, then ABD and CBD are both right angle triangles so that:

$\tan \theta_2 = AB/BD$ or $AB = BD (\tan \theta_2)$

$\tan \theta_1 = BC/BD$ or $BC = BD (\tan \theta_1)$

and the total tree height $AC = BD (\tan \theta_2 + \tan \theta_1)$

Hypsometers using the trigonometrical principle measure vertical angles. If they are for use from fixed horizontal distances then their scales may be graduated to read height directly, or they may be graduated to read the angles in degrees, or tangents of angles, or the tangents as a percentage, i.e. in Figure 9 ((AB/BD)100). All such hypsometers using the trigonometrical principle are basically similar.

An efficient hypsometer should embody the following characteristics:

FIGURE 8 Christen's hypsometer

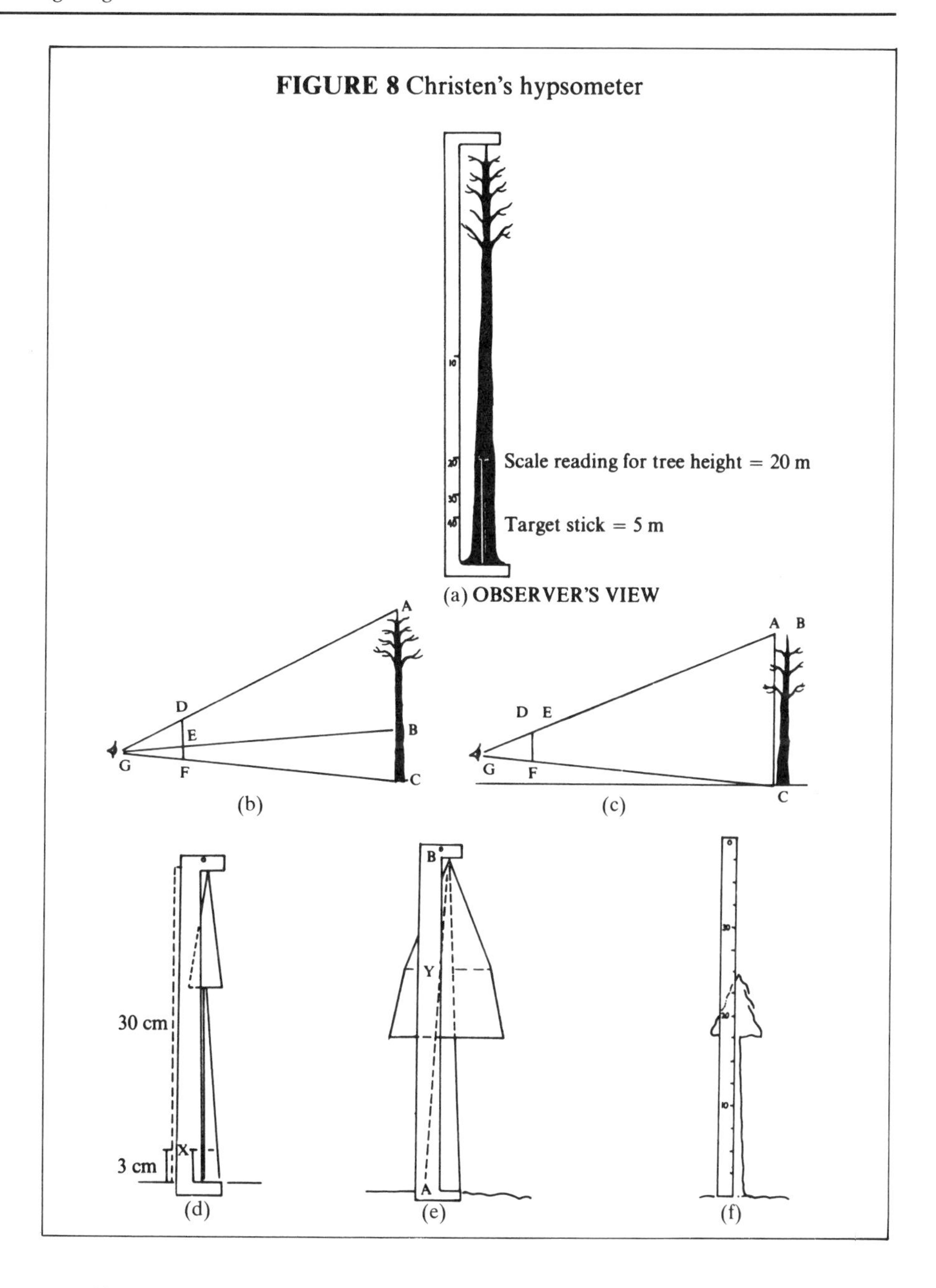

- Cheapness
- Graduated with several scales for direct reading from different fixed distances
- Have a scale of tangents expressed as percentages
- Incorporate a range finder for establishing the fixed distances
- Be lightly damped
- Have the scales visible when taking readings so that their oscillations may be observed. Readings are then taken when the oscillations have ceased

FIGURE 9 The trigonometry of tree height measurement

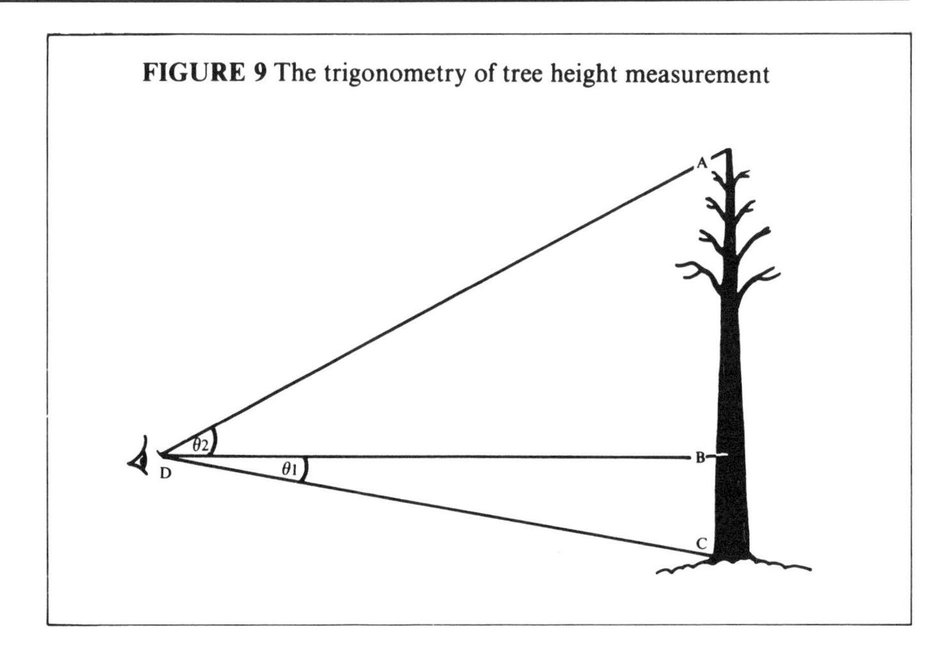

- Have scale identities visible when taking readings so that they are not misidentified

Types in common use are:

- Suunto hypsometer
- Blume-Leiss hypsometer
- Haga hypsometer
- The hypsometer incorporated in the Spiegel relaskop

None of the four named instruments has all the desirable characteristics. Their advantages and disadvantages are summarized below:

Suunto
Cheap
Average damping in oil bath
Small, light, robust 15 m and 20 m scales only
Range finder incorporated
Scales visible when in use but identities masked
Incorporates a single angle gauge of 1/50 radian

Blume-Leiss
Medium/high price
Very heavily damped
Medium size and weight, robust 15, 20, 30 and 40 m scales and degrees
Range finder incorporated
Scales not visible when sighting – care needed to select correct scale as all are visible

Haga
Medium priced
Heavily damped
Medium size and weight
15, 20, 25 m and tan % scales

Spiegel relaskop
Expensive
Lightly damped
Compact and robust but heavy
20, 25 and 30 m scales only on the normal model*

Scales not visible when sighting – single scale selected and exposed in viewing window before sighting	Scales visible when sighting but not identified Incorporates sophisticated angle gauges Multi-purpose instrument

*The wide angle Spiegel relaskop has a larger number of height scales

When the top and base of the tree cannot be seen from any of the fixed distances incorporated in the hypsometer being used, the tree height from any distance may be calculated using as a correction factor for the instrument readings the ratio of actual distance to distance assumed for the scale readings, e.g. tree height = scale reading (actual distance/scale distance).

EXAMPLE 10 Correcting a scale reading on a hypsometer

A tree is measured from a point 11 m away using the 20 m scale.

Reading to top of tree +13 m
Reading to bottom of bole −3 m

Total readings 16 m

$$\text{Tree height} = 16\left(\frac{11}{20}\right)$$
$$= 8.8 \text{ m - rounded to 9 m}$$

N.B. The sign attached to the hypsometer reading indicates whether the line of sight is above (+) or below (−) the horizontal. Normally the readings are totalled disregarding the sign. Only if the eye of the observer is below the base of the tree will both readings be positive, in which case the reading to the base of the tree must be *subtracted* from the reading to the top.

Conclusion

The instruments based on trigonometrical principles are more accurate than the more simple ones employing geometrical principles; however, the latter are much cheaper and can be made locally.

2.3.3 Errors in Height Measurements

The sources of the major errors in height measurement other than human errors are:

- failure to measure correctly the horizontal distance from the observer to the tree
- wind sway
- lean
- non-linearity of the relationship of tree height and angle of sight

One of the commonest human errors arises from failure to identify correctly the points of measurement at the top and bottom of the tree (Figure 10).

Romesberg & Mohai (1990) summarize practical measures to improve the precision of tree height estimates.

Errors in the measurement of the horizontal distance from the observer to the tree

If the distance from the observer to the tree is not measured horizontally, the observer will stand too near the tree, and the height will be overestimated by the direct reading on the instrument scale.

Wind sway

Wind causes tree tops to sway and this can be a very serious hindrance in tree height measurement and cause serious errors. Accurate readings cannot be made in high winds. The errors may be reduced by averaging readings taken at the extremes of the sway towards and away from the observer.

Leaning trees (see Figure 11)

If a tree, BD, is leaning away from an observer at A, then the height recorded on a fixed scale instrument appropriate to the distance AB is FB. However, if the observer moves to the diametrically opposed position C so that CB = AB, then the height recorded is EB. The average of the two readings FB and EB is close to but not exactly equal to the vertical height of D above the ground, that is DB′. The true length of bole BD can only be calculated if the angle of lean θ is measured. (This may be done by placing the base bar of a pair of calipers along the bole and reading the angle of lean with an inclinometer.)
Then

$$BD = DB'/\cos\theta$$

In most field mensuration studies DB′ is estimated and BD is not calculated. Some authorities recommended measuring leaning trees from points at right angles to the plane of the lean, or measuring the distance from the perpendicular from the top to the ground, i.e. from B′ to C in Figure 11, instead of from the base of the tree.

Non-linearity of the relationship of tree height and angle of sight

The smaller the angle of the sight the easier it is to define the highest point in the crown; *but* the nearer the angle of sight to 45° the smaller is the error caused by an inaccurate reading of that angle. The best compromise between these two conflicting considerations is to select the observation point so that the angle of sight lies between 30° and 45°, i.e. the observer should stand between one and one-and-a-half times the tree height away from the tree. Angles greater than 45° must be avoided as the probability of mistaking a side branch for the top of the tree is unacceptably high.

Further reading

ON ERRORS IN HEIGHT MEASUREMENT:

Loetsch, F. *et al.* (1973). *Forest Inventory*, Vol. II. BLV, Munich, pp. 43–44.

FIGURE 10 Errors in height measurements from failure to identify correctly the top of the tree

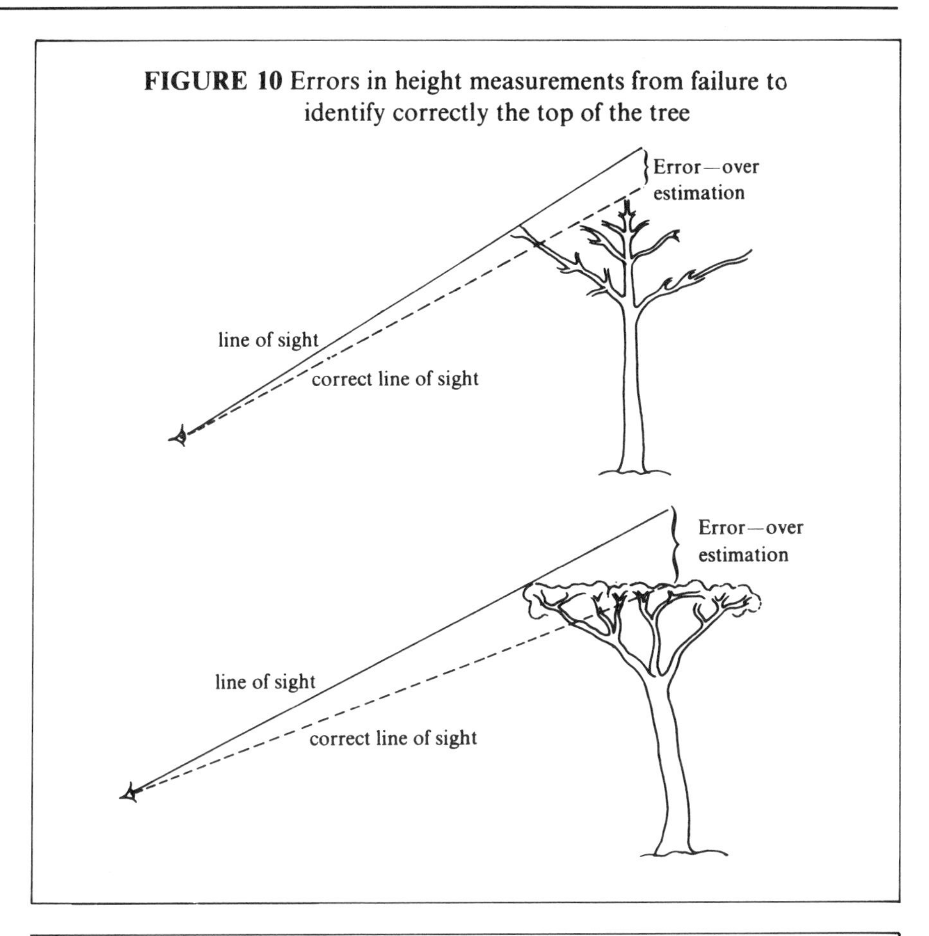

FIGURE 11 Height measurement on leaning trees

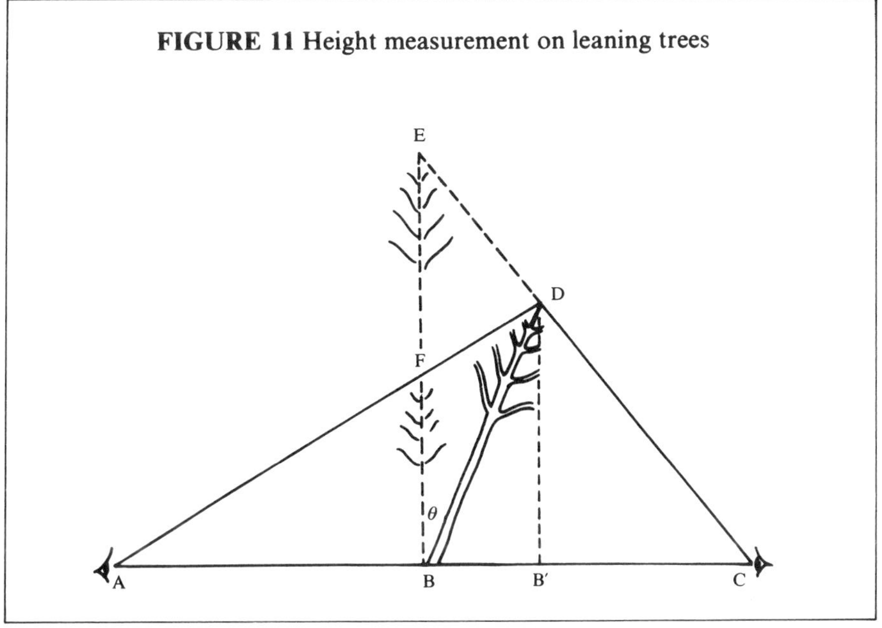

2.3.4 Standardization of Measurement Procedures

Great care is needed when measuring height as there are many sources of human error that must be eliminated if accuracy and precision are to be achieved. The measurer must:

- select a viewing point from which the top and bottom (upper and lower measuring points) of the tree can be seen clearly
- take an average of at least three readings that differ by not more than 0.5 m
- measure the horizontal distance to the centre of the base of the tree accurately. N.B. range finder targets should not be hung on the nearside of the tree, but at the side or vertically below the tip
- use an angle of sight between the upper and lower measuring points of between 30° and 45°
- read the correct scale
- if the height of a particular tree is required, then the effect of lean must be minimized by taking readings from diametrically opposed points and calculating their average, or at right angles to the plane of the lean
- if the change in the height of a particular tree is required and measurements are to be repeated in later years, then the points of viewing must be marked and used on successive occasions
- normally heights measured indirectly with a hypsometer are rounded to the nearest half metre (N.B. if recorded as 0.5 m, this last digit is not a significant decimal figure)

When measuring total height, the point of measurement is defined as the highest growing point. If the base of the tree is obscured by a rise in the ground, then a point should be marked on the bole a known distance above ground level and used as the lower measuring sight. Then the height calculated from the hypsometer readings has to be adjusted by adding the height above ground level of this lower sighting mark.

The horizontal distance from the observer to the tree must be measured or calculated. Step chaining, i.e. holding the measuring tape horizontal above sloping ground, or correcting for slope by the cosine of the angle of slope, or ensuring a horizontal line of sight for a range finder is essential. Normally the distance is measured from the centre of the tree bole at ground level.

2.4 THE FORM OR SHAPE OF THE TREE (*f*)

The outline of the cross-section of tree boles has already been discussed in Section 2.2 and height measurements in Section 2.3. This section considers the root system and the profile of the tree from ground level to the top of the crown.

The nature and extent of the root system, the form of the bole and the structure of the crown result from interactions of genotype and environment. On loam soils where roots can penetrate easily and deeply some species develop tap roots, e.g. teak (*Tectona grandis*), neem (*Melia azedarach*) and many pines, whereas on clay soils with a seasonally high water table the same species will develop a root plate. Also the extent of stem buttresses is, to some limited extent, a function of the site although age and genotype are also important.

Bole development also is governed by genotype, age and environment. Young trees tend to show strong apical dominance and a straight stem. As the tree ages and particularly when the reproductive phase is reached, the crown expands to provide flowering sites and apical dominance tends to be lost. The dominance of the apical meristem tends to be strong in most colonizing or pioneer or clasmophytic species such as *Betula pendula, Maesopsis eminii, Broussonetia payrifera, Musanga cecropoides* and *Trema guineensis.* (A clasmophyte is a species that colonizes gaps in more stable vegetation types, caused by destruction of the canopy by wind throw or wind break, for example.) Most members of the genus *Pinus* have strong apical dominance that persists until the tree is fully grown. Generally the faster the height growth the stronger is the apical dominance. Some species lose apical dominance of the terminal shoot at the end of a period of rapid height growth when a whorl of cluster of branches or an inflorescence develops. Then, after a period of little or no height growth, a lateral bud assumes apical dominance and a new period of height growth ensues. This type of periodicity of growth is common in many species of *Terminalia,* in *Alstonia boonei* and in *Cordia abyssinica.*

In savanna and miombo many trees have short boles and spreading crowns – either rounded like *Sclerocarya caffra* and *Brachystegia spiciformis,* or flat-topped like many species of *Acacia, Albizia,* and *Brachystegia longifolia.*

Further reading

Horn, H.S. (1971) *The Adaptive Geometry of Trees.* Monographs in population biology, No. 3. Princeton University Press, Princeton.

2.4.1 The Distribution of Dry Matter in Trees

Until recently almost all tree mensuration concerned the bole of trees because that was the portion harvested. Some research concerned crown size and root systems but this work was done to support physiological or silvicultural studies. Now the patterns of harvesting are changing; the branches and twigs may be utilized, leaves used as fodder and roots extracted and used in pulp production. Consequently the field of mensuration must be enlarged to encompass the estimation of volume and weight of these parts as well as the bole and include them in growth prediction models.

Various studies of the relative amounts of dry matter in trees have been done. Crude generalization provides the following pattern:

	Forest trees (%)	*Savanna/Woodland trees* (%)
Twigs and leaves	10	10
Branches	15	30
Bole	30	30
Roots > 5 cm diam.	45	30

Nevertheless the dry weight root/shoot ratio varies both between species and between individuals, and is affected by the site – both by physical soil characters and the nutritional status of the tree. The variation from tree to tree is of interest to the tree breeder, for example when selecting for wind-exposed sites, for wind breaks, or for clonal plantations such as poplar (*Populus* spp.) or rubber (*Hevea brasiliensis*) for example.

Further reading

ON THE DISTRIBUTION OF DRY MATTER IN TREES AND FORESTS

Nye, P.H. and Greenland, D.J. (1960) *The Soil and Shifting Cultivation.* Technical Communication No. 51, Commonwealth Bureau of Soils, Commonwealth Agricultural Bureaux, Farnham Royal, England.

2.4.2 The Tree Bole

Although the limits of utilization are extending to include tree roots and crowns, the bole of the tree is still of major importance to the forester as it contains between 30 and 50% of the dry matter, more of the generally harvested product and, because of its function in supporting the crown, its structure is designed to support weights and stand stress. Therefore the wood of the bole is well suited to use as a structural material and has a high strength to weight ratio. Also its size – both diameter and length – allows cheaper handling and processing compared to the roots and branches.

The bole consists of three parts:

- the part near the ground where the shape is affected by roots, root swell and buttress
- the major portion of the length from that part near the ground to below the live crown. It is usually relatively free from live side branches and usually consists of one or a few main stems
- the upper or crown wood part. If apical dominance has been maintained, this upper part consists of a stem and relatively simple side branches, but where apical dominance has been lost it is replaced by crown branch wood

All parts of the bole are more or less irregular and are imprecisely described by any regular geometric figure; nevertheless, in order to estimate volume, the tree mensurationist may assume a degree of regularity and base his estimates on one or a number of geometric figures belonging to the family described as 'solids of revolution'.

FOR THE ADVANCED STUDENT

2.4.2.1. Solids of Revolution

A solid of revolution is a solid three-dimensional figure defined by establishing a relationship between two rectangular coordinates representing the two dimensions, radius (y) and distance from the end or apical bud (x), and rotating the plane of this relationship a full 360° around the x coordinate axis (Figure 12).

The general formula describing the relationship of x and y for a solid of revolution is:

$$y = kx^r$$

Where: k and r are constants defining the rate of taper and the shape of the solid respectively.

Hence the cross-sectional area g_b at x_b

$$g_b = \pi(kx_b^r)^2$$

FIGURE 12 The definition of a solid of revolution

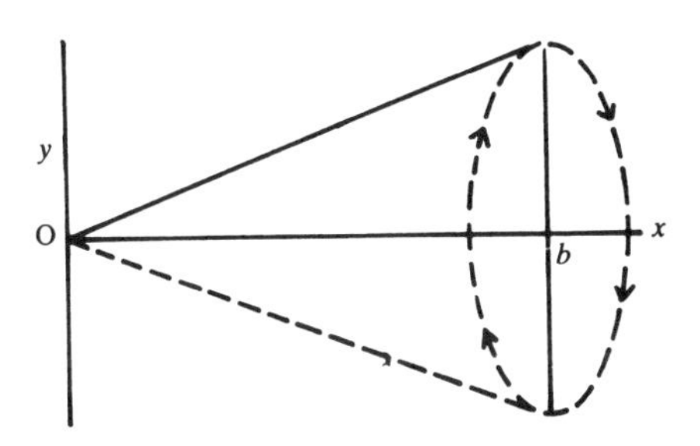

FIGURE 13 The volume of a cone ($r = 1$)

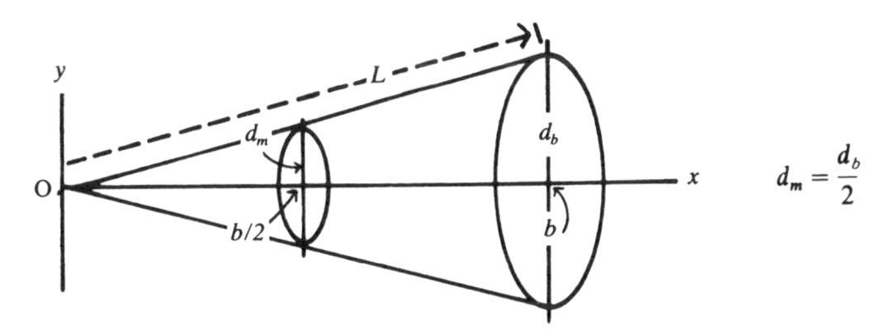

$y = k'x$ $\quad x_b$ = height of cone $\quad d_b$ = diameter of base of cone

g_b = cross sectional area of base of cone

d_m = diameter at half height $\quad k'$ = taper coefficient

$g_i = \pi y_i^2$ and $y_i^2 = (kx_i^r)^2$—as in the general formula

but $y_i^2 = (k'x_i)^2$—in the diagram above therefore $r = 1$ and $g_b = \pi k^2 x_b^2$

$$\text{then } v_b = \pi\int_0^b (kx)^2\, dx = \pi\left[\left(\frac{k^2x^3}{(2+1)}\right)\right]_0^b = \pi\left[\left(\frac{k^2x^3}{3}\right)\right]_0^b = \tfrac{1}{3}(\pi k^2 x_b^2)x_b$$

$= \frac{1}{3}$(basal area)(height) i.e. form factor = 0.33

The volume of the solid is the sum of the volumes of a series of infinitesimally thin cross-sectional slices. Using the symbols of the differential calculus,

$$V = \pi \int_0^b (kx^r)^2 \, dx$$

$$= \pi \Big|_0^b \left[\frac{k^2 x^{(2r+1)}}{(2r+1)} \right]$$

The ratio of the volume of the solid to that of a cylinder with the same base cross-section and height is called the form factor.

For a cone $y = kx$
The profile of a cone is a straight line, i.e. $y \propto x$, and the radius is proportional to the relative distance from the apex. The radius at the base of a cone is twice that of the radius at half height.

For a quadratic paraboloid
The cross-sectional area is proportional to the distance from the origin – i.e. the apex (or tip of the tree) – $y^2 \propto x$. The cross-sectional area at the base of a quadratic paraboloid is twice that of the sectional area at half height.
For a quadratic paraboloid $y^2 = k'x$

For a cubic paraboloid $y^3 \propto x$ so $y^2 = k'x^{2/3}$
The cube of the radius is proportional to the distance from the origin.

For a neiloid (semi-cubic paraboloid) $y \propto x^{3/2}$ and $y^2 = k'x^3$
The form factors and powers of r for the simple solids of revolution commonly met within tree mensuration are summarized below:

Solid of revolution	Form factor	Power of r
Cylinder	1	0
Cubic paraboloid	3/5	1/3
Quadratic paraboloid	1/2	1/2
Cone	1/3	1
Neiloid	1/4	3/2

The symbol k' in these formulae along with the power r govern the rate of change of diameter with distance from the apex. The greater the value of k', the greater is the rate of change in diameter.

FIGURE 14 The volume of a quadratic paraboloid ($r = 0.5$)

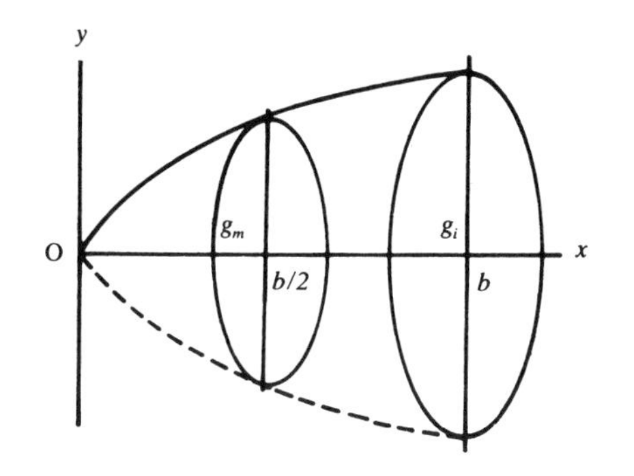

$$g_m = \frac{g_i}{2}$$

$y^2 = k'x$

$g_i = \pi y_i^2 \quad y_i^2 = (kx_i^r)^2 \quad$ so $\quad y_i^2 = (k'x_i) = (kx_i^r)^2$

therefore $\quad r = 0.5 \quad$ and as $g_i = \pi k'x \quad$ so $\quad k' = k^2$

then $\quad v_b = \pi \int_0^b (kx^r)^2 \, dx = \pi \Big[\frac{1}{(2r+1)} (k^2 x^{(2r+1)}) \Big]_0^b$

$= \pi[(\frac{1}{2}k^2x^2)] = \frac{1}{2}(\pi k'x_b)x_b = \frac{1}{2}$(basal area)(height)

i.e. form factor = 0.5 (Symbols as for cone)

FIGURE 15 The volume of a cubic paraboloid ($r = 0.33$)

$g_i = \pi y_i^2 \quad y_i^2 = (kx_i^r) \quad$ but also $y_i^2 = (k'x_i^{2/3})$

therefore $\quad r = 1/3 = 0.33; \quad$ and $\quad k' = k_2$

Then $\quad v_b = \pi \int_0^b (kx^{1/3})^2 \cdot dx$

$= \pi_0^b[(3/5)k^2x^{5/3}] = (3/5)\pi k^2 x_b^{5/3} = (0.6)\pi(k'x_b^{2/3})x_b$

$= (0.6)$(basal area)(height) Form factor 0.6

(Symbols as for cone)

FIGURE 16 The volume of a neiloid ($r = 3/2 = 1.5$)

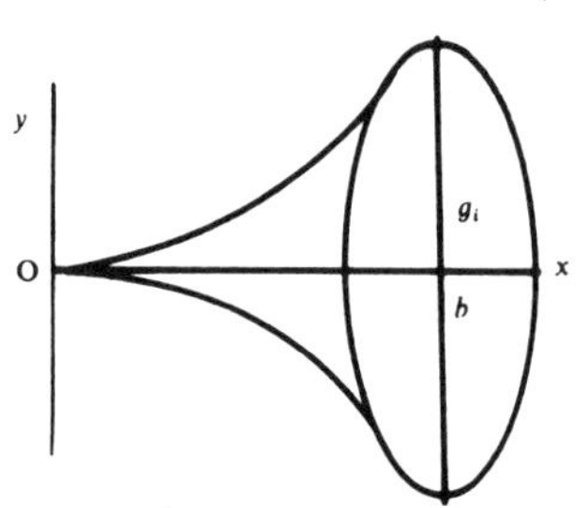

$g_i = \pi y_i^2 \quad y_i^2 = (kx_i^r)^2$ but $y_i^2 = k'x_i^3$

therefore $r = 3/2$ and $k' = k^2$ then $v_b = \pi\int_0^b (kx^{3/2})^2 \, dx = \pi \left[\tfrac{1}{4}k^2x^4\right]_0^b$

$= (\tfrac{1}{4})\pi k^2 x_b^4 = (\tfrac{1}{4})\pi k^2 x_b^3 x_b = (0.25)\pi k' x_b^3 x_b = 0.25$(basal area)(height)

form factor = 0.25 (Symbols as for cone)

FIGURE 17 Solids of revolution

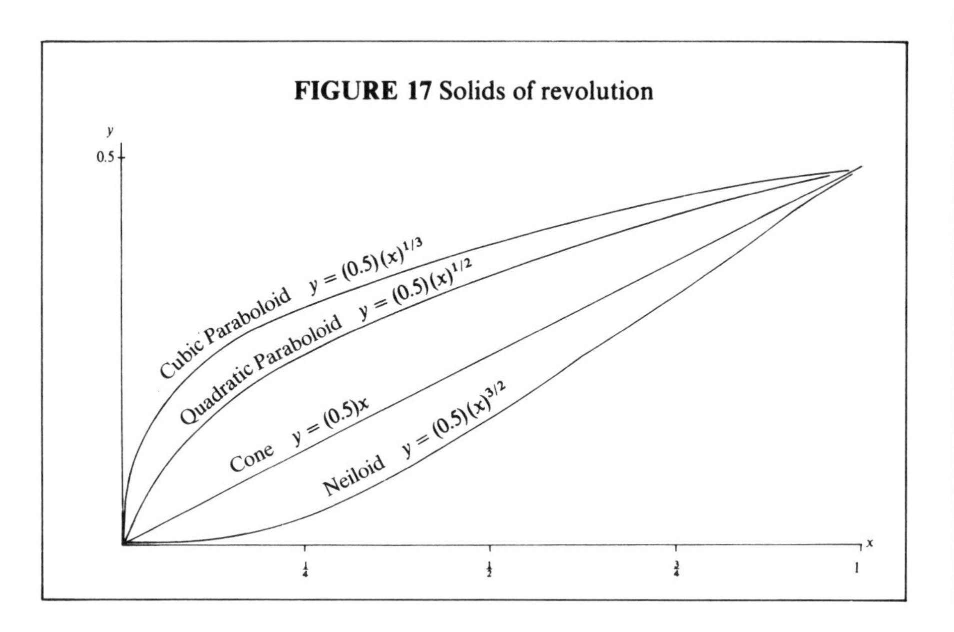

2.4.2.2 *The Form and Taper of Trees and Logs*

The best and shortest definition of the form of a tree or log is its shape. The shape may be regular, as for a solid of revolution, or – more commonly – irregular. The form may be measured by the form factor, that is the ratio of the volume of the tree or log to that of a cylinder of equal basal cross-sectional area and height. Even if a tree or log has an estimated form factor of 0.5 – equal to that of the quadratic paraboloid – the conclusion that its form is that of a quadratic paraboloid is unjustified as probably it is irregular. (N.B. for trees, the equivalent basal area is usually taken at 1.3 m above ground level, that is at breast height; in some circumstances the cross-sectional area at (0.9)(total height) from the growing point may be used so the form factors of trees of different total heights may be compared on a rational basis.) Taper is the rate of change of diameter over a specified length or height. Figure 18a shows two logs of the same form but different tapers. They are both conoid, have the same cross-sectional areas at the base and top but have different lengths (L_1 and L_2). Figure 18b illustrates two logs of the same average taper – that is the same end cross-sectional areas and the same length – but with different forms; (c) is conoid and the other (d) is paraboloid.

FIGURE 18 Log shape and tapers

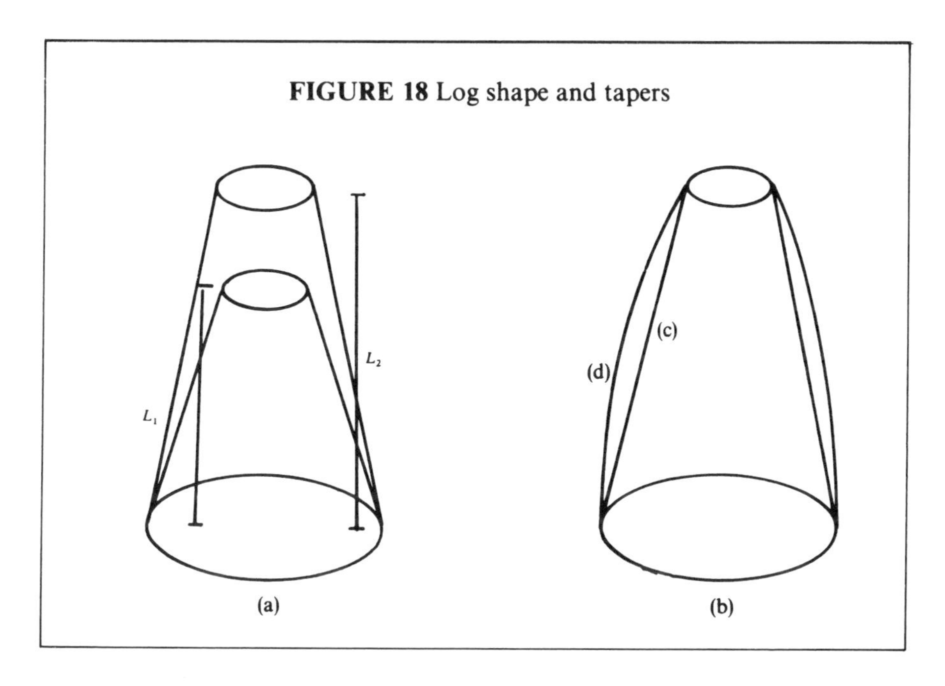

Irregularity in form complicates accurate prediction of tree form factors and volumes. Common causes of irregularities are:

- abrupt change of diameter at a node
- deformity after injury to the cambium
- abrupt change of diameter associated with heart rot
- swelling from occlusion of branches
- influence of root swell, buttress or stem flutes

Many research workers have expressed the profile of trees and logs by means of complex splined functions, often constrained by a diameter at breast height and total height. The literature on these techniques is extensive (see M'Hirit & Postaire, 1985; Newberry & Burkhart, 1986; Ormerod, 1986; Capelewski & McClure, 1988; see also Section 2.7.4.6).

2.4.2.3 *Definitions of Tree Form Factors*

When calculating form factors, the cross-sectional area and the height of the reference cylinder and the definition of the volume of the tree must be declared. Different authorities use differing declarations depending on the use of the calculated factor. Three form factors have been defined in the classic European work on tree mensuration and a fourth is commonly used in the field. They are:

Classic studies

- Absolute form factor $= \dfrac{v_t}{(g_{gl})\,(h_t)}$
- Artificial form factor $= \dfrac{v_t}{(g_{bh})\,(h_t)}$
- Normal (or Pressler's) form factor $= \dfrac{v_t}{(g_{0.9})\,(h_t)}$

Field practice

- Form factor $= \dfrac{v_m}{(g_{bh})\,(h_t)}$

Where:

v_t = total volume over bark
v_m = merchantable volume over bark, defined by a stated top diameter
h_t = total height
g_{gl} = cross-sectional area at ground level
$g_{0.9}$ = cross-sectional area at 0.9 of total height measured from the tip, i.e. 0.1 from ground level
g_{bh} = cross-sectional area at breast height, i.e. 1.3 m above ground level

The last form factor has the advantage that it yields the measure usually wanted by the practising forester – merchantable volume – when used in conjunction with g_{bh} and total height, both of which are unambiguously defined and easily measured in the field; however, the artificial form factor is independent of changing limits of merchantable volume and is, therefore, most useful. Pressler's form factor is useful in research in comparing trees of different heights. Form factors normally lie within the range of 0.25 and 0.50.

Apart from the variables used in its calculation, form factor has the following sources of variation:

- species and genotype
- age
- stocking
- crown size
- site factors – especially wind exposure

Some species have cylindrical boles – notably the emergents and upper canopy layer species in tropical high forest at low altitudes, e.g.

- *Entandrophragma* spp.
- *Celtis* spp.
- *Cephalosphaera usambarensis*
- *Tetraberlinia tubmaniana*
- many members of the Dipterocarpaceae

Such trees have high form factors. In even-aged crops the form factor tends to increase after the closure of the canopy before levelling out at about 0.45 and thereafter altering but slowly. The change in form factor at the time of canopy closure follows changes in the depth of live crown. Tall open grown trees with deep live crowns have low form factors whereas trees grown in dense stands with shallow crowns and trees that have been high pruned tend to have high form factors. Part of the bole below the live crown tends to be of paraboloid form and that within the crown tends to be conical. (N.B. Young trees below 4 m total height tend to have high form factors because the reference cylinder volume is small, being based on a cross-sectional area that is relatively high up the tree.)

In some areas exposure to wind has the effect of reducing the increase in height per unit of volume as the tree gets older and increasing the form factor. this effect has not been investigated thoroughly but is touched upon in Blyth's work in Scotland (Blyth, 1974).

2.4.2.4 Variation in Taper

Taper, that is the rate of change of diameter with height, or length, varies enormously. It is small along the lower part of the bole but large both where the bole is affected by root swell and in the crown.

Form quotients are ratios of two diameters, one of which is usually diameter at breast height. Form quotients using a diameter at a specified proportion of total height such as 0.1 or 0.5 of total height from ground level, or a diameter at a fixed height from the ground, such as 3, 4 or 5 m, are common. Form quotients may be grouped into classes – such as 0.775 to 0.825 – and referred to as the 0.8 form class. However, in the USA the Girard form class is defined as the ratio of an under bark diameter at the top of the butt log – $d_{17.3}$ – to that over bark at breast height, expressed as a percentage.

Many sawyers reckon on an average taper of 1 in 100 for saw logs, i.e. a reduction in diameter of 1 cm for every metre of length. This tends to be an overestimate especially in tropical high forest.

2.4.2.5 Implication of Form

Trees and logs are irregular, tapering solids with only approximately circular cross-sections; this has many practical implications:

- Opposite ends of logs have different cross-sectional areas; the longer the log the larger the difference tends to be
- When stacked on the ground or on a lorry with the smaller ends all on one side, the top of the stack will slope

- On sawing, the width of the lumber is limited by the small end diameter. The longer the log the greater is the change in diameter and the greater is the waste of wood in the butt end of the off-cut (that is the piece of wood and bark removed when 'squaring' the piece)
- When the log is sawn parallel to the central axis of the log the grain runs out on the surface of each board. This is the usual method of sawing as it is easier than sawing parallel to the profile of the log which leaves a tapered central core
- Volume estimation is imprecise unless many diameter measurements are made along the length of the tree or log

Also, for a tree exhibiting annual rings, the number of rings on the two ends of a log will differ – the upper or distal end having fewer than the lower end.

Further reading

ON FORM AND TAPER

Busgen, M., Munch, E. and Thomson, T. (1929) *The Structure and Life of Forest Trees.* Chapman and Hall, London.

Gray, H.R. (1956) *The Form and Taper of Forest Tree Stems.* Imperial Forestry Institute Paper No. 32, Oxford.

Larson, P.R. (1963) *Stem Form Development of Trees.* Forest Science Monograph No. 5.

Newnham, R.M. (1965) Stem form and variation of taper with age and thinning regime. *Forestry* 38, 218–224.

2.4.3 Branch Wood

Branch wood is used for making charcoal, pulp, small domestic items such as brooms and fire beaters and as fuel. Traditionally in the developed world only the larger pieces over, say, 7 cm in diameter are used and measurement is either by weight or stacked measure. However, in some areas where firewood is in short supply and is used for cooking, very small pieces may be harvested and used. (N.B. Small very dry sticks are employed to create a very hot fire in order to boil water quickly, whereas larger pieces are needed to maintain heat over a longer time. People who cook with firewood become very selective in what they choose; different woods or dung are not simple substitutes.)

Measurements by weight need standardization to an oven-dry base in order to be comparable. For example, in a household survey of firewood consumption throughout the year, green weight data collected at different seasons must be standardized before consumption in different months can be totalled into an annual figure. Samples are taken, weighed green, dried and the moisture content calculated by subtraction; a simple coefficient of the ratio of oven-dry to green weight can then be used to estimate the oven-dry weight of the green material in a particular month.

Stacked volume is a simple measure to employ. The branch wood is cut into standard lengths, for example 1 m pieces; these are stacked between uprights – say 1 m apart and 1 m tall – and the stacked volumes recorded in stacked cubic measure. The solid volume in the stack is less; if all pieces are piled parallel and are all cylindrical, then the theoretical volume will be 78.5% of the stacked volume i.e. ($\pi/4$). In practice this theoretical figure is unlikely and around 65–70% is more probable. An appropriate figure for a particular situation can be calculated either:

- by estimating the volume of each piece in sample stacks

or

- by photographing the face of a stack and estimating the area of solid wood compared to the area of air space on the photograph; this can be done by superimposing a rectangular grid over the photograph and counting the number of grid intersections that fall on wood compared to the number falling on air spaces

2.5 Bark

Often bark has to be removed before the wood is used; in many cases the bark is wasted although it is a possible source of energy: it may be burnt and the heat transformed into electricity. Bark is rarely thought of as more than a by-product of wood production even when it is processed and turned into a higher value product such as a horticultural growing medium, a use for which it is suited as it is relatively rich in plant nutrients. Bark contains the following percentage of the total mineral content of the above ground portions of Scots and Corsican pine growing on sand dunes in north Scotland (Wright & Will, 1958):

P 20–29%
Ca 20–41%
K 17–23%
N 18–25%
Mg 18–27%

The thickness of bark and its percentage of volume of the tree or log are important parameters in mensuration because most measurements on standing trees have to be made 'over bark'. The volume of bark has then to be deducted if the net volume of wood is needed. Bark thickness varies with:

- species
- age
- genotype
- rate of growth
- position in the tree

Some species have very thick bark, e.g. *Sequoiodendron gigantea* and, especially, many savanna trees or woodland species where the bark protects the cambium from fire and is an adaptation to the savanna conditions, e.g. *Erythrina abyssinica* and *Cussonia arborea*. Bark thickness also varies from individual to individual, but rarely as much as between species and genera.

Examples of trees with thick bark grown in East Africa

Savanna and woodland species
- *Pterocarpus angolensis*
- *Cussonia arborea*
- *Erythrina abyssinica*

High forest species
- Indigenous
  - *Phyllanthus discoideus*
  - *Entandrophrangma angolense*
  - *Maesopsis eminii*
- Exotics
  - *Tectona grandis* – also occurs in woodland and savanna
  - *Eucalyptus paniculata*
  - *Gmelina arborea*
  - *Pinus caribaea*

Examples of trees with thin bark

High forest species
- Indigenous
  - *Cynometra alexandri*
  - *Antiaris toxicaria*
  - *Newtonia buchananii*
- Exotics
  - *Eucalyptus deglupta*
  - *Terminalia superba*
  - *Aucoumea klaineana*

Bark is generally thickest near the base of the tree and thinnest in the crown. The thickness largely depends upon the nature of the bark cambium and the method of exfoliation. Exfoliation tends to be slow on slow growing trees and thick layers of corky bark may accumulate; in contrast, fast growing vigorous trees cast the old bark more rapidly and accumulate less. The exposure of inner bark along cracks in the outer bark is a sign of rapid expansion and diameter increment.

The volume of bark is a function both of bark thickness and tree diameter. Also the volume percentage of bark on a log varies with the position of the log in the bole and with species. In most species log bark percentage by volume decreases with height up the bole, but in some species the percentage is almost constant and in others the highest bark percentage occurs in logs from the base of the live crown. *Pinus patula* has a low bark percentage on logs with the immature papery type of bark, but an increasing percentage in that part of the stem with the thick fissured bark that develops after 15 years or so. Consequently the bark percentage in a lorry load of round timber may be very variable and vary with the species, age of tree, tree diameter and position of log in the tree. Commonly bark comprises between 10 and 20% of the over bark volume of a tree.

2.5.1 Measurement of Bark Thickness

The bark thickness of living trees may be measured with little damage to the trees using a Swedish Bark Gauge (see Figure 19).

This instrument consists of a blunt gauge with a shank graduated in millimetres. The gauge is thrust through the bark to the surface of the wood. The depth of penetration is measured on the shank at the end of the masking sleeve attached to a broad flange that is displaced as the gauge penetrates the bark. (The origin of the scale on the shank coincides with the end of the masking sleeve when the flange is in line with the distal end of the gauge.) When the gauge is forced through the bark, the flange with its attached sleeve is displaced an equal distance along the scale.

Alternatively, the bark either:

- has to be cut so that its thickness may be measured directly with a scale

or

- has to be removed at diametrically opposed points so that the diameter under the bark may be measured with calipers

or

- a band of bark completely encircling the bole has to be removed so that the girth under the bark may be measured with a tape

2.5.2 Calculation of Bark Percent

The bark percent of a tree or log may be found using the formula

$$\text{Bark percent} = \frac{(V_{ob} - V_{ub})\,100}{V_{ob}}$$

(N.B. As a % of total over bark volume)

Where:
V_{ob} = volume over bark
V_{ub} = volume under bark

FIGURE 19 The Swedish Bark Gauge

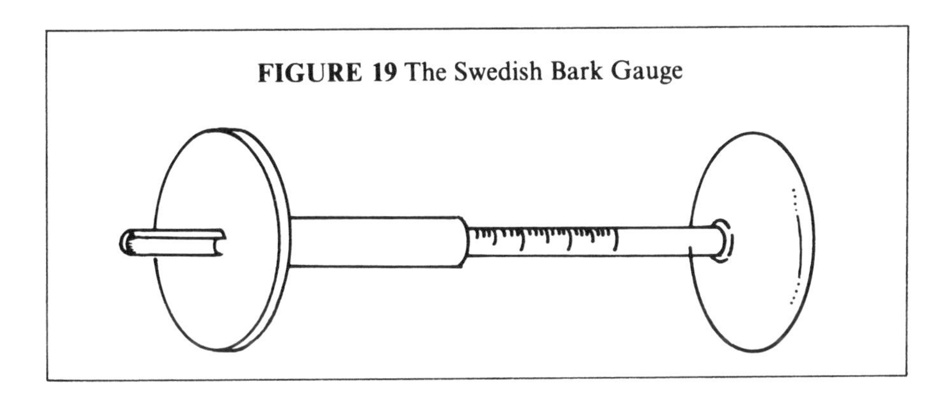

Assuming the form of a quadratic paraboloid, the bark percent of a log may be calculated using the formula

$$\text{Bark percent} = \frac{(d_{ob}^2 - d_{ub}^2)\,100}{d_{ob}^2}$$

Where:
d_{ob} = mid-diameter over bark
d_{ub} = mid-diameter under bark

(N.B. Expressed as a % of volume over bark.)

2.6 The Crown

The tree crown is the organ which supports the photosynthesizing tissue absorbing and employing radiant energy in the living processes. One of its main functions is to organize the position of the current photosynthetic area and provide for its renewal. Therefore one may expect that trees of a particular species with large crowns will grow faster than others of the same species with small crowns. In fact almost all the parameters of tree size and growth rate are closely correlated, i.e. diameter, height, crown size, bole volume, etc.

The commonest dimensions of crowns used in tree and crop mensuration are:

- Crown diameter (K) and crown area
- Crown depth or length
- Crown height – to lowest green branch or to lowest complete whorl of branches
- The ratio of the crown diameter (K) to the bole diameter (d), i.e. K/d
- Crown surface area
- Crown volume

The first four of these dimensions are illustrated in Figure 20.

2.6.1 Crown Diameter (K)

The task of estimating the vertical projection of the crown is quite as complex as estimating the cross-sectional area of the bole at breast height. Crown outlines are almost always irregular and branches of neighbouring trees may interlock, as for example in plantations of *Pinus patula* and *Burrtdavya nyassica*. With other species the crowns may be mutually antagonistic, as for example in plantations of *Cupressus lusitanica, Betula pendula* and most species of *Eucalyptus*. In part the antagonism appears to arise from sensitivity to injury caused when neighbouring crowns touch when blown in the wind.

As with boles, the area of a crown may be estimated employing the assumption that it is equivalent to the area of a circle defined by some average crown diameter. This average may be calculated in many ways, for example:

- the average of the maximum and minimum diameter

FIGURE 20 Crown dimensions

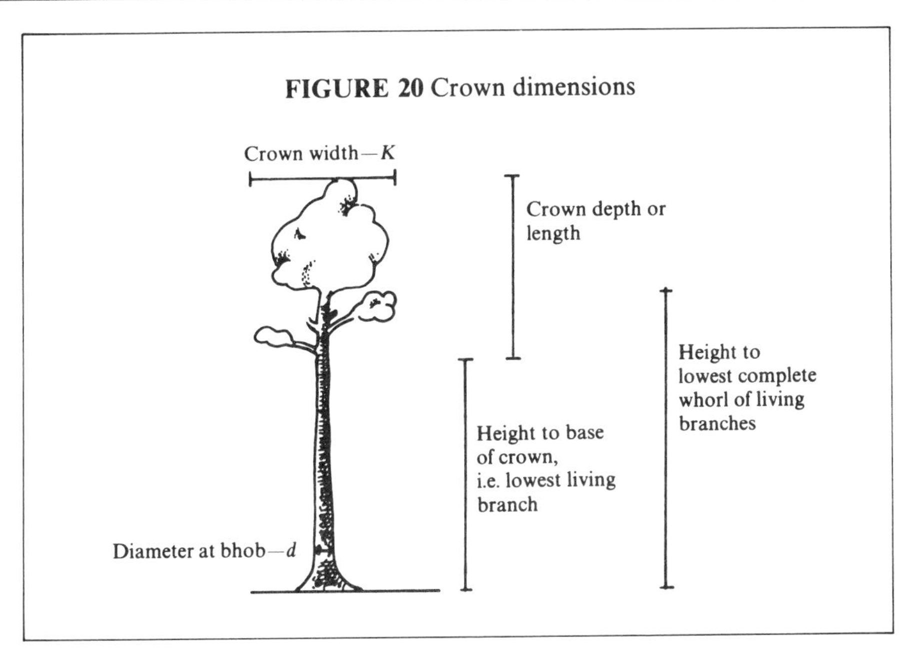

- the average of twice the maximum and twice the minimum radius from the centre of the bole to the periphery of the crown
- the average of the maximum diameter and the diameter at right angles to that maximum
- a single diameter chosen either at random or systematically with reference to a compass bearing
- the average of two diameters at right angles, the first chosen either at random or in some pre-determined direction, for example either towards the north or towards the plot centre

The vertical projection of the edge of a tree crown down to the ground is usually made with the aid of a mirror mounted on a gimbal with a sight that ensures a vertical reflection (Noordhin in Ayhan, 1977, unpublished). The projection is marked with pegs and the measurements taken with a tape. Also, similar measurements may be made on aerial photographs. However, the projections observed on the photograph and from the ground may not be identical because of masking by neighbouring crowns either from above or below. In some instances a negative bias has been attributed to measurements made on photographs (Dawkins, 1963). Measuring the diameter of individual trees may be impossible on small-scale photographs but the average crown size may be matched against a scale of circles of various diameters and, in this way, the average crown diameter estimated. Care must be taken to avoid personal bias.

2.6.2 Crown Depth and Crown Height

These parameters are normally measured on the standing tree using a hypsometer; individual workers have defined the terms in different ways and differently for conifers and hardwoods. In most definitions crown depth is the vertical distance from the highest growing point to lowest live foliage in the crown, although often epicormic shoots and occasional branches with a small amount of shade tolerant foliage are ignored. For conifers the lowest complete whorl of live branches is frequently used to define the lower margin of the crown. Crown height is the vertical distance from the lower margin of the crown to ground level.

2.6.3 The *K/d* Ratio

This ratio has been investigated by many research workers, often yielding different results for different species and in different parts of the world. *K* is usually measured in metres and *d* in centimetres. Interest in the ratio centres on the deduction of limiting basal areas for a given site, species and age that a rigid *K/d* ratio implies, and on its corollary of a predetermined number of trees of a given mean diameter.

Dawkins (1963) investigated 18 tropical forest species and deduced that for any given species the ratio was remarkably stable regardless of site, age or stocking. Other workers, e.g. Curtin (1964), found that for some species there was more variation. In part these different conclusions are a result of investigating different species, but in part they are a question of the accuracy of the measurements, the precision of the estimates and the percentage contribution to the variance that the particular worker deemed significant. However, all investigators agreed that the crown diameter and the breast height bole diameter are closely correlated for a given species and that one may be predicted from the other relatively precisely. The *K/d* ratio for any species, site, age and stocking is an important parameter that reflects many characteristics of the species. For example, Dawkins (1963) rated *Aucoumea klaineana* a more promising species for pure even-aged plantations than *Maesopsis eminii* because of the low *K/d* ratio of the former and, therefore, the feasibility of higher stocking in terms of basal area per hectare and the probability of higher volume per hectare growth rates.

Investigation of *K/d* ratios provides a useful means of checking the validity of yield tables and other growth models.

2.6.4 Crown Surface Area

As older leaves contribute little to the total net assimilate formed in a crown, many workers have deduced that the crown surface area will be correlated with that part of the total photosynthetic surface active in producing the net assimilate, and that therefore crown surface area should be a useful parameter in growth prediction. Usually the crown has been modelled as a cone whose surface area is given by the formula

$$a_c = \frac{\pi d_b L}{2}$$

Where:
a_c = crown surface area, m^2
d_b = diameter at base of crown, m
L = length of crown margin, m (as in Figure 13)

2.6.5 Crown Volume

A similar line of thought has led workers to conclude that crown volume should also be a useful predictor variable for individual tree growth. The crown volume of conifers and young hardwoods – especially *Eucalyptus* spp. – is estimated using the model of a cone:

$$v_c = \frac{\pi d_b^2 h_c}{12}$$

Where:
v_c = crown volume, m^3
d_b = diameter at base of crown, m
h_c = crown depth, m

2.6.6 Foliage

In some parts of the world tree foliage is an important product from the forest for livestock feed in months of scarcity. Its value depends upon various parameters including

- palatability
- digestible protein content
- energy content
- tannin and alkaloid content

The values for these parameters vary with species, age and season as well as between individuals. The ranking of species for value as a fodder is complex and must involve *in vivo* and *in vitro* experiments. Different fodders may be used by farmers to feed different types of stock or stock in different physiological conditions; for example, some fodders may promote milk production while others may assist in terminating lactation.

The harvesting of foliage is normally recorded in terms of green weight and has to be standardized to dry weight equivalents – as with other weights, see Section 2.4.3 (Wormald *et al.*, 1983).

2.6.7 Crown Position and Crown Form

Crown position and crown form (along with *dbh*) have been used to predict the diameter increment of individual tree growth in moist tropical forest (Wadsworth *et al.*, 1988). Dawkins (1958), reproduced with illustrations in Alder & Synott

(1992), described a crown position and crown form classification and scoring system:

Score	*Position*	*Form*
5	Emergent	Perfect
4	Full overhead light	Good
3	Some overhead light	Tolerable – distinctly asymmetrical or thin
2	Some side light only	Unsatisfactory with dieback and assymetry
1	No direct light received	Very poor, degenerating, dead

2.7 LOG AND TREE BOLE VOLUME

The volume of trees may be estimated either standing or felled. When the bole is subdivided into short lengths in a manner similar to that employed when the parts are destined for a saw mill, the parts are termed logs even when the subdivision is imaginary and the tree is still standing.

Foresters cannot easily measure the volume of the bole of a tree because it is irregular in both cross-section and profile. In fact the only feasible method of measurement is by water displacement in a xylometer. This is used only in special research projects and then usually for barked wood. The volume by displacement of rough-barked logs is usually considerably less than that estimated by other methods which give gross volume including that of the fissures in the bark. When a xylometer is employed the log volume is calculated from either the weight of the water displaced or the difference in the weight of the log measured in air and in water.

2.7.1 Log Volume by Direct Measurement

Volume estimation is based on measurements of diameter and length. These measurements may be made most accurately when the logs are separate and accessible to the measurer. Piles of logs cannot easily be measured accurately.

Logs are neither cylindrical nor often of any regular geometrical shape. Usually the model of a frustum of a quadratic paraboloid is adopted and the following formulae used to calculate volume:

$$v = \frac{\pi L(d_1^2 + d_2^2)}{8} \qquad \text{Smalian's formula}$$

$$= \frac{L(g_1 + g_2)}{2}$$

$$v = \frac{\pi L d_m^2}{4} \qquad \text{Huber's formula}$$

$$= L g_m$$

$$v = \frac{\pi L(d_1^2 + 4d_m^2 + d_2^2)}{24} \quad \text{Newton's formula}$$

$$= \frac{L(g_i + 4g_m + g_2)}{6}$$

Where:
d_1 = diameter at base of log, m
g_1 = cross-sectional area at base of log, m^2
d_m = diameter at mid-length of log, m
g_m = cross-sectional area at mid-length of log, m^2
d_2 = diameter at top of log, m
g_2 = cross-sectional area at top of log, m^2
L = log length, m
v = volume of log, m^3

All three formulae give the correct result for a frustum of a quadratic paraboloid and a cylinder. Newton's formula also applies for a frustum of a cone. (A frustum is that part of a solid of revolution bounded by two parallel planes at right angles to the axis of rotation.) However, if the log is not a frustum of a quadratic paraboloid and not a cylinder, then the use of either Smalian's or Huber's formula will introduce errors. For a conoid, i.e. a frustum of a cone, Smalian's formula overestimates the volume; Huber's formula underestimates the volume by half the amount of the overestimation with Smalian's. The errors given by both formulae are proportional to the length of the log and the square of the difference between the diameters at the two ends.

$$\text{Error} \propto L,(d_1 - d_2)^2 \text{ (see Appendix 2)}$$

i.e. the longer the log and the greater the taper, the greater the systematic error. (N.B. This relationship holds only for solids of revolution other than the cylinder and quadratic paraboloid.) Logs are rarely regular in shape so this relationship is only an indication of the expected errors.

Volumes of logs are normally estimated employing Huber's formula using logs that are as short as is practical. *N.B. Trees need not be cross cut before measurement.*

The choice of the positions to cross cut trees into logs depends upon the bole shape and taper and the markets. Conifers with a regular profile are cut so that the small end diameter and length suits the buyers. When the profile is irregular as in many hardwoods, large branches cause sudden changes in shape and taper; these breaks are called 'stops'. The stops dictate the points for cross cutting and the subdivisions for volume estimation.

When the volume of a single log has to be estimated accurately, Huber's formula should be employed and two diameters taken using calipers at the mid-length. When large numbers of logs are being measured then one diameter reading at the mid-point of each log is sufficient. When the mid-sectional area is calculated from a girth measurement, the tape has to be passed under the log. Care is needed to ensure that the tape is not twisted and is in contact with the log and has not included branches or other debris lying beneath the log. A 'measuring sword' is helpful in pulling the tape under heavy logs; it consists of a rigid but curved steel prong ending in a hook. The sword is pushed under the log, the tape attached to the hook and the whole withdrawn back under the log.

FIGURE 21 Cross cutting and timber measurement

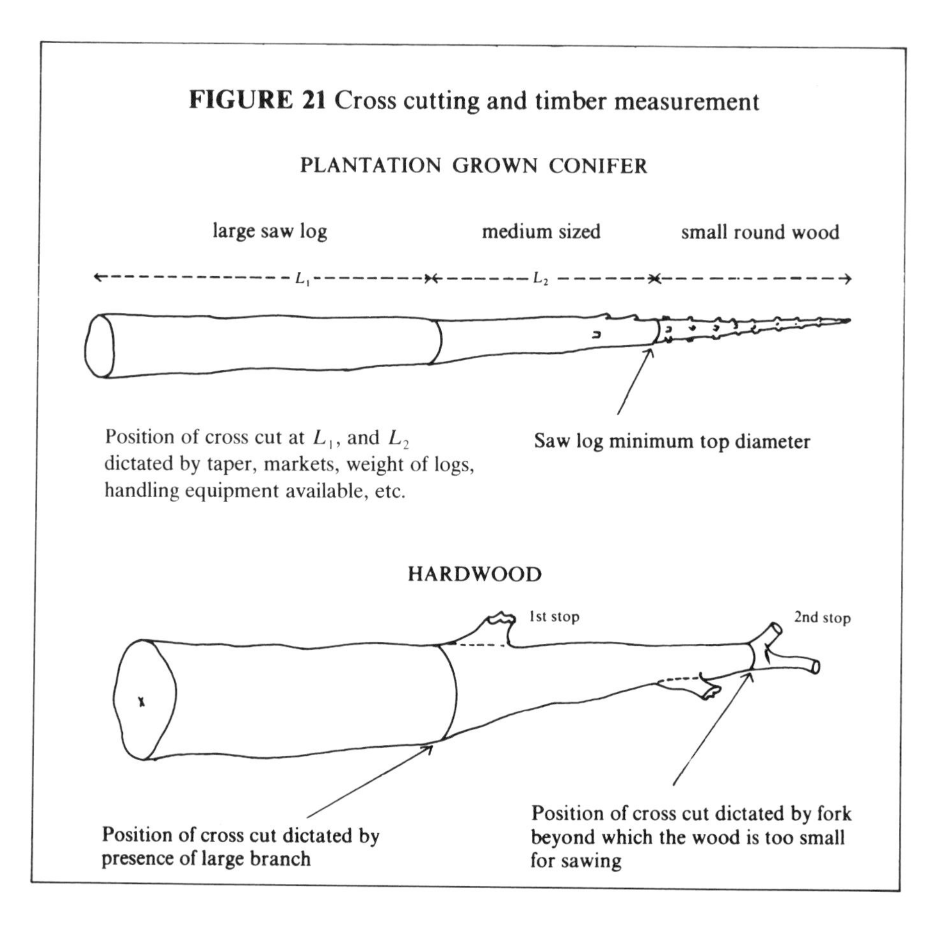

2.7.2 Tree Bole Volume Estimation by Direct Measurement

The volume of trees can be estimated either felled or standing.

2.7.2.1 Volume Estimation of Felled Trees

The bole volume of felled trees can be estimated by summing the volumes of its constituent logs. The tree bole need not be cut, but merely marked with chalk so that the length and mid-sectional diameters or girths of each section may be measured. Sometimes, especially in hardwoods, part of the bole may be unusable as, for example, when the heart rot has entered through a large branch stub. Either the gross value including the unusuable rejected portion or the net volume of usable timber may be required. An example is illustrated in Figure 23.

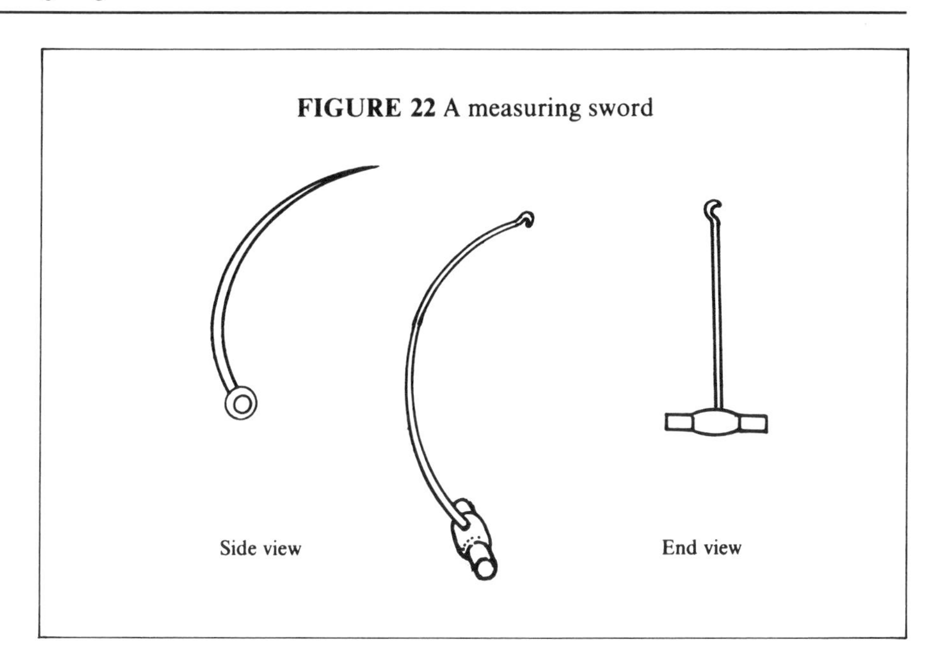

FIGURE 22 A measuring sword

FIGURE 23 Gross and net usable volume measurement

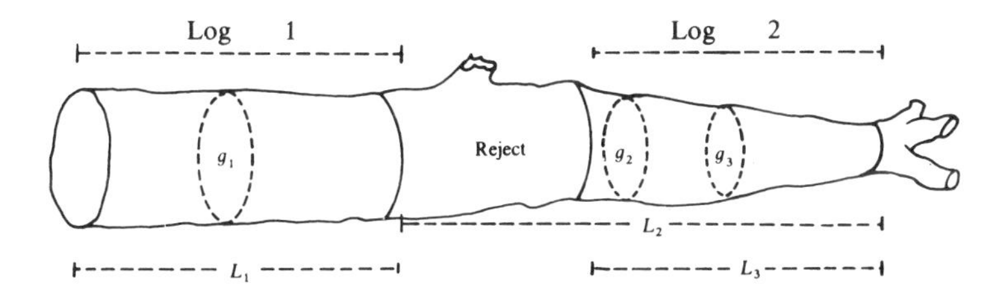

Gross volume $= g_1L_1 + g_2L_2$
Net volume $= g_1L_1 + g_3L_3$

g_1, L_1 = mid-cross-sectional area and length of log 1

g_2, L_2 = mid-cross-sectional area and length of remainder of bole

g_3, L_3 = mid-cross-sectional area and length of second log

When there is some irregularity of the bole at the mid-section:

- the average of the diameters or girths at equal distances on either side of the desired point of measurement avoiding the irregularity may be used; or
- the subdivision of the bole may be altered and shorter sections measured avoiding the irregularity; or
- measurements of diameters or girth may be taken at the ends of the log and Smalian's formula used.

Newton's formula is rarely used in practice.

2.7.2.2 Volume Estimation on Standing Trees by Direct Measurements

The girth or diameter at different points up the bole of a standing tree can be measured directly either by climbing or with a Barr and Stroud dendrometer, Spiegel relaskop, or other similar instrument.

2.7.2.2.1 Direct measurement of diameters by climbing

If a tree can be climbed, then its volume can be estimated by measuring the length and mid-sectional diameter of girth in exactly the same manner as when the tree has been felled. Even if only the lower part can be climbed, the volume of this part can be estimated using Huber's formula and the volume of the remainder estimated using either Smalian's formula or by employing the model of a cone. The most valuable log usually is the butt log and in a plantation-grown tree 40 cm dbh and 25 m tall, 80% of the bole volume is contained in the lowest 10 m. Therefore the error introduced by the assumption of a cone for the form of the unclimbed upper part of the tree may not be serious.

When Huber's formula is used for part of the bole and Smalian's formula for the remainder, then extreme care is needed to ensure that the cross-sectional area at the top of the last log measured by Huber's formula (that is the base of the first log measured by Smalian's formula) is observed. Smalian's formula is rarely used for the butt section of a standing tree because of the difficulty in estimating the cross-section at or near ground level and because it is less accurate than Huber's – particularly over the butt flare.

2.7.2.2.2 Direct measurement of diameters by an optical instrument

When climbing is impracticable and the outline of the bole is clearly visible, then diameters may be measured optically using either a Barr and Stroud dendrometer, Spiegel relaskop or the newer Telerelaskop. The Spiegel relaskop is described in Section 3.1.4. Detailed descriptions of the Barr and Stroud and Telerelaskop can be found in *Forestry Inventory* by Loetsch *et al.* (1973). These two latter instruments contain high quality optics and moving parts and are very expensive but, under ideal light conditions, are capable of high accuracy. The Spiegel relaskop and the wide angle adaption are less expensive but capable of reading only to values of the order of ± 1 cm diameter.

EXAMPLE 11 Calculation of the difference in the estimated volumes of a tree from measurements taken

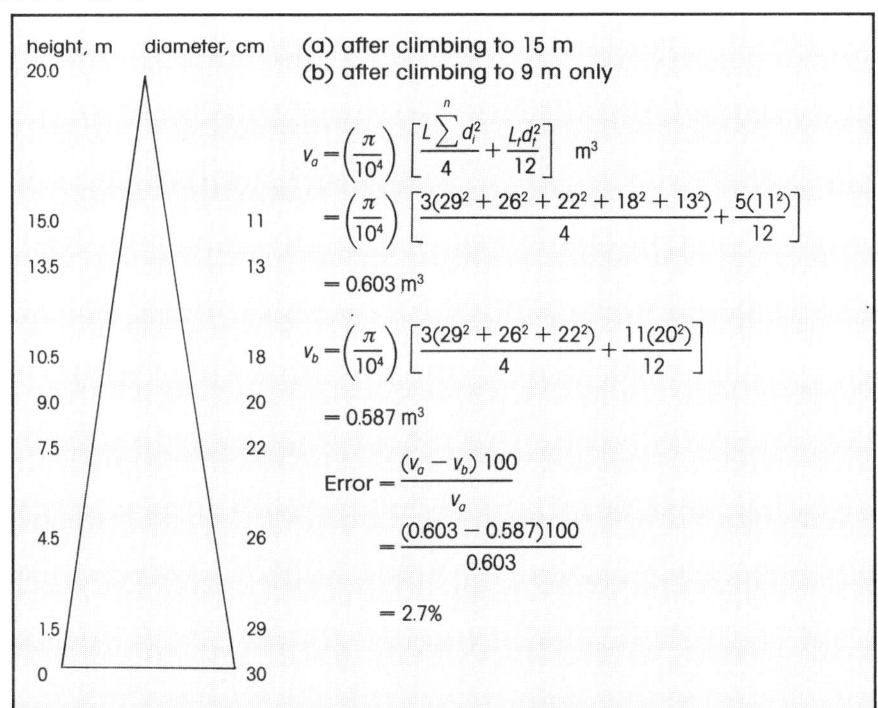

N.B. These calculations are based on values expected from a large number of measurements. Also the factor 10^{-4} is included to convert the basal area from cm^2 to m^2 because the diameters are in cm.

Diameters can only be measured with these instruments where the bole is clearly visible and not obscured by branches and foliage. Therefore the section lengths employed in the volume calculation will vary and care must be taken in determining the mid-length positions. As in climbing to a proportion of total height, the volume of the bole above the last section may be estimated using Smalian's formula. In conifers the position of the upper merchantable limit, e.g. 7 cm dob, may be estimated from an empirical formula such as:

ht to 7 cm = (Total ht − 3 m)

Further reading

ON OPTICAL INSTRUMENTS

Bitterlich, W. (Undated) *Spiegel-Relaskop and Telerelaskop.* FOB, Postfach 12, A-5035 Salzburg, Austria.

Bitterlich, W. (1984) *The Relascope Idea.* CAB International, Wallingford.

Findlayson, W. (1969) *The Relascope.* FOB, Postfach 12, A-5035 Salzburg, Austria.

FIGURE 24 Booking sheet for volume measurements

Name: A N GUNDI Forest: LWEYO Compt.: 19 Date: 28/2/82 Species: PINUS PATULA

Tree No.	Diam. at bh cm	Total Height m	Height to dob			Secn. No.	Length m	Ht to mid. pt. m	dob at mid. pt. cm	Bark Thick-ness mm	Mid Secn. Area ob m²	Mid Secn. Area ub m²	Vol. ob m³	Vol. ub m³	Bark	Tree Form Factor	Remarks
			7 cm m	14 cm m	24 cm m												
1	30.0	22.5	19.5	14.5	7.5	1	3	1.5	29.1	12	0.0665	0.0560	0.200	0.168	16		
						2	3	4.5	26.3	8	0.0543	0.0479	0.163	0.144	12		
						3	3	7.5	21.7	7	0.0370	0.0324	0.111	0.097	13		
						4	3	10.5	16.9	6	0.0224	0.0194	0.067	0.058	13		$\sum^{4}$ logs =
						5	3	13.5	12.5	5	0.0123	0.0104	0.037	0.031	16		ob 0.541
						6	4.5	17.2	9.2	4	0.0066	0.0055	0.020	0.017	15		ub 0.467
									TOTAL ob to 7 cm dob				0.598	0.515	14	0.365	
									Volume of 4 lower logs				0.541	0.467			
						5	2.5	13.2	12.6	5	0.0125	0.0106	0.031	0.026	16		
									TOTAL volume ob to 14 cm dob				0,572	0.493	14		
									Volume of lower log				0.200	0.168			
						2	4.5	5.2	24.5	7	0.0471	0.0419	0.212	0.189	11		
									TOTAL volume ob to 24 cm dob				0.412	0.357	13		

N.B. diameters read to nearest mm; height read to nearest 0.1 m; bark thickness read to nearest mm; basal area to 4 decimal places; volumes calculated to 3 decimal places.

2.7.2.3 *Booking Measurements of Felled or Standing Trees for Volume Estimation*

A suitable booking sheet is shown in Figure 24. The aims of using such a form are common to most such operations, namely:

- to facilitate the field work and minimize errors, e.g. to ensure that the diameters are measured at the correct locations
- to facilitate checks on the accuracy of the field work
- to ensure that all the required information is recorded
- to facilitate calculations and minimize the likelihood of errors in calculation, copying and preparation of data for computer input
- to ensure that the data can be identified

FOR THE ADVANCED STUDENT

2.7.2.4 *Tree Volume by Height Accumulation*

The volume of a tree can be calculated from the lengths of logs whose end diameters differ by a constant amount or taper step. The measurements may be taken conveniently with a Spiegel relaskop or similar instrument. The observer stands at a convenient distance from the tree so that he may measure heights easily. He selects a convenient taper step that he can perceive and measure. For example, standing 10 m distance a taper step of 5 cm is easily measured with the relaskop. The measurer views the bole of the tree and selects the lowest point on the bole where the diameter is an exact multiple of the taper step; for example if breast height diameter is 32 cm he selects the point where the bole is 30 cm. The height above ground to this point is recorded; the readings are repeated for the point at each taper step up the bole, i.e. at 25 cm, 20 cm, 15 cm, 10 cm, 5 cm, so that a total of six heights are recorded. These data form the basis of the calculation of volume. (See Appendix 3 for derivation of formulae.)

Then

$$v = \left(\frac{\pi T^2}{4}\right) \sum^{n} \left(i^2 L_i + iL_i + \frac{L_i}{k}\right)$$ excluding the cone above T m top diameter ob

or

$$v = \left(\frac{\pi T^2}{2}\right) \left(\sum^{n'} a_j + \frac{\sum^{n} L_i}{2k}\right)$$ excluding the cone above T m top diameter ob

$$v_t = \left(\frac{\pi T^2}{2}\right) \left(\sum^{n} a_j + \frac{\sum^{n} L_i}{2k} + \frac{L_0}{6}\right)$$ including the cone above T m top diameter ob

Where:

v = volume of tree to T m top diameter ob

v_t = total volume of tree including cone above T m top diameter ob

n = number of logs measured, excluding the cone
n' = number of taper steps to a top diameter of T
L_i = length of ith log, m $i = 1 \ldots n$
T = taper step, m
a_j = accumulated height to top of jth log, m $j = 1 \ldots m$
a_1 = ground level to top of first log
$a_2 = a_1$ + ground level to top of second log
$a_3 = a_2$ + ground level to top of third log, etc.
k = constant (= 2 for paraboloid, 3 for conoid, 4 for neiloid)
L_0 = length of cone

EXAMPLE 12 Calculation of volume by height accumulation

j	i	i^2	L_i	i^2L_i	iL_i	a_j – top diameter 5	15 cm
4	1	1	5	5	5	54	37
3	2	4	6	24	12	33	27
2	3	9	3	27	9	17	17
1	4	16	7	112	28	7	7
Total			21	168	54	111	88

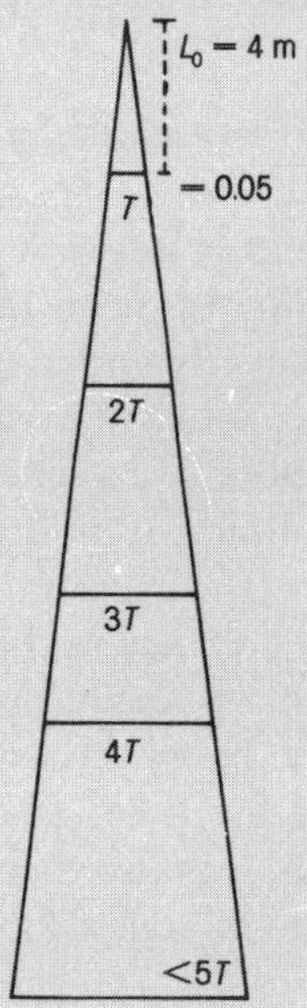

$$v = \frac{\pi(0.05)^2}{4}\left(168 + 54 + \frac{21}{3}\right) \quad \text{or} \quad v = \frac{\pi(0.05)^2}{2}\left(111 + \frac{21}{6}\right)$$

$$= (0.7854)(0.0025)(229.0) \qquad = (1.5708)(0.0025)(114.5)$$

$$= 0.4494\ \text{m}^3 \qquad = 0.4494\ \text{m}^3$$

$$v_0 = (1.5708)(0.0025)\left(114.5 + \frac{4}{6}\right)$$

$$= 0.4520\ \text{m}^3$$

N.B. $n = 4$
$n' = 4$

When the top diameter limit to volume is more than one taper step, for example 15 cm top diameter or $3T$, then a full set (n') of accumulated heights – a_j – are still needed, but the measurement 'ground level to top of the log' remains a constant.

Consequently in the example above, the calculation of the a_i for volume to a top diameter limit of 15 cm is:

a_1 = ground level to top of first log = 7
$a_2 = a_1$ + ground level to top of second log = 17
$a_3 = a_2$ + ground level to top of third log = 27
$a_4 = a_3$ + ground level to top of fourth log= 37

N.B. The distance 'ground level to top of third and fourth logs' remains 10 m as in a_2. L_1 is the length of bole between a diameter of T and $2T$ near the top of the tree, but a_1 is the accumulated height from ground level to the top of the first log.

$$V_{15} = \frac{\pi(0.05)^2}{4}\left(139 + 37 + \frac{10}{3}\right) \quad \text{or } V_{15} = \frac{\pi(0.05)^2}{2}\left(88 + \frac{10}{3}\right)$$

$$= (0.7854)(0.0025)(179.33) \qquad = (1.5708)(0.0025)(89.67)$$

$$= 0.35\ m^3 \qquad = 0.35\ m^3$$

N.B. $n = 2$

$n' = 4$

Where:

V_{15} = volume to a top diameter of 15 cm ob.

For a group of trees, separate volumes need not be calculated for each tree, the field data being accumulated for all trees. Providing that the taper step is small and therefore the length of each log is short, then the use of the form of a conoid with $k = 3$ is commonly employed. A practical difficulty, especially in buttressed trees, is the estimation of the number of taper steps at or near ground level.

Loetsch *et al.* (1973) give tables of constants with different taper steps and height units, but these impose an unnecessary restriction on the flexibility of the method. Grosenbaugh (1964), who initiated the system, has published sophisticated computer programs and less sophisticated routines for programmable calculators incorporating the height accumulation method of volume calculation for use in forest inventories.

2.7.3 Log Volume Estimation Using Tables

The breadth of a piece of sawn timber and, therefore to a limited extent, its value is governed by the small end diameter of the log from which it is cut. Consequently saw mills frequently measure the logs entering a mill by length and small end diameter and use log volume tables based on these two measurements. The volume may be calculated assuming either one rate of taper for all logs, or the taper may vary with species, log size and form.

2.7.3.1 Log Volume Tables Based on the Form of a Frustum of a Cone

Many saw mills use this form of log volume table. The rate of taper assumed is frequently 1:100, i.e. one centimetre change in diameter for each metre of length. The volume calculation for the frustum of a cone is based on the small end diameter (d_1), the length (L), and the large end diameter (d_2) derived from the rate of taper.

$$v = \left(\frac{\pi L}{12}\right)(d_2^2 + d_2 d_1 + d_1^2)$$

N.B. d_1, d_2 in metres.

EXAMPLE 13 Volume calculation – 1

For a log volume table based on the assumptions:

- form is a frustum of a cone
- taper is 1:100

$d_1 = 15$ cm $= 0.15$ m; $L = 5$ m; $d_2 = d_1 + L/100$ m
$d_2 = 20$ cm $= 0.20$ m

$$v = \left(\frac{\pi(5)}{12}\right)(0.20^2 + (0.20)(0.15) + 0.15^2)\ \text{m}^3$$
$$= 0.121\ \text{m}^3$$

Such tables are biased and usually underestimate the true volume because the form factor of a log is usually greater than that of a cone. Where saw mills chip waste wood, more accurate tables may be required in order to predict the outturn of chips.

2.7.3.2 *Log Volume Tables Based on Other Log Forms*

Compatibility between the estimates of log volume based on mid-length diameter and those based on small end diameter is convenient as it ensures that either basis provides the same estimate of volume. One way to do this is to calculate the volume from a mid-length diameter derived through a constant taper applied to the small end diameter. Obviously precise volumes will only be estimated for logs with the assumed form and taper.

$$v = \left(\frac{\pi L}{4}\right) d_m^2 \qquad \text{(Huber's formula)}$$

EXAMPLE 14 Volume calculation – 2

For a log volume table based on a constant taper to mid-length only.

- taper is 1:100

$d_1 = 15$ cm, $L = 5$ m; $\Rightarrow d_m = 17.5$ cm $= 0.175$ m

$$v = \left(\frac{\pi(5)}{4}\right)(0.175)^2\ \text{m}^3$$
$$= 0.120\ \text{m}^3$$

Alternatively the constant taper rate may be used to deduce the large end diameter and the form of a frustum of a quadratic paraboloid may be assumed. In this case the volume may be calculated using Smalian's formula.

$$v = \left(\frac{\pi L}{8}\right)(d_2^2 + d_1^2)$$

EXAMPLE 15 Volume calculation - 3

For a log volume table based on an average rate of taper over the whole log length:

- form is a frustum of a quadratic paraboloid
- average taper is 1:100

$d_1 = 15$ cm $= 0.15$ m; $L = 5$ m, $\Rightarrow d_2 = 20$ cm $= 0.20$ m

$$v = \left(\frac{\pi(5)}{8}\right)(0.20^2 + 0.15^2)\ \text{m}^3$$

$$= 0.123\ \text{m}^3$$

These three examples illustrate the variation in the volumes that may appear in log volume tables based on the same average taper but different bases of calculation. They are a useful but often biased form of volume ready reckoner. More complex log and tree volume tables based on a study of stem taper are dealt with under the section on single tree volume ready reckoners.

Conclusion

When a log volume table is required for use in a saw mill, the model using Huber's formula is recommended.

2.7.4 Tree Volume Estimation Using Functions and Tables

The object of a single tree volume table is to predict accurately the total volume of a number of trees without felling them, using measurements that can be obtained accurately, easily and cheaply. The phrase 'single tree volume table' contrasts with 'stand volume table'. The former predicts volume per tree and the latter volume per hectare. A single tree volume table is always used to predict the volume of a number of trees because the precision of the prediction of one tree alone is bound to be poor. An extract from a very simple volume table is shown in the example below.

EXAMPLE 16 Volume table for *Pinus caribaea* var. *hondurensis*

Location Turrialba, Costa Rica

$v = -0.0100 - 20.2623g^2 + 0.5693gh$

(0.6647) (0.0316)

Residual mean square = 0.009015 $F = 3378$

$R^2 = 0.983$ $\bar{v} = 1.062\ m^3$

Standard error of $\bar{v} = 0.0074\ m^3$ using 164 samples

v = total volume ob, m^3
g = basal area, m^2
h = total height, m

N.B. 'Fox tails' excluded from the population.
Standard errors of the partial regression coefficients are quoted in parentheses.

		Total height, m			
dbh, cm		20	22	24	26
20	$v =$	0.327	0.363	0.398	0.434
22		0.359	0.399	0.438	0.477
22		0.393	0.435	0.479	0.477
23		0.427	0.454	0.522	0.569
24		0.463	0.514	0.566	0.617
25		0.499	0.555	0.611	0.666

Range of data used in the construction of the table:
d 8.3 to 52.6 cm
h 6.9 to 31.0 cm

The precision of prediction is improved by restricting the population to which the tables refer. One such restriction is that volume tables normally refer to one species only or, occasionally, to two or three closely related species, e.g. *Eucalyptus camaldulensis* (Zanzibar variety) and *E. tereticornis*. Another restriction that may be applied is to limit the area of the population to a single locality.

The definition of the volume predicted has also to be stated. In the past bole volume of a size suited to the saw mills was a common definition, i.e. > 18 cm top diameter over bark; when there is a demand for small round wood then total volume to 5 cm dob may be the prediction.

There are many types of volume table and they can be classified into many different and overlapping categories. For example:

- general or local volume tables
- volume tables prepared graphically (subjectively) or by reference to mathematical models (objectively)
- tables using one, two, or three predictor variables

- tables predicting volume and tables predicting form factor, volume being derived in a subsequent calculation from $v = (ghf)$
- tables predicting the relationship between diameter and the distance down from the tip of the tree, i.e. a taper table, volume being then calculated by integration of the solid of revolution; these may be compatible with companion volume tables, or without such compatibility
- tables predicting volume from measurements taken in the field and those using measurements taken on aerial photographs

2.7.4.1 General and Local Volume Tables

This is an outmoded division – a relict of the past when volume table construction was an arduous, subjective operation involving measurements from thousands of trees. General volume tables were made for a whole country, e.g. the Standard Volume Table for *Pinus patula* published by the Tanzanian Forestry Division. Once a forest project or a region has an adequate range of tree sizes, then local volume tables should be prepared. General or standard volume tables should *not* be used in a small area without first testing for bias (see Section 2.7.4.3.4). Even within a project tables may be biased when applied to parts of the whole population from which the tables were derived; e.g. the tables may be biased when applied only to small trees. Also tables may have to be revised if younger tree crops are a different phenotype from those used in the tables' construction; such phenotypic differences may either reflect genetic differences introduced by varying the seed source or by tree breeding, or be caused by differences in silvicultural treatments such as pruning, spacing, etc.

2.7.4.2 Subjective and Objective Methods of Preparing Volume Tables

The graphical method of volume table construction is unsatisfactory because

- of its subjectivity; different designers using the same data will construct different tables
- the error in estimated volumes cannot be predicted

The second objection is the more serious especially when the lack of a predicted error leads to a false sense of the table's reliability. It also renders the choice of model and predictor variables a matter of personal choice. Before the application of probability theory and regression analysis to the problems of volume table construction, graphical methods using data from very large numbers of trees had to be used; recently more efficient and more objective methods have been used with advantages of

- provision of an estimate of the precision of prediction
- an objective method of choosing between different mathematical models
- cheapness as fewer data are needed to achieve a given precision

These advantages have been obtained at the cost of introducing a small degree of mathematical and statistical complexity. In the following sections only objective methods will be discussed because their advantages far outweigh this one disadvantage.

2.7.4.2.1 Techniques in volume table construction

The work of constructing a volume table consists of three parts:

- measuring the volumes of selected trees in a sample representing the population
- establishing relationships between the measurements taken on the tree and its volume – usually using regression analysis techniques
- choosing the best model and verifying the accuracy of the table constructed

Measuring the volumes of selected trees

Before trees can be selected the population has to be defined without ambiguity. This definition normally specifies:

- species and sometimes provenance
- geographical location
- size or age of trees – because extrapolation of the results outside the population sampled can introduce serious bias
- the qualitative characters of the population, i.e. whether edge trees, forked trees, trees with damaged crowns, leaning trees or abnormally grown trees such as 'fox tails', are included in the population

The aim of sampling is to select objectively trees that represent the relationships of volume to the easily measured parameters such as dbh, height, taper, buttress height, bole length, etc. Sampling units should be selected from all size classes by some form of stratification (Demaerschalk & Kozak, 1974, 1975; Marshall & Demaerschalk, 1986); in plantations age is a useful basis for this division. In extensive natural forest for practical reasons often the only feasible means is to sample trees felled in normal commercial operations; nevertheless the sampling design must endeavour to ensure adequate representation of the unusually small and large trees. The aim of sampling is not to estimate accurately the mean volume of the sample trees and of the population, but to represent accurately the relationships of volume and the predictor variables. Consequently as great a degree of objectivity as possible must be attained and if, for practical reasons such as access difficulties, part or parts of the population remain unsampled, then this must be stressed in the presentation of the results.

The number of sampling units required cannot be predicted before:

- the precision required is stated
- the mathematical models are selected, and
- the residual variance is estimated from a pilot trial

Consequently a preliminary sample of, perhaps, 30 trees must be measured; additional sampling units will have to be selected and measured if the desired precision has not been attained. In uniformly grown, pure, even-aged plantations of conifers 50 sample trees are likely to provide confidence limits of the mean within ± 10% at a probability of $p = 0.05$.

An independent sample is needed to verify the volume table, but the data may be collected in one field operation. These data that are to be used for verification may be concentrated in selected parts of the population – for example independent sets of 20 sample trees may be selected and measured in each of three classes

- the small trees
- the medium-sized trees
- the large trees

Then separate tests to verify the model can be performed independently on each (see Section 2.7.3.4). The sample trees are usually felled and measured as in Section 2.7.2.1.

Establishing relationships

A degree of subjectivity is introduced through the selection of the models to be analysed. This effect should be minimized by

- scanning the literature to compile likely relationships
- preparing simple graphical illustrations of the data to show relationships between pairs of parameters

e.g.

$$v \text{ on } g \text{ (or } d^2)$$

$$v \text{ on } d^2h$$

$$log(v) \text{ on } \log(d) \text{ etc.}$$

Such illustrations assist in selecting likely models for analysis.

The techniques of regression analysis are detailed in standard texts (Snedecor & Cochran, 1967, for example) and will not be repeated here, but see Section 5.1.1.8 for formulae. However, the assumptions of the analyses must be recalled as commonly they conflict with the actuality of the data. The main assumptions in the usual least squares estimates of regression parameters are:

1. The sampling units (trees) are selected independently.
2. The parameters measured on each unit are independent.
3. The variance in volume of trees with identical predictor variables is constant and independent of those predictor variables.

Also, in the estimate of standard errors and confidence limits it is assumed that:

4. The measurements are made without error.
5. The differences between the estimated and measured volumes are normally distributed with a mean of 0 and constant variance – estimated by the residual mean square of the regression analysis of variance.

Tree data rarely conform with assumptions 2,3,4 or 5. When possible volume table data should be transformed to reduce the heteroscedasticity (inconstant variance); often logarithmic transformations achieve this. Alternatively the weighted least squares estimates of the regression parameters should be calculated (Cunia, 1964; Snowdon, 1985).

Whatever the model fitted, illustrations of the residuals, that is volume measured − volume predicted, plotted against the measured volume should be displayed, so that any systematic pattern or bias may be detected and the extent of the heteroscedasticity revealed. Displays of this type are most effective in revealing bias (consistent under- or overestimation) in subdivisions of the whole population (see Figure 27).

Choosing the 'best' model and verifying the results

The interpretation of the results of linear regression analyses is not simple. Models must comply to basic conditions before being included in further analyses:

- the variance ratio (F) must be significant at the chosen level of probability
- a plot of residuals must exhibit
  - — no bias
  - — constant variance

The judgement of these conditions is usually subjective although objective statistical tests are feasible. Models that are accepted may be ranked by:

1. Their goodness of fit as measured by r^2 or R^2.
2. The standard error of the mean.
3. The sum of the mean squared differences between the measured and estimated volumes divided by the degrees of freedom for the residual variance.
4. Furnival's Index (Furnival, 1961) – F.I.

F.I. = (residual mean square)$^{0.5}$/geometric mean of y

If the models have been derived from the same data set, have the same predicted variable and the same number of degrees of freedom for the residual variance, then all criteria provide the same rankings. Otherwise criterion 3 and 4 should be used.

The model with the lowest index value is chosen. A short list of the most useful models should be assembled and each verified using the independent data sets collected at the time of the field sampling. Then the relative efficiency of the satisfactory models may be compared (see Section 2.7.4.3).

2.7.4.2.2 Merchantable volume tables

Volume tables are often derived from measurements taken from trees felled in the course of normal commercial harvesting operations, and the volumes recorded have been those derived from the current felling practices. These practices change as the technology of wood processing advances; for example, the development of particle board has encouraged the utilization of logs of very small top diameters. Presently the trend is towards constructing volume tables predicting total volume and deriving from these tables providing volumes to particular top diameters.

If the limit of utilization is a small top diameter, then the reduction from total volume is the volume of a small cone that is independent of tree size.

$$MV = TV - a$$

Where:
MV = merchantable volume
TV = total volume
a = volume of the cone above merchantable top diameter

When the limit for utilization of a particular product, such as a veneer or saw log, is defined by a relatively large top diameter, then such a simple relationship is inappropriate. Two strategies for developing volume tables in such cases are:

- to derive the volume tables through a taper function – see Section 2.7.4.5
- to derive merchantable volume from a function of total volume

Bruce *et al.* (1968) used complex multiple regression equations based on powers of diameter and height. If such a procedure is employed for more than a single table, care must be taken that they are logically compatible so that the merchantable volume to, say, 10 cm top diameter is consistently larger than that to 12 cm top diameter. An alternative approach is to fit a more flexible function such as

$$\frac{MV}{TV} = k + c(1 - e^{-a(x - x_0)})$$

Where:
MV = merchantable volume
TV = total volume
c, k = constants to be fitted
x = dbh
x_0 = value of dbh when $MV = 0$

that will produce a family of compatible curves as in Figure 25. (See also Section 2.9.)

Further reading

ON TECHNIQUES OF VOLUME TABLE CONSTRUCTION

Alder, D. (1980) *Forest Volume Estimation and Yield Prediction,* vol. 2. Yield prediction. FAO Forestry Paper No. 22, Rome.

2.7.4.3 One, Two and Three Parameters for Predicting Volume

The general formula for the volume of a tree is:

$$v = ghf = (\pi/4)d^2hf \text{ m}^3$$

Where:
d = diameter at breast height, m
g = basal area, m^2
h = total height, m
f = form factor

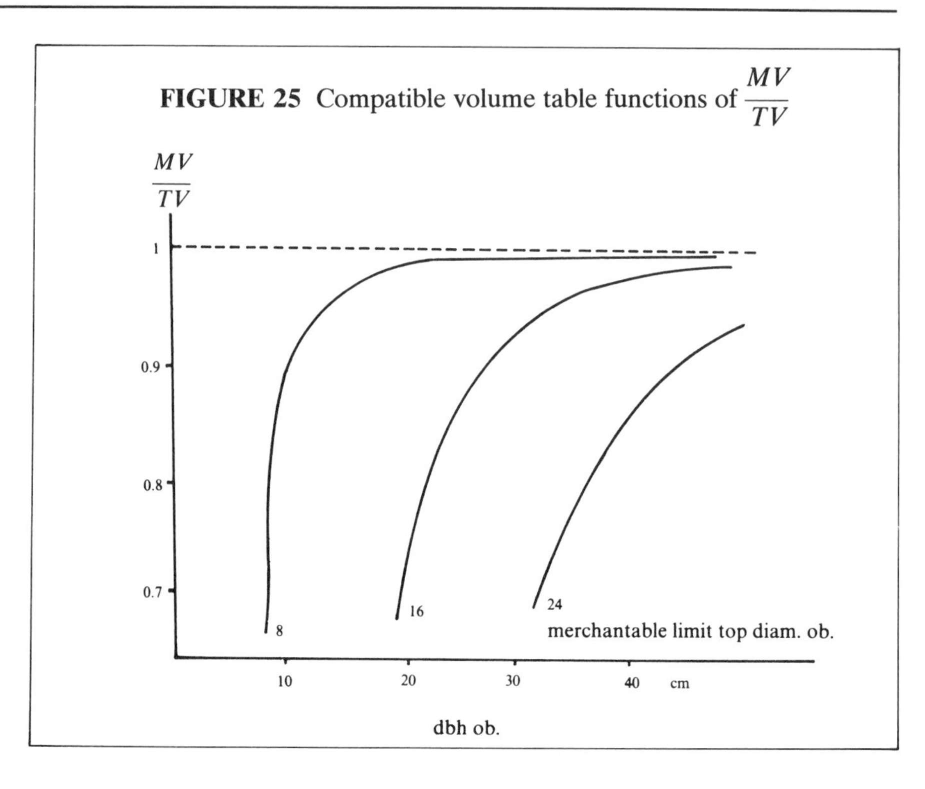

FIGURE 25 Compatible volume table functions of $\frac{MV}{TV}$

When using a single predictor variable or parameter, then g or d^2 is the obvious choice; diameter is a cheap, easy and precise measurement. For two predictor variables, diameter or basal area is allied to height – either total height, or timber height, or bole length. A second parameter will normally reduce the residual variance of the model; for example, two trees of the same dbh may vary greatly in volume because they have different form heights, but two trees of the same dbh and total height will have different volumes only if their forms are different. As the form factor can neither be measured nor calculated without first measuring the volume, form itself cannot be used as a third parameter. Its place is usually taken by some measure of the rate of taper over the lower and more accessible part of the bole. Trees of the same dbh, height and average taper from, say, dbh to 3 m above ground level may still have different volumes as they are irregular in shape, but this residual variance will be small and the estimate of the total volume of a given number of trees will be more precise than if only one or two parameters are used. However, the time taken for the measurements and therefore their cost will be greater than if fewer parameters are used. (N.B. The reduction in the residual variance with additional parameters depends upon the correlation of these additional variables with the residual variance in volume.)

2.7.4.3.1 One parameter volume tables

A frequently used mathematical model of a tree bole is a simple, linear relationship of volume on basal area:

$$v = a + b(g) \text{ or } v = a + b(d^2) \text{ the volume–basal area line}$$

Where:
v = volume over bark to 7 cm dbh diameter, m^3
g = basal area over bark, m^2
d = diameter ob at breast height, cm
a, b are constants to be estimated

This model is illustrated in Figure 26. It implies that

- trees of zero basal area have volumes to 7 cm of a, m^3. (N.B. a is often negative)
- as the general formula for volume is $v = ghf = a + b(g)$, then $v/g = hf = a/g + b$, then as $a \rightarrow 0$ so b approaches a constant form height; therefore, as height tends to vary with basal area, form must vary inversely with basal area, i.e. as trees increase in basal area and height the form factor decreases

Both these conditions imply bias in the table.

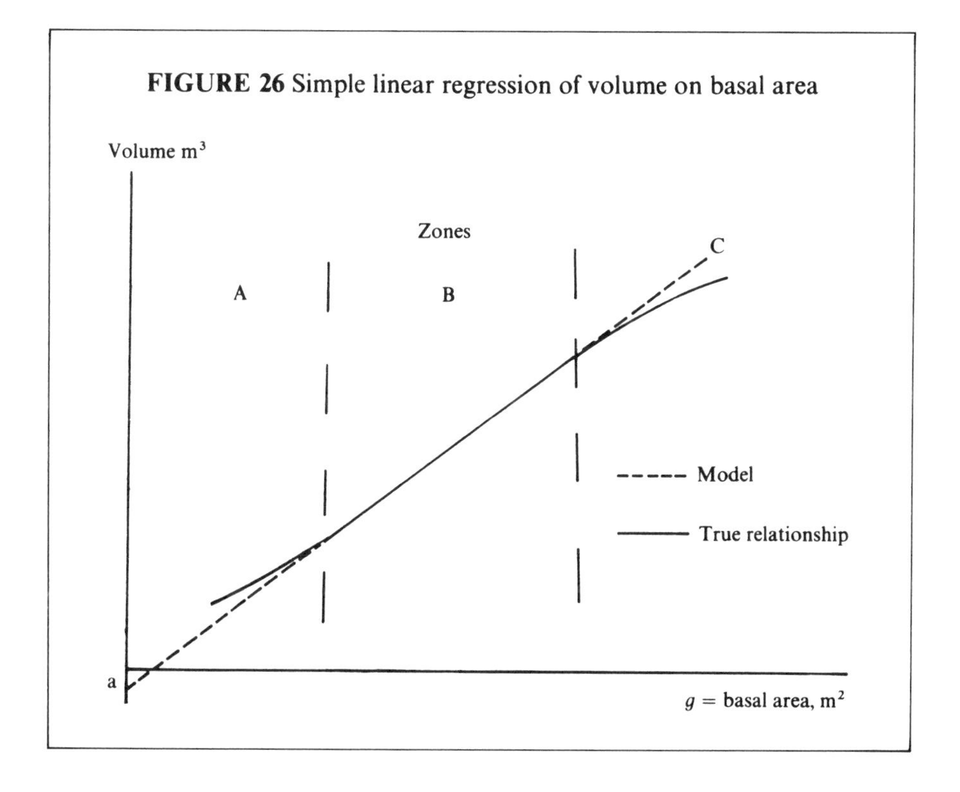

FIGURE 26 Simple linear regression of volume on basal area

These implications of the model are a severe constraint on its application; if used over the entire range of tree sizes normally found in a softwood plantation from the stage of first thinnings to the final felling, considerable bias will be introduced. The prediction of volume for small trees (zone A in Figure 26) is often the source of the bias which is introduced by the arbitrary definitions of the height above ground for the cross-section g. On very small trees breast height is a greater proportion of total height than on larger trees causing an underestimate of volume; an arbitrary definition of volume, as for example to 7 cm top diameter, means that small trees can have no volume and have to be excluded from the population. As a result, the constant a is frequently found to be negative, increasing the bias in the estimate of small trees (zone A). This bias is then reflected in balancing biases in zones B and C. Zone C reflects the situation when height growth slows and the crown of the tree develops to provide flowering and fruiting sites. For these reasons such a simple model has to be restricted to a small range of diameters as in a specific stand at a specific age.

An alternative and much used model is:

$$v = ad^b \text{ or } \log(v) = \log(a) + b\log(d)$$

It is usually the best single parameter model for conifers. More complex models for a one parameter table are quite feasible; for example, a polynominal relationship of volume on increasing powers of diameter:

$$v = a + b_1 d + b_2 d^2 + b_3 d^3 \ldots$$

In spite of their limitations, provided that the constraints are understood and accepted and the possibility of bias kept in mind, one parameter volume tables are extremely useful and, allowing for costs of measurement, relatively very efficient.

2.7.4.3.2 Two parameter volume tables

These are probably the commonest form of volume table. Loetsch *et al.* (1973) list eleven common models; they fall into three main categories:

- The simple combined variable model:

$$v = a + b(d^2h)$$

- Multiple regressions with powers of d, h and gh:

$$v = a + b_1(d^2) + b_2(h) + b_3(d^2h) + b_4(h^2) + b_5(d^2h^2) \ldots$$

- Logarithmic models:

$$v = ad^{b_1}h^{b_2} \text{ or } \log v = \log(a) + b_1(\log d) + b_2(\log h)$$

Where:
d = diameter at breast height, m
g = basal area, m^2
h = height, m
v = predicted volume, m^3

Models from all three categories have been used frequently; more recently the logarithmic models have proved to be useful for many species from miombo woodland (Temu, 1980). When calculating a logarithmic model (and given v_m = measured volume, in m^3), the ordinary least squares solution gives a slight bias as the sum of the deviations ($\log v_m - \log v)^2$ is minimized and therefore the sum of the deviations ($v_m - v$) is not zero. Usually this bias is too small to be of practical consequence. However, the curvi-linear model can be used and the powers b_1 and b_2 calculated by an iterative process on modern computers if the bias is serious.

Similar models can be used for log volume tables on the log length and diameter at the smaller end.

Example 17 illustrates the form of one and two parameter volume tables. If the extra parameter eliminates bias and enables one model to represent a wide range of tree sizes, two parameter tables may be better than one parameter ones. Their disadvantage is in the higher cost of application because heights have to be measured. The high cost may result in a lower relative efficiency – that is a higher cost of obtaining an estimate with a certain precision compared with the cost using a one parameter table.

2.7.4.3.3 Three parameter volume tables

The common models are similar to the two parameter ones:

$$v = a + b_1 d^2 h + b_2 q$$

or

$$\log v = a + b_1(\log d) + b_2(\log h) + b_3(\log d_i)$$

Where:
h = height, m
d = diameter at breast height, m
d_i = diameter at height i above ground level, m
q = d_i/d

EXAMPLE 17 Calculation of volume tables

(a) *One parameter*

Model: $v = a + b(g) = 10.00(g) - 0.04$*
Species: *Pinus patula*
Population: Trees 15-35 cm dbh in Kiwira Forest Project in Tanzania

dbh ob cm	g m^2	v to 7 cm dob m^3
15	0.018	0.14
16	0.020	0.16
17	0.023	0.19

When:
d = 15 cm
g = 0.018 m^2
v = 10.00g − 0.04 = 0.14 m

Example 17 continued

(b) *Two parameters*

Model: $v = 0.03 + 0.41(gh)$
Species: *Cupressus lusitanica*
Population: Trees 15-40 cm dbh in Meru Forest Project, Tanzania

dbh ob cm	g m^2	Volume to 7 cm dob, m^3 Total height, m 10	12	14	16
15	0.018	0.10	0.12	0.13	0.15
16	0.020	0.11	0.13	0.14	0.16
17	0.023	0.12	0.14	0.16	0.18

*If the model $v = a + b_1 d^2$ is used when d is entered in centimetres then

$$b_1 = \frac{\pi b}{(4)(10^4)} = (0.00007854)b$$

When:
$d = 15$ cm
$g = 0.018$ m^2
$h = 10$ m

$$v = 0.03 + (0.41)(0.018)(10) = 0.10 \text{ m}^3$$

Volume increases $(0.41)(g)$ for every metre increase in height.
When $g = 0.018$ m^2, volume increase = 0.007 for each metre increase in height.

Loetsch *et al.* (1973), quoting Schmid's work on spruce *Picea abies,* suggest that the second diameter should be taken as high as is practicable – at least 5 m from the ground. This raises the cost of applying the tables and usually reduces their relative efficiency unless more simple unbiased models cannot be designed. The tables are calculated for different q values that are taken as the mid-point of a form class. For example $q = 0.85$ may be taken as the mid-point of a form class covering the range $0.825 < q < 0.875$. These tables are not common but have been used in inventories of tropical high forest in order to avoid making separate tables for each species.

FOR THE ADVANCED STUDENT

2.7.4.3.4 Bias in and efficiency of volume tables

The usefulness of a volume table is measured by two important characteristics – bias and efficiency. A volume table should not be biased. Whatever the size of the trees whose volume is to be predicted, the expected values of their predicted volume and their measured volumes should be the same. The more restricted the definition of the population to which a particular table is applicable, the less likely is bias to be serious; this is true whether the bounds of the population are defined by tree volume, location or site quality. If a bias is found to be consistent, then an appropriate correction can be applied to the predicted values. In practice a bias of less than 2% is rarely worth correcting whereas one of 5% may be. The decision depends upon the penalty of a faulty estimate. A 5% shortfall in revenues from a forest could well affect both the profitability of the forest and any industry receiving its produce and cause misallocation of additional capital.

A test for bias is made by subdividing the population and examining the deviations between the measured (v_m) and predicted values (v) within each sub-population separately. Such a test must be done on independently collected samples, and not on the data from which the volume table was originally constructed.

Obviously v_m must be measured accurately on felled trees. Regularity in the pattern of deviations, as shown in Figure 27, is an indication of bias.

FIGURE 27 Graph of $(v_m - v)$ on v_m

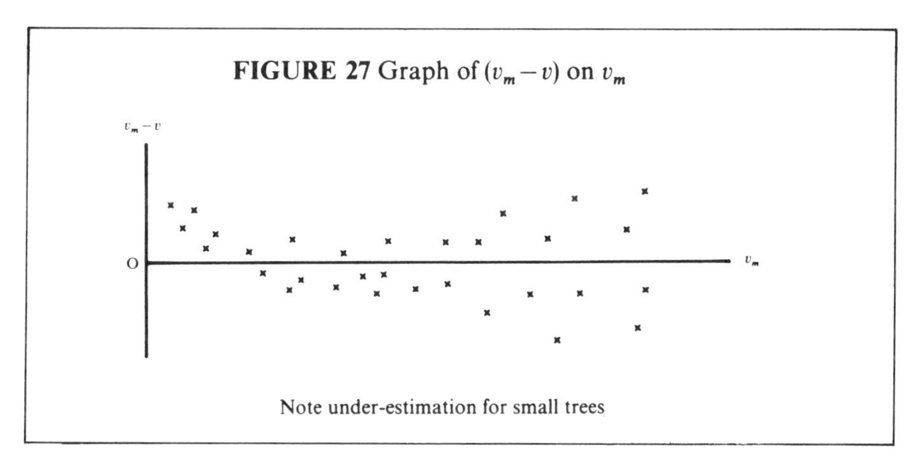

Note under-estimation for small trees

Figure 27 also shows that the variance of the deviations is not homogeneous. This is commonly the case and it makes the paired t test inappropriate. The χ^2 test is also inappropriate, it ignores the sign of the differences although it does scale the differences. Also the observations and estimates are not samples of a frequency distribution but refer to a single individual in the population. The more appropriate tests are non-parametric – for example, either the sign test or, more precisely, the Wilcoxon paired observations, rank sign test (Dawkins, 1975; Daniel, 1978). More sophisticated tests examine the independence of the deviations.

For the Wilcoxon test, the measure $(v_m - v)$ is calculated for each tree in a set of independent data. These residuals are ranked regardless of their sign, starting with the smallest. When two or more are identical, disregarding the signs, they are given the same rank – that is the average of the ranks they would have received if different but less than the next largest residual. In the example below, 0.004 occurs twice after rank 7; therefore they are ranked equal at 8.5 which is the mean of the two ranks 8 and 9. The ranks are then assigned the same sign as the corresponding residual.

The null hypothesis of the test is that the sum of the positive ranks less the sum of the negative ranks equals 0. This is tested by z calculated as shown, which for large samples is normally distributed.

EXAMPLE 18 The Wilcoxon paired observations rank sign test of bias

$$H_0: \sum^{N} \text{(positive ranks)} - \sum^{N} \text{(negative ranks)} = 0$$

V_m	V	$V_m - V$	Rank
0.127	0.126	0.001	2
0.624	0.625	−0.001	−2
0.209	0.208	0.001	2
0.166	0.163	0.003	5.5
0.418	0.421	−0.003	−5.5
0.756	0.753	0.003	5.5
0.173	0.170	0.003	5.5
0.685	0.681	0.004	8.5
0.780	0.784	−0.004	−8.5
1.185	1.191	−0.006	−10.5
0.191	0.197	−0.006	−10.5
0.656	0.647	0.009	12
0.583	0.572	0.011	13
0.864	0.876	−0.012	−14
0.848	0.834	0.014	15.5
1.124	1.138	−0.014	−15.5
0.514	0.529	−0.015	−17
0.889	0.910	−0.021	−18.5
0.872	0.851	0.021	18.5
1.073	1.045	0.028	20.5
0.996	0.968	0.028	20.5
1.055	1.024	0.031	22
1.300	1.268	0.032	23.5
1.155	1.123	0.032	23.5
1.236	1.199	0.037	25.5
1.114	1.077	0.037	25.5
1.175	1.135	0.040	27.5
1.257	1.217	0.040	27.5
1.064	1.018	0.046	29
23.089	22.750	0.339	

Difference = 23.089 − 22.750 = 0.339

$$\text{Bias} = \frac{0.339 \cdot 100}{23.089} = 1.5\%$$

$$T = \sum^{29} x_i = 102.0$$

$$N = 29$$

Where:

x_i = ranking of the ith difference in the smaller of the two sums of ranks

$$z = \frac{T - \frac{N(N+1)}{4}}{\sqrt{\frac{N(N+1)(2N+1)}{24}}}$$

$$= -2.50$$

The probability of obtaining a $|z|$ of $|2.50|$ under the terms of the null hypothesis is less than 0.025, i.e. so low that we reject the hypothesis and conclude that the table is biased.

Given that a one and a two parameter table are both unbiased, then their relative efficiency has to be judged by the costs of using them to obtain estimates of equal precision, i.e. the better table is that which being unbiased gives the cheaper estimate of volume of a given precision or within given confidence limits (CL).

EXAMPLE 19 Comparison of the eficiency of two unbiased volume tables - *A* and *B*

	A	B
Residual variance of volume table	50	20
No. of trees to be measured to provide CL of ± 10% of the mean v	75	30
Time taken to measure the predictor parameters per tree and estimate its volume	1 min	3 min
Total time for prediction with CL of ± 10% $C_A N_A$, $C_B N_B$	75	90
Relative efficiency ($C_A N_A$: $C_B N_B$)	5	6

Where:
C_A, C_B = cost per tree for Table A, B
N_A, N_B = numbers of trees needed to achieve CL of ± 10% of the mean v
N.B. The relative efficiency of 5:6 means that A is more efficient than B

2.7.4.4 Volume Tables Derived from Predictions of Form

One of the assumptions of the least squares solution for the estimation of regression parameters is that the residual variance is homogeneous and therefore is independent of the prediction. This assumption is rarely valid when volume is the predicted variable; a procedure for weighting the individual deviations is recommended (Cunia, 1964). Evert (1969) suggested that the variance of form factor is homogeneous and independent of the volume prediction and advocated its use as the dependent or predicted variable. Common models predicting volume can be converted by dividing both sides by gh thus:

$$v = a + b(g)$$

$$v/gh = a/(gh) + b/h = f$$

Evert (1973), in a test on *Pinus resinosa,* found that the introduction of the third variable of form class increased the precision of estimates. The equation with the smallest standard error of estimate in his test was:

$$f = a + b_1(d_i/d) + b_2(d_i/d)^2 + b_3(h/(h - k))^2 + b_4(d_i/d)(h/(h - k)) + b_5(d_i/d)^2(h/(h - k))^2$$

Where:
d_i = diameter at height i above ground level
d = diameter at breast height
h = total height
k = breast height

Volume has to be calculated and tabulated using the formula

$$v = ghf$$

This tabulation is done for mid-points of diameter, height and form classes as for the normal three parameter tables. The availability of computer programs for weighted least squares regression estimates has reduced the need for and use of this type of model.

FOR THE ADVANCED STUDENT

2.7.4.5 Mathematical Models Relating d and h with Volume Calculated by Integration

A very flexible system of volume estimation is obtained by developing a mathematical model reflecting the relationship between diameter and height. Once this has been established then the volume of the whole bole or any part of a bole defined by limits of either diameter or height can be calculated by integration of the solid of revolution. The limiting diameters and heights are expressed as quotients of dbh and total height from the tip to ground level. As with logarithmic models, the tables are slightly biased as the tree with the average profile is not identical with the tree of average volume. The extent of this bias has to be investigated before a table is put into use.

An often used taper function is

$$(d_i/d)^2 = a + b_1(h_i/h) + b_2(h_i/h)^2$$

Where:
h_i = height i metres above ground
d_i = diameter at height i metres, cm
h = total height, m
d = diameter at breast height, cm
(Kozak *et al.*, 1969)

Often d_i is an under bark measurement. In this model a must be set equal to $(-b_1 - b_2)$ to ensure that $d_i = 0$ when $h_i = h$, so

$$\left(\frac{d_i}{d}\right)^2 = -b_1 - b_2 + b_1\left(\frac{h_i}{h}\right) + b_2\left(\frac{h_i}{h}\right)^2$$

$$= b_1\left[\left(\frac{h_i}{h}\right) - 1\right] + b_2\left[\left(\frac{h_i}{h}\right)^2 - 1\right]$$

The constants b_1 and b_2 can be estimated using the least squares technique. Kozak *et al.* (1969) give the derivatives and the formulae for estimating volumes of logs to specified diameters and lengths, a diameter for a given height, or a height for a given diameter.

EXAMPLE 20 Calculation of a taper equation and derived volumes (much simplified)

Given the following independent observations made on a random sample of eight trees from a population

x_1 dbh cm	x_2 h m	x_3 d_i cm	x_4 h_i m	x_5 x_4/x_2 Rel.ht	x_6 x_3/x_1 Rel.dia.	x_7 $(x_6)^2$
36	22	9.0	1.1	0.05	0.25	0.06
13	15	4.4	2.1	0.14	0.37	0.12
34	23	17.0	5.3	0.23	0.50	0.25
25	20	15.5	8.0	0.40	0.62	0.38
19	18	14.2	9.9	0.55	0.75	0.56
15	17	13.0	13.9	0.82	0.87	0.76
40	26	40.0	26.0	1.00	1.00	1.00
29	22	26.1	19.8	0.90	0.90	0.81

Regression analysis

$$\left.\begin{array}{l} x_5 = -0.007 + 1.055x_7 \\ x_7 = 0.002 + 0.957x_5 \end{array}\right\} \begin{array}{l} r^2 = 0.99 \\ r = 0.99 \end{array}$$

N.B. Owing to sampling errors these lines do not pass through either the origin or the point (1,1).

$$\text{'Relative } v\text{'} = \frac{\pi}{4}\int_0^1 \left(\frac{d_i}{d}\right)^2 \delta\left(\frac{L_i}{h}\right)$$

$$= 0.7854 \int_0^1 \left(0.957\left(\frac{L_i}{h}\right) + 0.002\ \delta\left(\frac{L_i}{h}\right)\right)$$

$$= 0.7854 \left[\frac{0.957}{2}\left(\frac{L_i}{h}\right)^2 + 0.002\left(\frac{L_i}{h}\right)\right]_0^1$$

$$v = \frac{0.7854}{10^4}\left[\frac{0.957}{2}\left(\frac{L_i}{h}\right)^2 + 0.002\left(\frac{L_i}{h}\right)\right]_0^1 d^2h \qquad \text{m}^3$$

Example 20 continued

Where:

v = total volume, m^3
d = dbh, cm
h = height, m
L_i = distance i from top of tree, m
d_i = diameter at L_i, cm

The total volume of a tree 25 cm dbh and 20 m total height is:

$$v_t = (0.785)(0.478 + 0.002)(0.0625)(20)\ m^3$$
$$= 0.47\ m^3$$

The volume of the same tree between 5 and 15 m from the top is:

$$v_{(5-15)} = (0.785)[((0.478)(0.75)^2 + (0.002)(0.75)) - ((0.478) \times (0.25)^2 + (0.002)(0.25))]$$
$$= (0.785)[((0.478)(0.75^2 - 0.25^2)) + ((0.002)(0.75 - 0.25))]$$
$$= (0.785)(0.239 + 0.001)$$
$$= 0.19 \text{ in relative terms}$$

or in units of m^3

$$= (0.19)(0.0625)(20)\ m^3$$
$$= 0.24\ m^3$$

The volume of a tree between two stated diameters may be obtained by calculating the corresponding (L_i/h) using the first equation:

$$(L_i/h) = (1.055(d_i/d)^2 - 0.007)$$

The volume of the same tree as above between 12 and 21 cm diameter is:

$$v_{(12-21)} = (0.785)[((0.478)(\{1.055(21/25)^2 - 0.007\}^2 - \{1.055(12/25)^2 - 0.007\}^2]) + [(0.002)(\{1.055(21/25) - 0.007\} - \{1.055(12/25) - 0.007\}])]$$
$$= (0.785)[(0.478)(0.543 - 0.062) + (0.002)(0.737 - 0.250)]$$
$$= (0.785)[(0.230 + 0.001)]$$
$$= 0.181$$

or in terms of m^3

$$= (0.181)(0.0625)(20)\ m^3$$
$$= 0.22\ m^3$$

However, the regression taper equations as used in the example are subject to sampling errors. Consequently estimates of volume obtained using both the diameter and the height regressions will not be identical. In this example when

$(d_i/d) = 1$ then (L_i/h) is predicted to be 1.048; similarly when
$(L_i/h) = 1$ then (d_i/d) is predicted to be 0.959.

Many authors have used complex series of equations to represent the profile of the tree, including splined and conditioned equations to ensure conformity between the different sections described by the different equations. Recent references are M'Hirit & Postaire (1985), Newberry & Burkhart (1986), Ormerod (1986), Cao *et al.* (1986).

2.7.4.6 Compatible Volume Tables and Taper Functions

Compatible volume and taper tables give the same estimate of the total volume of a tree of a given diameter at breast height and height.

Demaerschalk (1972, 1973) illustrated the means of converting several commonly used expressions of volume to compatible taper functions that gave the best possible fit for taper, subject to the condition that the total volume estimates be identical with those given by the existing volume table.

EXAMPLE 21 Derivation of a compatible taper equation for the volume function $v_t = a + b(d^2h)$

Under the condition that the volume predicted by the volume function and that from the integration of the taper function must be equal:

$$v_t = \left(\frac{\pi}{4}\right)\int_0^h (d_i)^2\, \delta L = a + b(d^2h)$$

$$= \left(\frac{\pi}{4}\right) \left[(d_i^2\, L)\right]_0^h = a + b(d^2h)$$

$$= \left(\frac{\pi}{4}\right) d_i^2\, h \qquad = a + b(d^2h)$$

$$d_i^2 = (4a + 4b(d^2h))/(\pi h)$$

or

$$(d_i/d)^2 = (4a/d^2 + 4b/h)/(\pi h)$$

Where: d_i is a specific diameter at which this relationship satisfies the constraint of equal volumes from both functions. This specific function can be made general for any d_i by including in each term of the function a first derivative

$$\left(\frac{L_i}{h}\right)^{(rk+1)}$$

Where: r and k are independent parameters calculated by minimizing the residual variance in $(d_i/d)^2$.

$$(d_i/d)^2 = \left(\frac{4a}{\pi}\right)(q+1)L_i^q h^{-(q+1)} d^{-2} + \frac{4b}{\pi}(p+1)L_i^p h^{-p}$$

Example 21 continued

Where:
v_t = total volume, m^3
d = diameter at breast height, m
h = total height, m
L_i = distance i from top of tree
d_i = diameter at L_i
p, q = exponents chosen to minimize the residual variance in predicted d_i

The advantage of compatibility between the two systems can be gained by converting an existing volume table into a compatible taper equation. Then volume tables to any desired merchantable limit may be derived. Though more complex and indirect, this approach is preferred to deriving a taper equation from field data and then integrating this equation to provide volumes. This is because a totally unbiased taper equation does not exist. However, a volume table based system, if derived from an unbiased equation, will give satisfactory results for both total volume estimation and volume to different merchantable limits (Munro and Demaerschalk, 1974; Gor-kesiah & Demaerschalk[1]).

The recent literature on taper functions and compatible volume tables is considerable; as an introduction readers are referred to Cao *et al.*, 1986; Reed & Green, 1984; Byrne & Reed, 1986; Amateis *et al.* 1986; Valenti & Cao, 1986; McTague & Bailey, 1987; McClure & Czalewski, 1986; Kozak, 1988; Newberry *et al.*, 1989.

EXAMPLE 22 Calculations employing a compatible volume and taper equation (after Munyuku, 1980)

N.B. Units are inches, feet and cubic feet.

Volume table $v = ad^bh^c$

v = total volume, ft^3
d = dbh, in
h = total height, ft
a = 0.0007347
b = 1.715
c = 1.432

Taper table constrained to give the same total volume

$$d_i = 10^{b_1}d^{b_2}h^{b_3}L^{b_4}$$

Where:
b_1 = −0.2184
b_2 = 0.8575
b_3 = −0.6415
b_4 = 0.8576

If d = 12.7 and h = 80.0:

1 Gor-kesiah, J.O. and Demaerschalk, J.P. Unpublished paper based on Gor-kesiah's MSc thesis, Faculty of Forestry, University of British Columbia.

By volume table

$$\log v = -3.1339 + 1.715 \log d + 1.432 \log h$$
$$= 1.4843$$
$$v = 30.5 \text{ ft}^3$$

By integration of taper function

$$v = \frac{\pi}{(4)(12)^2} \int_{L-0}^{L-H} d_i^2 \ \delta L$$
$$= (0.005454) \int_0^H (10^{b_1} d^{b_2} h^{b_3} l^{b_4})^2 \ \delta L$$
$$= (0.005454)(10^{(-0.4368)} d^{(1.715)} h^{(-1.283)} \left[\frac{L^{2.7152}}{2.7152}\right]_0^H$$

or

$$\log v = (-2.2632 - 0.4368 + 1.715 \log (d) - 1.283 \log (h)$$
$$+ 2.715 \log (L) - 0.4338)$$

When $d = 12.7$ and $h = 80.0$

$$\log v = 1.485$$
$$v = 30.5 \text{ ft}^3\text{, i.e. the models are compatible}$$

Using the taper table to estimate the diameter 25 ft above ground, i.e. $L = 55$

$$\log (d_i) = (-0.2184 + 0.8575 \log (d) - 0.6415 \log (h)$$
$$+ 0.8576 \log (L))$$

When $d = 12.7$ and $h = 80.0$

$$\log (d_i) = 1.0028$$
$$d_i = 10.1 \text{ in}$$

Similarly, if $d_i = 9.0$ in

$$\log (L) = \frac{0.2184 - 0.8575 \log (d) - 0.6415 \log (h) - \log (d_i)}{0.8576}$$
$$= \frac{1.4469}{0.8576}$$
$$= 1.6872$$
$$L = 48.7 \text{ ft from the tip or}$$
$$31.3 \text{ ft above ground level}$$

N.B. This model has not been constrained to ensure that the diameter at 1.3 m above ground level = dbh.

2.7.4.7 *Continental-type Tariff Tables*

In parts of Europe where the forests have been under controlled forest management for more than a century, and, especially, in irregular, uneven-aged forests in the mountains, local one parameter volume tables – called tariff tables – are employed by forest managers. In such crops the correlation between diameter and height is close so that a one parameter volume table is both convenient and precise and, being local tables, any bias is likely to be slight and relatively unimportant because they are not used to predict the basis of sale values, but only to calculate growth and potential production. Both growth and outturn are first calculated using the tariff tables, but sale lots are re-measured after felling.

A single average tariff table may be used for several species in one forest for the sake of simplicity. After felling and measuring the true volume, separate factors to convert the tariff prediction to true volume may be calculated for each species and used in succeeding years. In such instances the unit of volume employed in the tariff tables is not the cubic metre; in France where this type of forest, forest management system and tariff tables are common, the unit of the tables is called a 'silve'. There is no general conversion factor from silves to cubic metres, but each forest and species within the forest has its own conversion rate; these may vary from year to year with the size and composition of the annual outturn. In years where the outturn is mainly from thinnings the conversion rate may be very different from years following wind blow when the outturn is from salvage clearances.

2.7.4.8 United Kingdom-type Tariff Tables

In an even-aged plantation of one species the variation in height is usually quite small; this is especially so in regularly thinned plantations. The variation in diameter is much more marked. Also, in one plantation of say age 20 years, a suppressed tree may have a breast height diameter of 20 cm and a height of 15 m. In another younger plantation a 20 cm tree may be dominant with a height of only 10 m. Consequently among a number of even-aged plantations the correlation of diameter and height is much weaker than in uneven-aged forests. Thus, European-type tariff tables are unsuitable, being both imprecise and inaccurate in even-aged stands.

Hummel (1955) adopted a different approach but unfortunately used the same name – Tariff Tables – for the ready reckoners that he constructed to estimate the standing volumes of low value thinnings from the state forests in the United Kingdom. Now the tables are widely applied to estimate standing volumes for sales in both state and privately owned forests in the UK.

Hummel's research established that for plantations of conifers in the UK the following assumptions were generally valid:

- In an even-aged conifer plantation of one species a precise linear relationship existed between the basal area at breast height and the volume of single trees

  $$v = a + bg$$

  Where:
  v = volume over bark to 7 cm top diameter
  a = basal area at breast height over bark
- The regressions, or volume–basal area lines as Hummel termed them, for different plantations of the same species have a common point of inter-

section. In the recent publications this point is located at coordinates *x*, *y* of (0.004, 0.007); this approximates to the basal area and average volume of a tree of 7 cm dbh.

These assumptions are not absolute truths, but are sufficiently accurate to provide a basis for a useful and simple system of volume estimation. Having accepted these two assumptions, Hummel constructed an *artificial* series of such regression lines, all with a common point of intersection at (0.004, 0.007) and with equal intercepts of volume for a given basal area, i.e. at the same basal area the difference in predicted volume for any pair of adjacent lines was a constant. Each line was given an identity number in a continuous series from that with the smallest gradient or slope to that with the greatest. The range of the lines was such as to cover the range in tree volumes found in the plantations. These lines are not calculated from field data but are an entirely artificial construction forming a family of regression lines of volume on basal area.

The Tariff Tables were then calculated and tabulated to provide in columns – one column for each regression line – the volumes of trees for each centimetre of diameter.

The tables are applied following a form of destructive sampling which has to be repeated every time a volume estimate is required. This is because as the plantation increases in age and height, the slope of the regression line of volume on basal area increases, i.e. the serial number of the corresponding tariff table increases. Hamilton (1975) in *A Mensuration Handbook* (Forestry Commission Booklet No. 39) details the recommended procedures for the use of these tables. Similar tables are in use in Zambia and a simple single test was done in Tanzania (Philip *et al.*, 1979). More extensive tests are likely to show that this type of volume ready reckoner is applicable in even-age pure conifer plantations in the tropics.

2.8 Tree Growth

Tree growth is the increase in its size with time. Growth takes place simultaneously and independently in different parts of a tree and can be measured by many parameters, for example growth in diameter, in height, in crown size, in bole volume, etc.

In Europe, trees tend to grow fastest in spring with no growth in the cold winter; there it is common and meaningful to speak of the annual increment, i.e. the change in size in the one growing season each year. In the tropics growth patterns are less regular; in areas with marked wet and dry seasons growth may stop and restart more than once in a single year, whereas in the humid tropics growth may be almost continuous. In areas with a monsoon-type climate and marked potential soil moisture deficits, trees may shrink in diameter in dry seasons and grow by both release of moisture tensions and cell division and vacuolation in the rainy periods. The cross-section of the bole of some tropical high forest trees shows annual rings – teak (*Tectona grandis*) and *Pterocarpus angolensis* for example – whereas other trees may be less regular; some growth rings may be crescent shaped when the cambium is active around only part of the circumference of the bole at certain seasons.

FIGURE 28 A simplified illustration and example of UK tariff tables

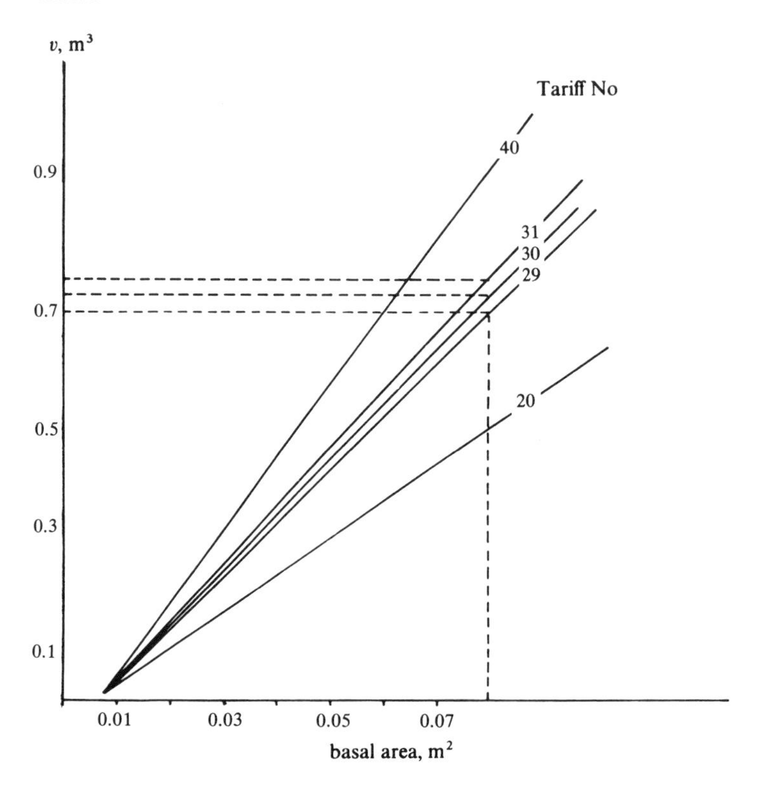

Extract from Tariff Tables
v = volume, m³

dbh, cm	Tariff Number 29	30	31	32	33
30	v = 0.61	0.63	0.65	0.68	0.70
31	0.66	0.68	0.70	0.72	0.75
32	0.70	0.73	0.75	0.77	0.80
33	0.75	0.77	0.80	0.82	0.85
34	0.80	0.82	0.85	0.88	0.91

The Tariff Number applicable to a crop is calculated by felling a representative sample of trees, and measuring

*dbh

*height

*diameter at half height

A minimum number of 20 sample trees is recommended. The volume of each tree is calculated using Huber's formula and its Tariff Number determined by scanning the line in the tables corresponding to the dbh and selecting the column and Tariff Number providing a volume nearest to that calculated. For example using the extract quoted above

sample tree	dbh, cm	volume, m^3	Tariff Number
3	32	0.76	31

The average Tariff Number from the sample is calculated and that Tariff Table is then applied to the whole crop as a one parameter volume table. Further information about these tables may be found in Forestry Commission publications, notably Forestry Commission Booklet No. 36 (Hamilton, 1971) and Booklet No. 39 (Hamilton, 1975).

The total volume increment of a tree is derived from the change over time in the three parameters:

- g basal area
- h total height
- f form factor

Volume increment has seven components, the first three of which are by far the most influential. Allowing for the difficulties in estimating increment, these first three alone are normally considered to comprise the total increment.

1. Relative increase in basal area ($\Delta g/g$)
2. Relative increase in total height ($\Delta h/h$)
3. Relative increase in form factor ($\Delta f/f$)
4. The product of ($\Delta g \Delta h/gh$)
5. The product of ($\Delta g \Delta f/gf$)
6. The product of ($\Delta h \Delta f/hf$)
7. The product of ($\Delta g \Delta h \Delta f/ghf$)

These elements of increment are additive.

EXAMPLE 23 The components of increment in volume (see example in Section 2.2.5)

In 1975 the dimensions of a tree were:

$d = 0.560$ m
$g = 0.25$ m^2
$h = 20$ m
$f = 0.5$
$v = (0.25)(20)(0.5) = 2.5$ m^3

Between 1975 and 1980

$\Delta d = 0.060$ m or 10% relative increase $\frac{\Delta d}{d} = 0.1$

$\Delta g = 0.05$ m^2 or 20% relative increase $\frac{\Delta g}{g} = 0.2$

$\Delta h = 2$ m or 10% relative increase $\frac{\Delta h}{h} = 0.1$

$\Delta f = 0$ i.e. f remained unchanged

In 1980

$v = (0.30)(22)(0.5) = 3.3$ m^3
$\Delta v = 0.8$ m^3

Example 23 continued

Alternatively, in terms of relative increases:

$$\left(\frac{\Delta v}{v}\right) = \frac{\Delta g}{g} + \frac{\Delta h}{h} + \frac{\Delta f}{f} + \frac{\Delta g \Delta h}{gh} + \frac{\Delta g \Delta f}{gf} + \frac{\Delta h \Delta f}{hf} + \frac{\Delta g \Delta h \Delta f}{ghf}$$

$$= 0.2 + 0.1 + 0 + 0.02 + 0 + 0 + 0$$

$$= 0.32 \text{ or } 32\%$$

$$\Delta v = (2.5)(0.32) = 0.8 \text{ m}^3$$

In this example the product of the changes in basal area and height amount to 0.02 out of 0.32 or approximately 6% of the change in the initial volume.

Increment is usually expressed either as a current or marginal rate of change – current annual increment (CAI) – in cubic metres a year (m^3 yr^{-1}), or as an average rate of change or mean annual increment (MAI) over the whole life of the tree. Growth frequently varies from year to year – for example because some years have more rain than others – therefore current annual increment is normally measured over several years and the result expressed as an average per year; this is sometimes referred to as the periodic mean annual increment (PMAI). Figure 29 illustrates the normal relationship between CAI and MAI. As long as the current growth rate is greater than average, the average or mean annual increment rises, but as soon as the current growth rate falls below the average growth rate, then the average must fall. The maximum average growth rate occurs when *CAI equals MAI.*

This simple picture illustrates the normal relationship between CAI and MAI, but the shape of the curves may vary very considerably for different species and growing space. An open grown teak tree, for example, in Nigeria has a volume CAI that falls from its maximum only very slowly; as a result the peak of the MAI curve is very broad. In contrast, a fast growing but short lived tree such as *Trema guinensis* has a volume CAI with a very sharp peak, and the MAI curve is also markedly peaked.

FIGURE 29 Current and mean annual increments; single trees

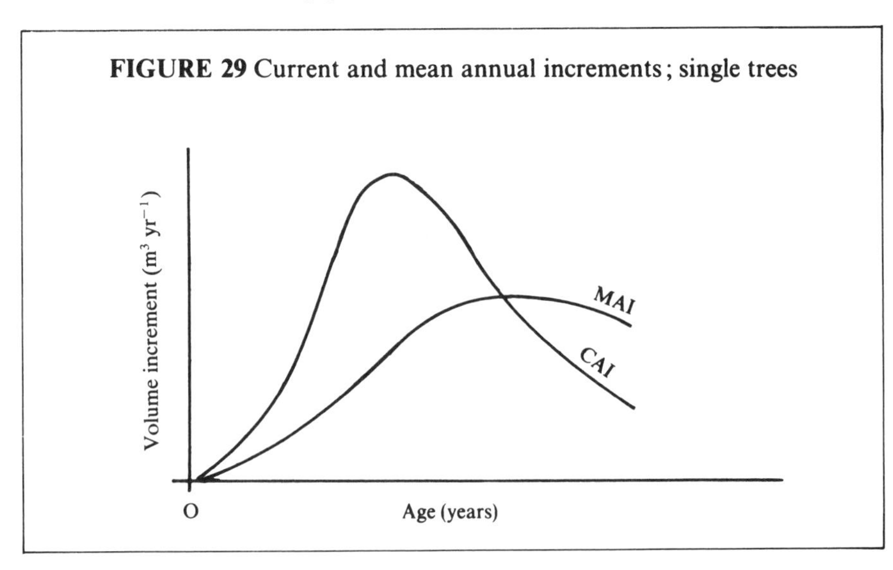

2.8.1 Simple Expressions of Growth Rates

There are three simple expressions of growth rates over short periods:

- The simple growth rate model (= constant rate)
- The compound growth rate model (= constant relative growth rate)
- The continuous growth model

In the simple growth rate model the change in size is expressed as a simple proportion of the size at some point in time – either the beginning, the middle or the end of the period of growth in question. If the period is more than one growing season then the average change in size per year is calculated and this change is held constant for each year of the period.

$$v_n = v_0 + ni \text{ or } i = (v_n - v_0)/n$$

In the compound growth rate model the change in size increases in each year of the period in question in accordance with a power or compound interest function.

$$v_n = v_0(1 + i)^n \text{ or } i = \sqrt[n]{\left(\frac{v_n}{v_0}\right)} - 1$$

In the continuous growth rate model there are no discrete annual increments:

$$v_n = v_0 \, e^{in} \text{ or } i = (\log_e v_n - \log_e v_0)/n$$

Where:
v_n = size in year n at end of period between measurements
v_0 = size at beginning of period between managements
n = number of years in period between measurements
i = rate of growth

2.8.2 Calculations of Growth at Constant Rates

The most used simple model of growth rate is the first or simple growth model. Using this model the mean annual increment of a tree is calculated by:

$$\text{MAI} = \frac{v}{\text{age}}, \text{ m}^3 \text{ per year, where } v = \text{volume of tree, m}^3$$

Current annual increment is normally measured over a period of several years and averaged for that period, i.e. in terms of bole volume

$$(i_v) = \frac{(v_n - v_0)}{n}, \text{ m}^3 \text{ per year}$$

Where:
v_n = volume after n years, m^3
v_0 = volume at the beginning of the period of n years, m^3
n = number of years between the measurement of v_n and v_0
i_v = CAI, m^3

The increment in volume may be expressed as a percentage of the average of the volume at the start and end of the period:

$$i\% = \frac{(v_n - v_0)200}{(v_n + v_0)n}$$

This is known as Pressler's formula. It may be modified for CAI% in terms of basal area and other parameters. The volumes v_n and v_0 may be measured directly or estimated from a volume table; in the latter case the same volume table must be used on both occasions. This method is particularly important in the tropics because many trees have no discernible annual rings. Sample trees must be identified and re-measured at intervals to provide reliable data on growth (see Section 4.2.2). As the change in height is more difficult to measure accurately than a change in basal area and form can only be derived, and because volume increment is strongly correlated with basal area increment, often only the latter is measured. Pressler's formula is then amended to:

$$(i_g)\% = \frac{(g_n - g_0)200}{(g_n + g_0)n} \quad \text{or} \quad \frac{(d_n^2 - d_0^2)200}{(d_n^2 + d_0^2)n}$$

Where:
g_n = basal area, m^2, after n years
g_0 = basal area, m^2, at the beginning of the period of n years
d_n = dbh after n years
d_0 = dbh at the beginning of the n years

In temperate countries diameter increment is often measured on cores extracted using an increment borer but, often, the identification of annual rings is far from simple. It is even more difficult and often impossible in the tropics. Then diameter increment must be measured on marked sample trees using a tape or caliper at a marked point of measurement.

Alternatively and more accurately, a vernier girth band may be attached to the tree and read directly without disturbing it (see Figure 30).

An alternative expression, Schneider's formula, expresses the CAI in basal area as a percentage of the current basal area.

$$(i_g)\% = \frac{400}{md} \text{ (see Appendix 4 for the derivation of this formula)}$$

Where:
d = current dbh (Schneider's formula), cm
m = number of rings in outer one centimetre of radius
i_g = CAI of basal area

The disadvantage of Schneider's formula for stands is that the average current annual ring width is derived from a different number of years m for each tree.

Volume increment may be estimated crudely from basal area increment using coefficients that vary with tree form height increment

$$\text{Volume increment \%} = k_i \text{ (basal area increment \%)}$$

Where k_i varies usually within the range of 1.30–1.45.

FIGURE 30 A vernier girth band

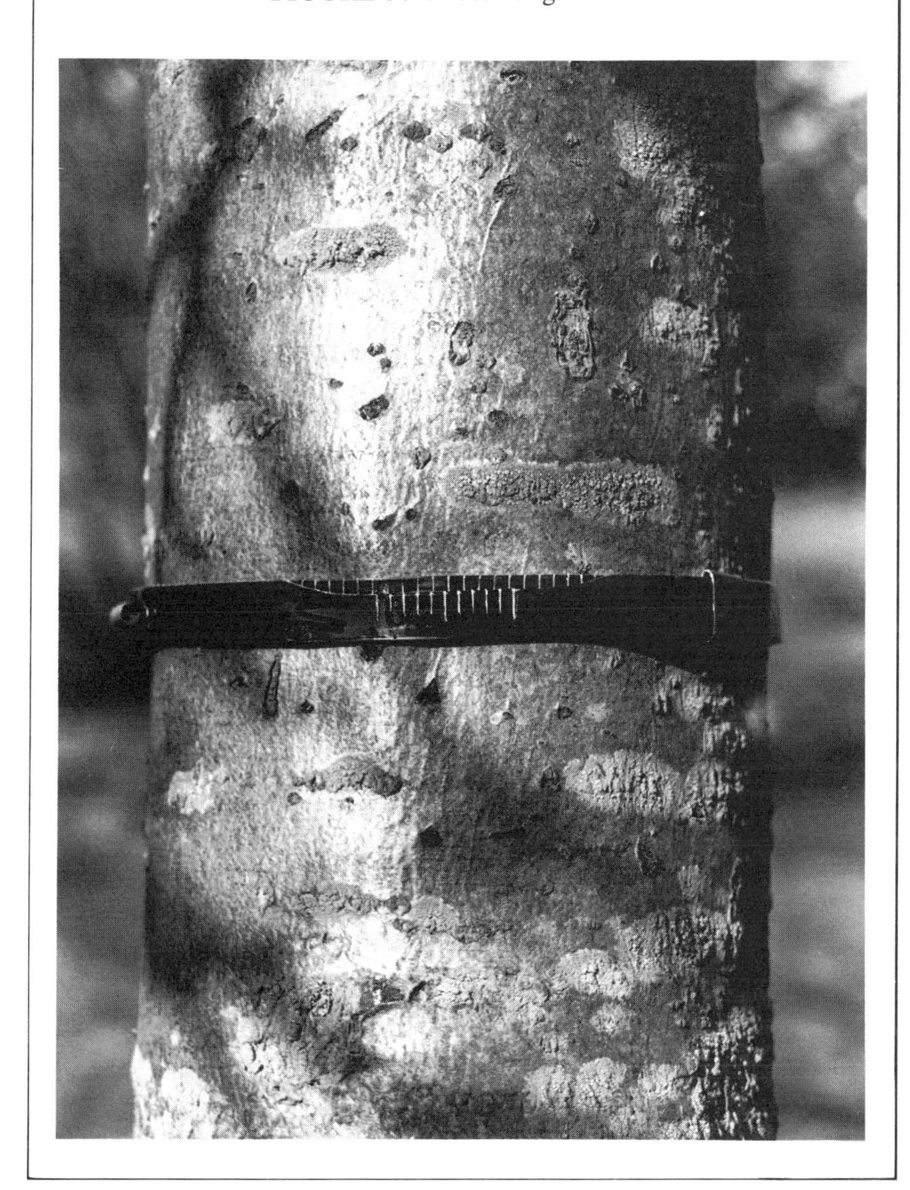

EXAMPLE 24 Estimation of increments

Pressler's formula

In 1975 $v_o = 0.25\ m^3$
In 1979 $v_n = 0.30\ m^3$

$$(i_v)\% = \frac{(0.30 - 0.25)200}{(0.30 + 0.25)4} = 4.5\%$$

Schneider's formula

$$d = 45\ cm$$
$$m = 3$$
$$(i_g)\% = \frac{400}{(3)(45)} = 3\%$$

If $k_i = 1.3$, then $(i_v)\% = (3)(1.3) = 3.9\%$

Symbols as in text

2.8.3 Other Expressions of Growth Rates

The growth pattern of most living organisms follows a sigmoid pattern, with slow initial and terminal growth rates, fastest growth during the middle of their life and a maximum final size. This is illustrated in Figure 31. This pattern is mirrored in two commonly used growth models – the logistic and the Gompertz functions (see Section 6.2.3.2.2).

The disadvantage of the logistic function is that there is no asymptote or maximum size implied by the model; also the curve cannot pass through the origin, i.e. there must be a positive intersection for the start of the growth model. This latter source of distortion is shared with the Gompertz function but can be eliminated in practice.

Nevertheless in spite of these distortions these two functions are useful models of growth for longer term studies, the distortions at the extreme sizes being relatively unimportant. Richards (1959) has also deduced other models. These models are dealt with in greater detail in Section 6.2.3.2.2.

2.8.4 The Pattern of Growth

The patterns of the growth of a tree in terms of the changes in diameter, basal area, height, form and volume are affected to varying degrees by the site, crop structure, competition and stocking. Even-aged crops of one species have very different growth patterns from uneven-aged crops of several species.

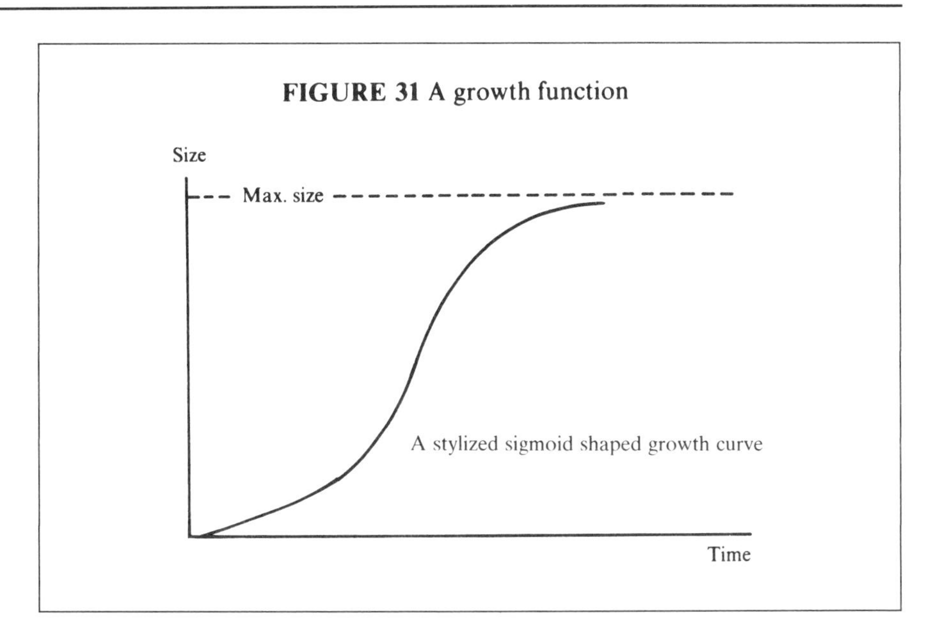

FIGURE 31 A growth function

2.8.4.1 *Diameter Growth*

The pattern of diameter growth in an even-aged plantation of one species is commonly marked by three stages:

- the early immature stage before canopy closure when the rate of diameter growth is little affected by competition
- the responsive middle stage; on canopy closure ring width decreases but responds quickly to treatment such as thinning and fertilizing
- the final (mature) stage when ring width is narrow and is not so markedly responsive to treatment

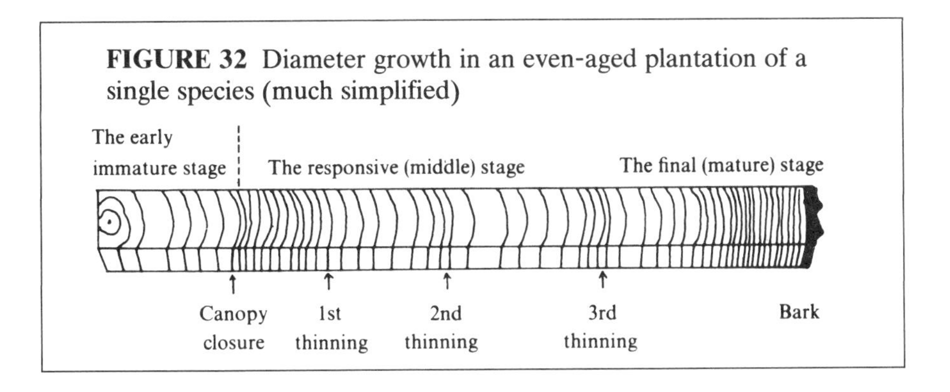

FIGURE 32 Diameter growth in an even-aged plantation of a single species (much simplified)

In contrast the pattern of diameter growth of a shade-tolerant tree regenerating under a canopy in a managed forest, growing slowly until released by the removal of the overwood, exhibits:

- an early stage of extremely slow diameter (and height) growth while the sapling is dominated by the overwood
- a middle stage when the tree is growing more rapidly but is still severely affected by its larger neighbours
- the final stage when the tree is a dominant with a large, free, well-developed crown and neighbouring trees of the same size are few and distant

FIGURE 33 Diameter growth in a managed uneven-aged forest

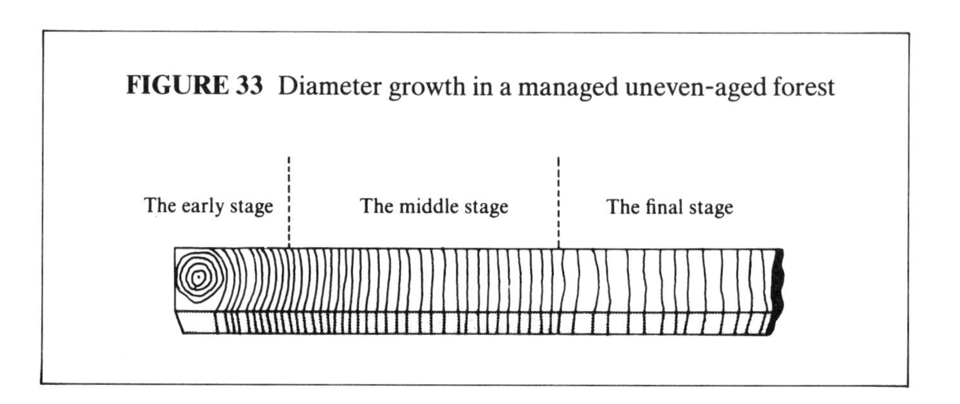

The contrast in the two patterns is accentuated by their wood properties. In the plantation grown tree a relatively large part of the bole volume is of low quality juvenile wood; in the uneven-aged forest grown tree the core of juvenile wood is small, and the larger part of the radius of the bole consists of high quality mature wood with an even growth rate. This again contrasts with the plantation grown tree in which the growth rate varies considerably.

These two contrasting patterns illustrate the fact that whatever the structure of the forest, the rate of diameter growth depends upon the degree of competition. In plantations through spacing and through thinning the forest manager can control the pattern of diameter growth to a large degree. Regular thinning with low stocking promotes fast diameter growth, and vice versa.

Diameter growth may not be regular along the whole of the bole length and there is no constant relationship between the diameter increment at one point along the bole and at another. It is obvious from the shape of the tree that diameter growth tends to be greatest towards ground level. After an initial stimulus to growth at the base of the crown, thinning tends to promote growth in the lower part of the stem, decreasing the form factor; dense stocking provides relatively fast growth just below the crown, thus increasing the form factor; green pruning of the crown has a similar effect. In very dense stands, diameter increment may be greater near the base of the crown than in the lower part of the bole (Larson, 1963).

2.8.4.2 Height Growth

In even-aged plantations height growth on age shows a typically sigmoid pattern with slow early and late stages and fast growth between. The rate of height growth is often almost constant from the time when the trees are well established and about to close canopy to the time when height is nearing its maximum and the tree crown is starting to develop a more complex branching system in order to provide flowering points for the reproductive phase. In even-aged plantations height growth is commonly modelled using the Gompertz, logistic or Schumacher functions (Alder, 1977; Nokoe, 1978).

In irregular forest and especially with shade-tolerant species the early rate of height growth may be extremely slow and only accelerates when the tree is released from overhead shade. In tropical high forest, shade-tolerant species such as *Strychnos mitis* and *Cynometra alexandri* may be only 2 m tall at 10 years of age; similar patterns occur in temperate consociations of beech *Fagus* spp. and *Abies* spp. Saplings of many climbers behave in the same fashion.

2.8.4.3 The Change in Form Factor with Age

As with diameter growth, the change in form factor and rate of taper in different parts of the bole of a tree depend upon the competition and site factors affecting a tree at a particular age. Much of the response of trees to external stresses such as wind takes the form of an adaption optimizing the distribution of net photosynthate to maximize the tree's mechanical resistance to the forces to which it is subjected. Often trees in sheltered sites have high form factors and trees in sites exposed to strong winds have low form factors. Immediately after thinning form factor tends to increase temporarily as crown development stimulates the production of crown wood; in contrast trees in regularly and heavily thinned plantations and open grown trees have low form factors, whilst trees grown in dense plantations have higher form factors. Pruning also affects tree form, usually in a manner similar to thinning.

Very young trees have a large proportion of their stem volume below breast height, and their breast height diameter is more representative of the mid-timber length diameter than their diameter at ground level. Such trees have an abnormally high form factor. In part the change in the relative level of breast height accounts for the change in form factor with age.

2.8.5 Stem Analysis

Some trees – either exotic or indigenous – such as Mninga (*Pterocarpus angolensis*), teak (*Tectona grandis*) and pines such as *Pinus patula*, growing in most parts of tropical countries with monsoon climates, have rings that provide an accumulated record of their growth that can be decoded and analysed. These rings may not be annual but associated with seasonality in the climate and patterns of growth.

An example of such a stem analysis

A teak tree 30 years old, 28 m in height and 42 cm dbh is felled in 1980. Cross-sectional discs about 2 cm thick are cut and removed at ground level, breast height and at 1 m intervals up the bole. The discs are examined in the laboratory and the boundaries of the annual rings are marked along two diameters at right angles (these diameters should be aligned in the same direction on each disc) (see Figure 34). The ring immediately below the bark on each disc grew in the current year and formed a continuous sheath from ground level to the current terminal shoot. By counting towards the pith from the cambium at the junction of the bark and wood, the year of formation of each ring on each disc can be determined. If the innermost (central) ring on a disc cut 14 m above ground level has only 20 rings, and there are 30 rings on the disc cut at ground level because the tree is 30 years old, then 10 rings are missing at a height of 14 m. This means that the tree had not yet reached 14 m tall at the end of the 10th year but had reached this height at the end of the eleventh year because the ring of that year is present. From the examination of each disc the rate and pattern of height growth can be determined.

FIGURE 34 Stem analysis

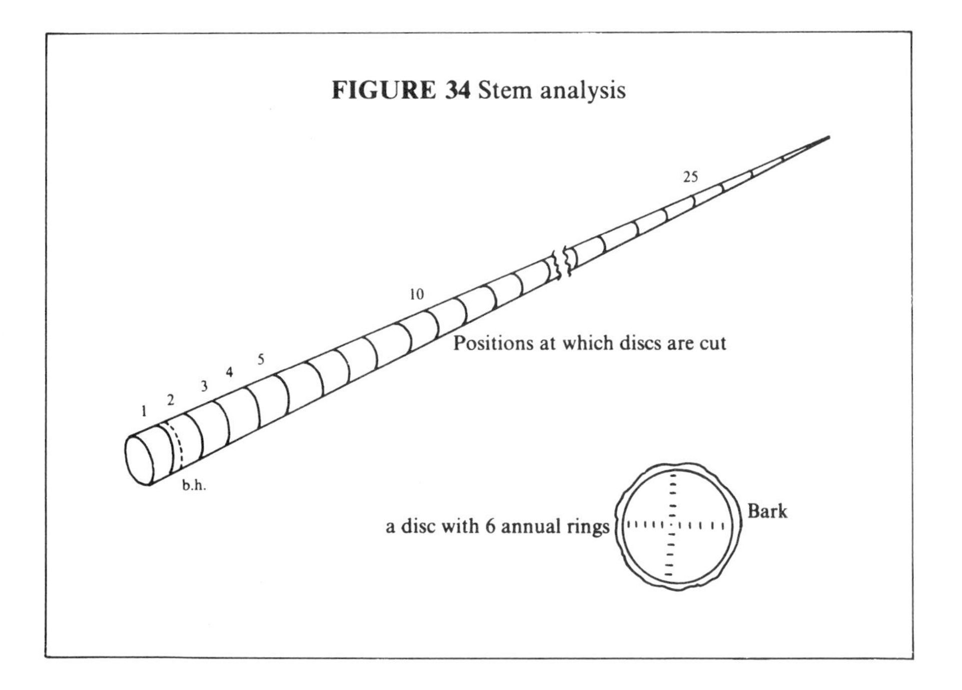

Similarly, by examining and dating each ring on the section cut at breast height, the rate and pattern of diameter growth at breast height can be determined. The volume under bark of each log 1 m in length can be calculated using Smalian's formula:

$$v = \left(\frac{\pi L}{4}\right)\left(\frac{d_1^2 + d_2^2}{2}\right)$$

In this way the under bark volume of each log in successive years can be determined from the measurements of the diameters in successive years marked on the discs. These volumes can be summed to determine the rate and pattern of volume growth. Similarly the under bark form factor can also be calculated for each year of the life of the tree.

A layout for the calculations is shown in the following example. Overbark volumes can be derived through a coefficient, k, from under bark diameters, where

$$k = \frac{\sum^{n} \text{dob}}{\sum^{n} \text{dub}}, \qquad \text{dob} = k(\text{dub}) \quad \text{and} \quad n = \text{number of discs}$$

Examples of booking and calculation sheets for stem analysis

1. Data for the *height* on *age* relationship. Measurements taken in 1980

Section (1)	Height section (2)	No. of rings (3)	Year in which the height recorded in col. (2) was reached (4)	Age of tree in year recorded in col. (4) (5)
Stump	0	30	1950	0
2	1	29	1951	1
3	breast height	29	1951	1

etc.

2. Data for the *diameter* on *age* relationship. Measurements taken on the section cut at breast height

Year (1)	Age (2)	Dub 1 (3) cm	Dub 2 (4) cm	Mean (5) cm	Dob (6) cm
1951	1	2.1	1.9	2.0	2.4
1952	2	2.9	2.9	2.9	3.4

etc.

$$\text{dob} = k(\text{dub})$$

$$k = \frac{\sum^{n} \text{dob}}{\sum^{n} \text{dub}}$$

$$n = \text{number of discs}$$

3. Calculation of log and tree volume

	1961	1962	1963	etc.
Section 9 m from ground level				
Dub	18.6	20.3	21.0	
Dob	20.3	22.2	22.9	
g	0.0324	0.0387	0.0412	
Section 10 m from ground level				
Dub	17.4	19.2	20.2	
Dob	19.0	21.0	22.1	
g	0.0284	0.0346	0.0384	

etc.

Volume of log ob in:

	1961	1962	1963
9th log	0.0304	0.0366	0.0398

Volume of tree ob in:

	1961	1962	1963

4. Summary of data of volume

Year (1)	Age (2) yr	dbh ob (3) cm	h (4) m	g @ bh (5) m^2	V (6) m^3	V cyl. (7) m^3	form factor

etc.

5. Summary of growth rates

Year (1)	Age (2) yr	Volume ob (3) m^3	Periodic increment (4) m^3	CAI (5) m^3	MAI (6) m^3
etc.					

Then these four patterns of growth:

- diameter at bh, or basal area on age
- height on age
- volume on age
- form factor on age

provide a complete picture of the tree's development. CAI and MAI can be calculated for each parameter of the tree. In contrast they tell *nothing about the crop* from which the tree was felled unless records of past fellings are available.

N.B. Stem analysis is only feasible if growth rings are clearly visible and can be dated.

2.9 Tree Biomass

The increase in rural populations in many developing countries, and the demand by interested groups in developed countries for a part in the decision making process of forest management, has led to a widening of the task of the mensurationist and the need to quantify many products of the forest that previously were ignored. (The assessment of non-quantifiable or intangible forest products is outside the scope of this book.) One response of the mensurationist has been to try to measure the whole tree – not just the bole; another has been to try to express the production by a tree or forest in units of energy or dry matter. The concept of using dry matter was adopted to permit comparison of species whose end uses may be very different. Nevertheless foresters must keep in mind that 'utility' is what the collector or purchaser considers: two species producing the same dry matter quantity may be valued very differently if the fodder value of their foliage is different, or their combustion properties differ, or the distribution of the dry matter between the main stem and branches differs.

Nevertheless a voluminous literature on the measurement of biomass has accumulated in the last decade. The principle of measuring biomass is easy:

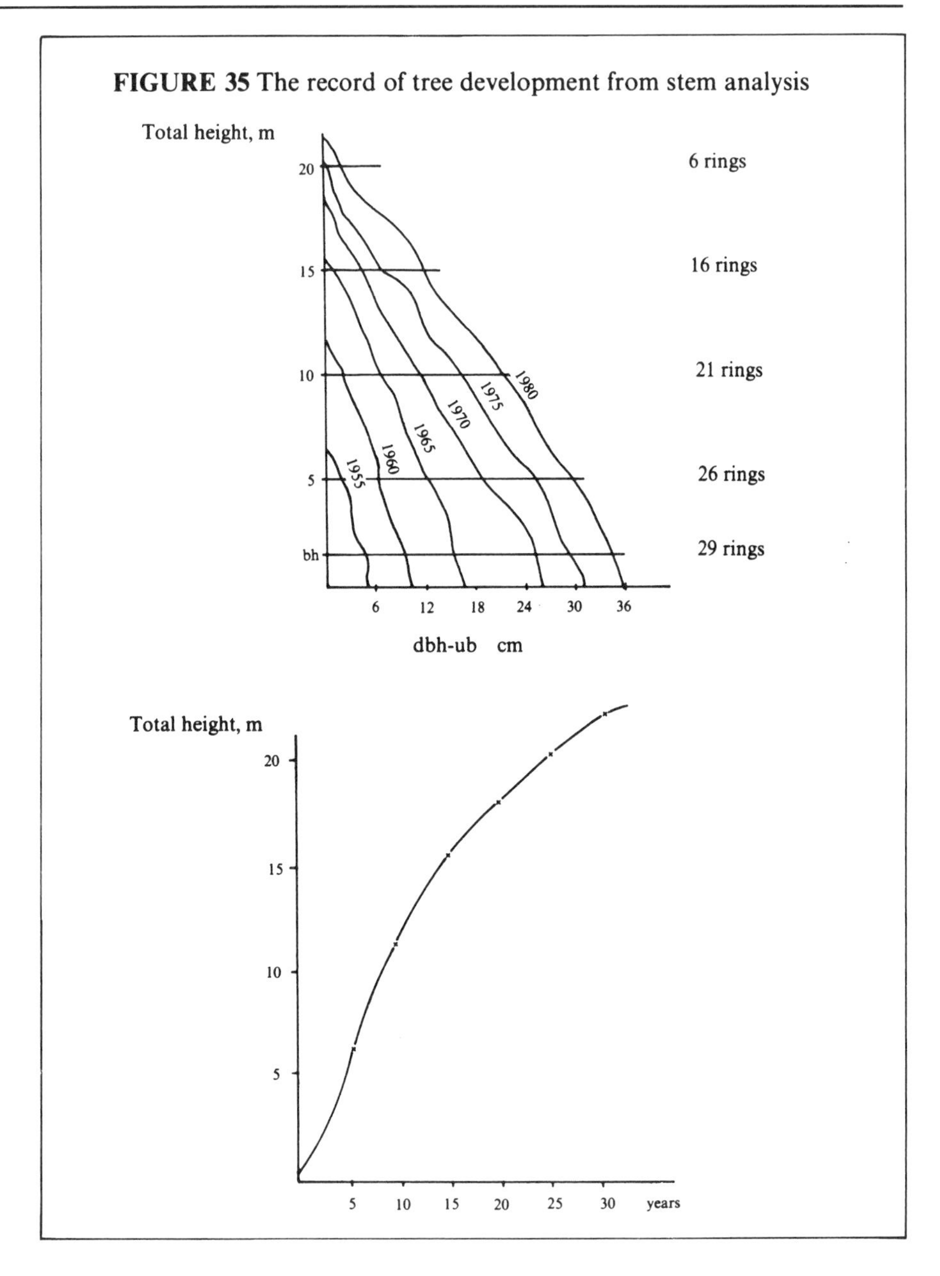

FIGURE 35 The record of tree development from stem analysis

- fell (or dig up) a representative sample of trees
- take small representative samples of the different parts of the tree – bole branches, leaves (and roots)
- weigh the samples green and oven dried
- bulk up from the samples to the whole and sum for the tree
- establish relationships with parameters easily measured on the standing tree and develop predictive mathematical models for application avoiding destructive sampling

The problems are in:

- selecting representative trees and parts of trees
- developing unbiased predictive models
- ensuring additivity of the parts to equal the whole, i.e. that the sum of the predicted parts equals the prediction of the whole

The techniques already described for volume tables are generally applicable to the construction of biomass tables; many authors have found that total biomass may be predicted satisfactorily from dbh.

$$\log(w) = a + b(\log d) \quad \text{or} \quad w = a' \cdot d^b$$

Where:
w = tree dry weight
d = diameter at breast height
a, a', b = constants

An introduction to the detailed literature may be gained from Cunia & Briggs (1984, 1985); Valentine *et al.* (1984); Campbell *et al.* (1985); Reed & Green (1985); Snowdon (1985, 1986); Applegate *et al.* (1988).

3 Measuring Tree Crops

The task of measuring forest crops is termed forest inventory. The techniques of forest inventory will be discussed in Chapter 4, but first the parameters used to describe crops will be detailed. There are many instruction manuals detailing methods of measuring forest crops under differing conditions such as Hamilton (1975).

Forests have different structures, that is the different species, ages and sizes of tree are grouped in different patterns in different forests. A common classification of forest crop is shown in Figure 36.

Examples of the six categories are:

1. Plantations of exotic conifers – *Pinus patula*.
2. Areas of miombo woodland dating from a time when a particular area of shifting cultivation reverted to woodland, or plantations of two or more species intimately mixed, or natural forest of mixed species managed on a coppice system for fodder and fuel.
3. Plantations of *Eucalyptus* or another species managed by coppice with standards for small round wood and saw timber.
4. Areas of bamboo, *Arundinaria alpina* with *Podocarpus* spp., *Crossopteryx febrifuga*, etc. as emergent forest trees, or areas of mixed forest managed on a coppice with standards system for a wide range of wood products.
5. Areas in mixed tropical high forest with consociations of *Cynometra alexandri*, *Strychnos mitis* or in temperate zones of *Fagus* spp., *Abies* spp., etc.
6. The common structure in tropical high forest or forest managed on the single tree selection system.

Uneven-aged forest may vary from two-storeyed high forest in which the two age classes are distinct, to forest with many age classes where regeneration occurs at scattered centres over a long time period, to an intimate mixture of species and age classes with no obvious boundaries between them.

A most important parameter used to describe blocks of forest is that of area. The techniques of area measurement are not covered in this book. Husch *et al.* (1982) and Loetsch *et al.* (1973) provide instructions and further references to this topic.

The description of even-aged crops is relatively easy because the average

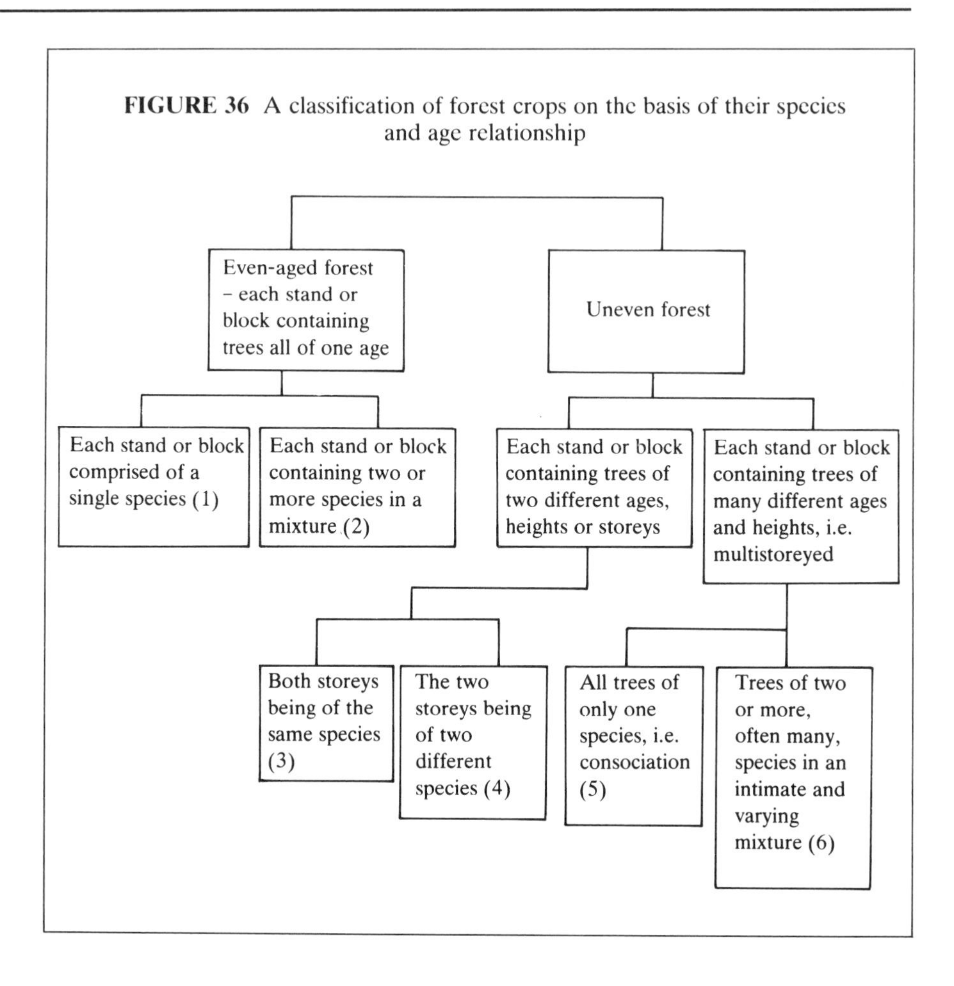

FIGURE 36 A classification of forest crops on the basis of their species and age relationship

value of a parameter, e.g. average height, average volume, etc. provides a meaningful measure and for the most part represents the crop satisfactorily from both the silvicultural and management points of view. This is because distributions tend towards the Normal and variances are either small or relatively easily estimated. The description of uneven-aged crops is not so simple because their distributions are less easily predicted and the average or mean is not a satisfactory basis of description or comparison.

3.1 Parameters Describing Even-aged Crops of One Species

The common parameters used to describe blocks of even-aged crops of a single species are:

A area, hectares, ha
t age, years
N number of trees per hectare
$\bar{d}$ mean diameter, cm

$\bar{g}$ mean basal area, m^2
$d_{\bar{g}}$ diameter of tree of mean basal area, cm
G basal area per hectare, $m^2\ ha^{-1}$
$\bar{h}$ mean height, m
h_d dominant height, m
h_L Lorey's mean or weighted height, m
V volume per hectare, $m^3\ ha^{-1}$
$\bar{v}$ mean volume per tree, m^3
$\bar{f}$ mean form factor
CAI current annual increment, $m^3\ ha^{-1}$
MAI mean annual increment, $m^3\ ha^{-1}$

3.1.1 Age

The age of plantation raised from transplants is normally taken from the year of planting; only if trees are raised by direct sowing on the plantation site is the age that of the plants themselves. Where early survival has been poor and beating up extensive or where there has been advance natural regeneration, an even-aged stand may contain a range of ages; stands are regarded as even-aged as long as the age range does not affect the application of silvicultural and management techniques normal for even-aged crops established in a single year. A maximum in the range of ages is normally considered to be one-quarter of the expected rotation length. In crops with a range of ages approaching this maximum, care in the use of average values and expected distributions of diameter, height and volume must be taken as they may not represent these crops adequately.

3.1.2 Numbers of Stems per Hectare

The number of stems per hectare is a useful description of a crop, but alone it does not define stocking density. With age, height or diameter it does reflect a picture of crop density. A crop standing at 500 stems per hectare if young and consisting of small trees may be quite open or lightly stocked, whereas a crop of 500 mature trees per hectare will be very dense or heavily stocked. A common spacing for new plantations of conifers grown for timber is 2.5 by 2.5 m giving 1600 stems per hectare; final crops at the end of the rotation often have between 200 and 500 stems per hectare, the balance having been felled in thinnings. In contrast, crops raised for energy or small fuelwood may have thousands of stems per hectare and remain unthinned.

3.1.2.1 Estimating the Number of Stems per Hectare

There are two techniques commonly used to estimate the number of trees in a stand:

- by measuring the number of trees in a small plot or plots of known area representing the stand
- by measuring the diameters of trees subtending an angle equal to or greater than that of a known angle or relascope scale at representative points in a crop. (This technique will be described in detail in Section 3.1.4.2.1 in the theory of the relascope.)

Estimating the numbers of trees per hectare by means of a sample of small plots is a common and efficient method. There are three variants:

1. At representative points in the crop demarcate small plots of known area a and count the number of trees in each, n_i.

Then $$N = \frac{\bar{n}}{a}$$

Where:
N = number of stems per hectare
$\bar{n}$ = average of the counts, n_i

2. At representative points measure the length – L_i – separating the nearest tree from its closest neighbour (see Figure 37).

Then $$N_i = \frac{A}{a_i} = \frac{10^4}{\left(\frac{\pi}{4}\right) L_i^2}$$

$$N_i = \frac{4 \cdot 10^4}{\pi L_i^2}$$

$$N = \frac{(1.273)10^4}{m} \sum^{m} \left(\frac{1}{L_i^2}\right)$$

Where:
A = 1 ha or 10^4 m^2
a = area of plot defined by L
L_i = distance between the two trees at the ith sampling point, m
m = number of sampling points
N_i = estimate of number of stems per hectare at the ith sampling point
N = estimate of number of stems per hectare in the population

3. A variation on 2 above; at representative points measure the distance from the sampling point to the nearest, 2nd nearest, 3rd nearest or nth nearest neighbouring tree

Then $$\frac{N_i}{(n - \frac{1}{2})} = \frac{A}{a_i} = \frac{10^4}{\pi k_i^2}$$

$$N_i = \frac{(n - \frac{1}{2})10^4}{\pi k_i^2}$$

$$N = \frac{(n - \frac{1}{2})10^4}{m\pi} \cdot \sum^{m} \left(\frac{1}{k_i^2}\right)$$

Where:
k_i = distance from sampling point i to the nth nearest neighbour, m

Methods **2** and **3** do in fact define small circular plots whose perimeter is defined by the distance to the centre of a neighbouring tree and containing a known number of trees. In the case of method **2**, measuring to the closest neighbour, then there is one tree within the plot made up of one-half of each of the two trees lying on the plot perimeter, as in Figure 37. In the case of method **3**, measuring from the sampling point to the 4th nearest neighbour, there are three complete trees in the plot and one half tree lying on the plot perimeter. Therefore $(n - \frac{1}{2})$ is used in the formula.

A rule of thumb suggests

- for distances up to 1 m measure to the nearest 0.01 m
- for distances from 1 m up to 3 m measure to the nearest 0.05 m
- for distances over 3 m measure to the nearest 0.1 m

These techniques are biased; the variances of the techniques are different and vary with the size of the sampling plot used; empirical tests suggest that an efficient technique is to measure the distance from the sampling point to the 4th nearest tree, or alternatively to count the number of trees in a small plot containing on average about five trees. However, Payandeh & Ek (1986) suggest that a larger number of trees should be included. They also refer to literature on the quadrant method used by ecologists. The efficiency and choice depend upon the pattern of variation in the stand and the ease of measurement. The cost efficiency of different methods to minimize the least squares estimate of N for a given cost is discussed in Pryag & Gore (1989).

3.1.3 Diameters

The distribution and frequency of different diameters in a crop may vary enormously with the crop's species, age and history. Even-aged crops that have been regularly thinned may have a small variation in diameter around the mean; in contrast, plantations that have not been thinned may have a much larger range in diameter. Usually even-aged crops tend towards a simple distribution, normal or slightly skewed, although in some instances thinning may produce bimodal distributions.

EXAMPLE 25 Calculation of the numbers of trees per hectare

Method 1. At 10 points in a plantation chosen systematically, the following data were collected on the number of trees - n - in circular plots of area 0.01 ha.

Data: n = 16, 14, 18, 13, 12, 9, 17, 15, 16, 14

$$\sum^{10} n_i = 144 \qquad \bar{n} = 14.4$$

$$N = \bar{n}/a = 14.4/0.01$$

$$= 1440 \text{ stems ha}^{-1}$$

Method 2. In a similar plantation and at a similar ten points chosen systematically, the distance - L_i - between the nearest two trees (n = 1) was measured and recorded to the nearest 0.05 m

Data: L_i = 2.25, 3.75, 1.95, 3.65, 2.75, 2.90, 3.10, 3.45, 3.60, 2.85

$$\sum^{10}\left(\frac{1}{L_i^2}\right) = 1.246$$

$$N = \frac{(1.273)10^4}{10} \cdot (1.246)$$

$$= 1586 \text{ stems ha}^{-1} \qquad \text{where } n = 1$$

Method 3. In a similar manner but in another older plantation the distance from the sampling point to the 4th nearest tree (n = 4) was measured and recorded to the nearest 0.1 m

Data: k_i = 4.8, 6.2, 5.4, 6.1, 5.7, 6.0, 5.8, 5.6, 6.2, 6.0

$$\sum^{10}\left(\frac{1}{K_i^2}\right) = 0.3045$$

$$N = \frac{(3.5)10^4}{(10)(3.14)} \cdot (0.3045)$$

$$= 339 \text{ stems ha}^{-1}$$

The distribution of diameters in forest crops is important because trees of different diameters may be used for different purposes and have different values per cubic metre of wood. Many silvicultural operations affect diameter distributions and the values and uses of the crops. Therefore studies of such operations as thinning and spacing must consider the effects both on rate of growth and volume production and on the diameter distributions. Similarly predictions of stand growth are most useful if both volume and diameter distributions are included.

3.1.3.1 *Diameter Distributions*

Actual diameter distributions may be calculated from data derived from forest inventories; such distributions incorporate irregularities due to particular sources of variation operating on the population at the particular time of the

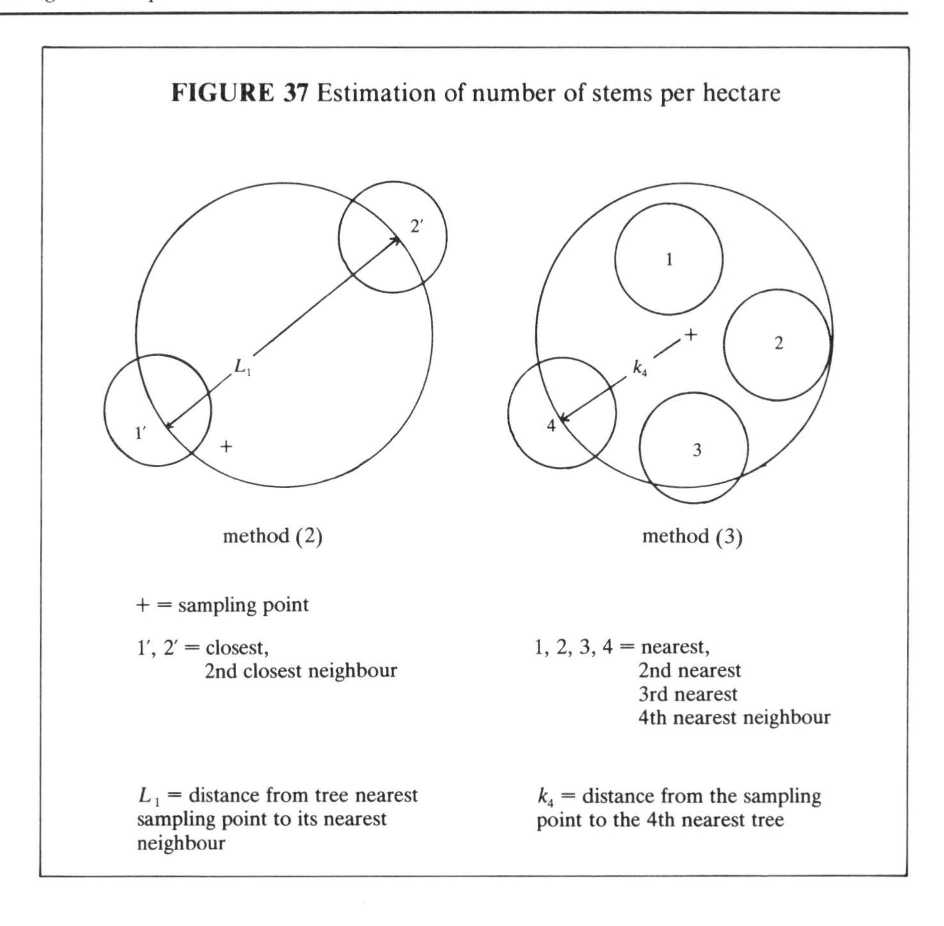

FIGURE 37 Estimation of number of stems per hectare

inventory. In many cases more useful predictions may be based on models that generalize and portray the expected distributions. Three models have been commonly used although many others, such as those modelling two-peaked (bi-modal) distribution, may be relevant. The common models are:

- the Normal distribution, e.g. as in an even-aged crop of a single species
- the Beta distribution, e.g. as in very young plantations and many components of tropical high forest
- the Weibull distribution

3.1.3.1.1 The Normal distribution

The Normal distribution is characterized by the 'bell shaped curve' with mean, mode and median all equal, and is defined by the mean and the variance. The distribution may be predicted using tables that show the percentage of the population expected to lie within bounds defined in terms of the normal deviate, on either side of the mean (see Snedecor & Cochran, 1967).

Example

A population is deemed to have a Normal distribution of diameters defined by the estimated mean of 50 cm with a standard deviation (s) of 5 cm. Hence the expected diameter distribution for a population of 900 stems per hectare is:

cm		Probability	N, stems ha^{-1}
<40	$(<\bar{d} - 2s)$	0.0228	21
40–45	$(\bar{d} - 2s)$ to $(\bar{d} - 1s)$	0.1359	122
45–50	$(\bar{d} - 1s)$ to $(\bar{d})$	0.3413	307
50–55	$(\bar{d})$ to $(\bar{d} + 1s)$	0.3413	307
55–60	$(\bar{d} + 1s)$ to $(\bar{d} + 2s)$	0.1359	122
>60	$(<\bar{d} + 2s)$	0.0228	21
		Total	900

FOR THE ADVANCED STUDENT

3.1.3.1.2 The Beta and Weibull distributions

The Beta function is

$$Y_x = k((x - a)^{\alpha}(b - x)^{\beta})$$

Where:

Y_x = frequency of trees of diameter x per hectare
k = scaling factor
a = lower limit of diameter distribution, i.e. smallest diameter in the stand
b = upper limit of diameter distribution, i.e. largest diameter in the stand
α, β = constants controlling the shape of the distribution

In contrast to the Normal, the Beta distribution has a finite range governed by the constants a and b. The constants may be determined by collecting data from actual stands or permanent sample plots, for example, and calculating values for α and β that minimize the sum of the squared deviations between the measured and the predicted values. The constants may also be calculated by several other methods (see Loetsch *et al.*, 1973, and Burkhart & Strub, 1971).

The Weibull distribution is

$$Y_x = cb^{-1}\left[\left(\frac{x - a}{b}\right)\right]^{c-1} \exp\left[-\left(\frac{x - a}{b}\right)^{c}\right]$$

Where:

Y_x and x are defined as above
a, b, c are constants controlling the scale and shape of the distribution

Alder (1977) used this function to model diameter distributions in conifer plantations in East Africa. He commented:

> The co-efficients of multiple determination were of the order of 0.3; indicating that the relationships were very poorly defined. Nonetheless,

the resultant predicted diameter distributions appear to change in a reasonable manner.

In practice, initial distributions are very variable, and not strongly determined by mensurational factors such as height, spacing, et cetera. It is probable that the variability is largely a result of heterogeneous spatial patterns of survival following planting and mortality in early growth. These patterns are themselves strongly influenced by microtopography and local edaphic variation.

Further reading

ON DIAMETER DISTRIBUTIONS

Alder, D. (1980) *Forest Volume and Yield Estimation.* FAO Paper 22.2, Rome.

Bailey, R.L. and Dell, T.R. (1973) Quantifying diameter distribution with the Weibull function. *Forest Science* 19(2), 97–104.

Burkhart, H.E. and Strub, M.R. (1971) A model for simulation of loblolly pine stands. In: Fries, J. (ed.) *Growth Models for Tree and Stand Simulation.* Royal College of Forestry, Stockholm.

Grey, D.C. (1989) Environmental factors and diameter distributions in *Pinus radiata* stands. *South African Forestry Journal* 149, 36–43.

Hafley, W.L. and Schreuder, H.T. (1977) Statistical distributions for fitting diameter and height data in even-aged stands. *Canadian Journal of Forestry Research* 7, 481–487. Quantifying diameter distributions with Weibull function.

Knuckel, H. (1953) *Planning and Control in the Managed Forest.* Oliver and Boyd, Edinburgh.

Little, S.N. (1984) Weibull diameter distributions for mixed stands of different conifers. *Canadian Journal of Forestry Research* 13(1), 100–107.

Lynch, T.B. and Moser, J.W. Jun (1986) A growth model for mixed species stands. *Forest Science* 32(3), 697–706.

Rennolls, K., Geary, D.N. and Rollison, T.J.D. (1985) Characterising diameter distributions by the use of the Weibull distribution. *Forestry* 58(1), 57–66.

3.1.3.2 *The Description of Crop Diameters*

Crop diameters have to be defined by at least three parameters

- the type of distribution
- the mean
- a measure or measures of scatter, i.e. variance, etc.

These parameters are normally estimated by measuring a representative sample and assembling a diameter frequency table that can be summarized by the distribution parameters. Normally a sample of 200 trees (Hamilton, 1975) gives an adequate picture in an even-aged plantation.

3.1.3.2.1 Estimating the mean diameter

When large numbers of trees are being measured diameters are normally recorded in classes rather than by individual records. Trees are rarely circular in outline and recording diameters to say the nearest 0.001 m is neither justified nor necessary. The mean diameter and the frequency of each diameter class are used to represent all the trees recorded in the class. This introduces small biases, particularly in classes with low frequencies and with wide class intervals. A larger error is introduced when the mean diameter of a class and the class frequency are used to convert the data to basal area because this involves squaring the diameter and the 'average of squares' is greater than the 'squared average'. Providing the diameter class widths are small this bias is not of practical significance.

EXAMPLE 26 The average of squares is greater than the squared average

Diam. class (cm)		Frequency (f)	$\bar{d}$	$\bar{d}^2$	$(f)(\bar{d}^2)$	$(f)(\bar{d})$
8.0	10.0	10	9.0	81.00	810.00	90.0
10.0	12.0	8	11.0	121.00	968.00	88.0
12.0	14.0	12	13.0	169.00	2028.00	156.0
		30			3806.00	334.0

$$\text{mean}\,(d^2) = \frac{3806.00}{30} = 126.87 \qquad (\bar{d})^2 = \left(\frac{334}{30}\right)^2 = 123.95$$

$\sqrt{\text{mean}\,(d^2)} = 11.26$ which is greater than $\sqrt{(\bar{d})^2} = 11.13$

N.B. $d_g = \sqrt{\frac{\sum^n d^2}{n}}$

The average or mean diameter is calculated weighting by frequency so that

$$\bar{\bar{d}} = \sum f\bar{d} / \sum f$$

where $\bar{\bar{d}}$ is the weighted mean diameter of the crop. However, the diameter corresponding to the tree of mean basal area $d_{\bar{g}}$ is more often used. A rule of thumb attributed to Weiss is that the tree of mean basal area is that whose relative frequency is 40% from the higher end of the diameter distribution – that is, if the diameters are ranked in a diameter frequency table, then the tree of mean basal area lies in the diameter class above which 40% of the population occurs.

3.1.3.2.2 Measuring the diameters in the field

In crop measurement only one diameter on each tree is needed, but the orientation of that diameter should be random with respect to any systematic pattern in the orientation of the irregularities of tree outline, e.g. on hill sides because of reaction wood, trees tend to have large diameters at right angles to the slope. The diameters must be measured accurately using a standard procedure so that check measurements can be made in a comparable manner; e.g.

- the measurements must be taken at 1.3 m above ground level on the highest side
- the point of measurement must be cleaned of dead and loose bark scales
- in a circular sample plot the arm of the caliper should be directed towards the plot centre

When diameter classes are used, then the class limits and class identity must be clarified, e.g. $\geqslant 10.00 < 12.00$ is the 11 cm class; or $\geqslant 11.00 < 12.00$ is the 11 cm class; or $\geqslant 10.50 < 11.50$ is the 11 cm class. Only then can the appropriate mean diameter of the class be identified ($\geqslant$: equal to or greater than).

Adlard (1990) gives excellent accounts of field measurement techniques in the context of sample plot measurement for recurrent forest inventory.

3.1.4 Basal Area

The cross-sectional area of a tree estimated at breast height is called the tree basal area, for which the internationally recognized symbol is g. It is normally expressed in m^2. The sum of the basal areas of all trees on an area of one hectare is symbolized by G m^2 ha^{-1}. It is usually measured over bark. Common values in young plantations are between 10 and 20, rising to a maximum of around 60 m^2 ha^{-1} in exceptional circumstances in older plantations, with higher figures in *Eucalyptus* stands in Australia. Dawkins (1958) gives a pan-Tropic average of 35 m^2 ha^{-1}.

Basal area is a useful measure by which to compare the stocking of two stands of the same species, age and height. In undisturbed, i.e. unthinned, forest or natural forest it is a good measure of site potential.

There are two common ways of estimating stand basal area:

- from an enumeration of diameters in representative small sample plots
- from counts of the number of trees subtending an angle equal to or greater than that of a gauge or relascope at representative sampling points in the stand – sometimes referred to as horizontal point sampling

3.1.4.1 Estimating Basal Area per Hectare Using Sample Plots

At representative points, using either a random or systematic layout, small sample plots are demarcated and a diameter frequency table constructed by measuring the dbh of all the trees in each plot and recording them – usually in 1 cm diameter classes. The total basal area for each plot is calculated by summing the basal areas in each diameter class, and the average basal area per hectare calculated by proportion to plot area

$$G = \frac{\sum^{n}\sum^{m_i} g_{ij}}{\sum^{n} a} \text{ m}^2 \text{ ha}^{-1}, \text{ and } \bar{g} = \frac{G}{N} \text{ estimated by } \frac{\sum^{n}\sum^{m} g_{ij}}{\sum^{n} m_i}$$

EXAMPLE 27 Estimating basal area per hectare using four plots

$a = 0.01$ ha

d class cm	f_1	f_2	f_3	f_4	g/tree m^2	f_1g m^2	f_2g m^2	f_3g m^2	f_4g m^2	Total for 4 plots m^2
11	1	–	2	–	0.009	0.009	–	0.018	–	0.027
12	1	1	–	1	0.011	0.011	0.011	–	0.011	0.033
13	2	3	2	2	0.013	0.026	0.039	0.026	0.026	0.117
14	5	4	4	4	0.015	0.075	0.060	0.060	0.060	0.255
15	3	3	4	5	0.018	0.054	0.054	0.072	0.090	0.270
16	–	1	–	2	0.020	–	0.020	–	0.040	0.060
17	2	1	2	2	0.023	0.046	0.023	0.046	0.046	0.161
18	–	–	1	–	0.025	–	–	0.025	–	0.025
19	1	–	–	1	0.028	0.028	–	–	0.028	0.056
	15	13	15	17		0.249	0.207	0.247	0.301	1.004

$$G = \frac{\sum^{n}\sum^{m} g_{ij}}{\sum_{n} 0.01} \text{ m}^3\text{/ha} \quad \text{and} \quad d_g = \sqrt{\frac{(1.004)(4)}{60\pi}}$$

$$= 1.004/0.04 \qquad d_g = 0.146 \text{ m}$$

$$= 25.1 \text{ m}^2\text{/ha} \qquad d_g = 14.6 \text{ cm}$$

N.B. 11 cm class $\geqslant 10.50 < 11.50$ cm

Where:
G = average basal area per ha
g_{ij} = basal area in the jth diameter class of the ith plot, m^2
m_i = number of diameter classes in the ith plot, $1 \ldots j \ldots m_i$
n = number of plots in stand, $1 \ldots i \ldots n$
a = area of each sample plot, ha

3.1.4.2 *Estimating Basal Area per Hectare Using a Relascope*

A relascope is an instrument used in the forest to discriminate between trees on the basis of whether or not the tree subtends an angle equal to or greater than that of the relascope when viewed from the sampling point. When held to the eye the gauge of the relascope provides a fixed angle. If the tree subtends a smaller angle than the gauge, the gauge will appear wider than the tree; if the tree subtends a larger angle than the gauge, the tree will appear wider than the gauge. Relascopes or angle gauges may be of many and varied types. They may be a simple notch at the end of a stick or chain or a sophisticated optical device employing either reflecting mirrors or narrow glass prisms. Figure 39 shows three common types. When using a relascope the observer stands with the gauge at a point in the forest and records the number of trees subtending an angle equal to or greater than a fixed and known angle. This record is multiplied by the 'relascope factor' to give a direct reading of basal area per hectare.

3.1.4.2.1 **The theory of the relascope**

Many variations on the theory of the relascope are given in the textbooks and literature (Grosenbaugh, 1952; Findlayson, 1969; Husch *et al.* 1982; Bitterlich, 1984). I prefer that given by Dr J.A. Petty (personal communication), which is as follows.

Given a random point in a stand of trees of varying diameters d such that a particular tree of diameter d_j exactly subtends the angle of the relascope, θ radians, when viewed from the random sampling point, tree d_j is distance L_j from the sampling point (d_j and L_j are measured in metres). The radius L_j describes a circle or sweep of area πL_j^2.

Then for small angles $\theta = \dfrac{d_j}{L_j}$ and $\theta^2 = \dfrac{d_j^2}{L_j^2}$

$$\text{Basal area of tree} = \frac{\pi}{4} d_j^2$$

This circle contains n_j trees of diameter d_j and all subtend angles greater than θ.

$$\text{Therefore} \quad \frac{\text{Basal area of } n_j \text{ trees}}{\text{Area of circular sweep}} = \frac{n_j \frac{\pi}{4} d_j^2}{\pi L_j^2}$$

$$= n_j \frac{\theta^2}{4}$$

But this is independent of the diameter d_j; therefore a count n of all trees subtending an angle equal to or greater than θ estimates the sum of the tree basal areas per m^2 of their respective circles

$$= n \cdot \frac{\theta^2}{4}$$

Or by proportion, the basal area per hectare is estimated by $G = n \cdot \left(\frac{\theta^2}{4}\right) \cdot 10^4$

or if $\left(\frac{\theta^2}{4}\right) \cdot 10^4$ is denoted as the relascope factor F, then $G = n \cdot F$

This proof indicates that trees exactly subtending the angle should be counted as ½; in practice such borderline trees are very rare and if the angle subtended by a tree is close to θ then the angle should be calculated by measuring d and L. If this precaution is not taken and all doubtful trees counted as a half a very considerable personal bias is likely. If the angle of sight of the relascope is not horizontal then the gauge angle has to be reduced by a factor of cos β, where β is the slope angle.

Proof

Given a point P_1, L_1 metres away from a tree of diameter d_1, from which the tree of diameter d_1 exactly subtends the angle θ_1 radians; what is the angle θ_2 which the tree will subtend if it is viewed from a point P_2 vertically above P_1 such that the angle of sight is $\beta°$ from the horizontal? See Figure 38.

Corollaries of this proof are that if a relascope with a fixed angle is used to

- measure the diameter of a tree at a point up to the bole subtending an angle of $\beta°$ from the horizontal, then the calculated diameter has to be adjusted by the factor (1/cos β).
- estimate the basal area per hectare on uniformly sloping ground, then the plan area of the sweep corresponding to a boundary tree of diameter d_j at horizontal distance L_j is the area of the ellipse $= \pi(L_j \cdot L_j \cdot \cos\beta)$. Therefore $G = nF/\cos\beta$.

The relascope can also be used to estimate the numbers of stems of a given size in a stand, and the total numbers of stems per hectare. This entails more measurement and calculation than the first method described in Section 3.1.2.1. If the trees subtending an angle equal to or greater than the relascope angle are also measured for dbh, then the number counted in the sweep area n_j of a given diameter d_j can be used to calculate the number per hectare N_j

$$\frac{n_j}{a_j} = \frac{N_j}{A} \Rightarrow \frac{n_j}{\pi L_j^2} = \frac{N_j}{10^4}$$

so

$$N_j = \frac{n_j 10^4}{\pi L_j^2} = \frac{n_j 10^4}{\pi\left(\frac{d_j^2}{\theta^2}\right)} = \frac{n_j 10^4 \theta^2}{\pi d_j^2}$$

but

$$10^4\theta^2 = 4F \quad \text{and} \quad \pi d_j^2 = 4g_j$$

$$\Rightarrow N_j = n_j \frac{F}{g_j}$$

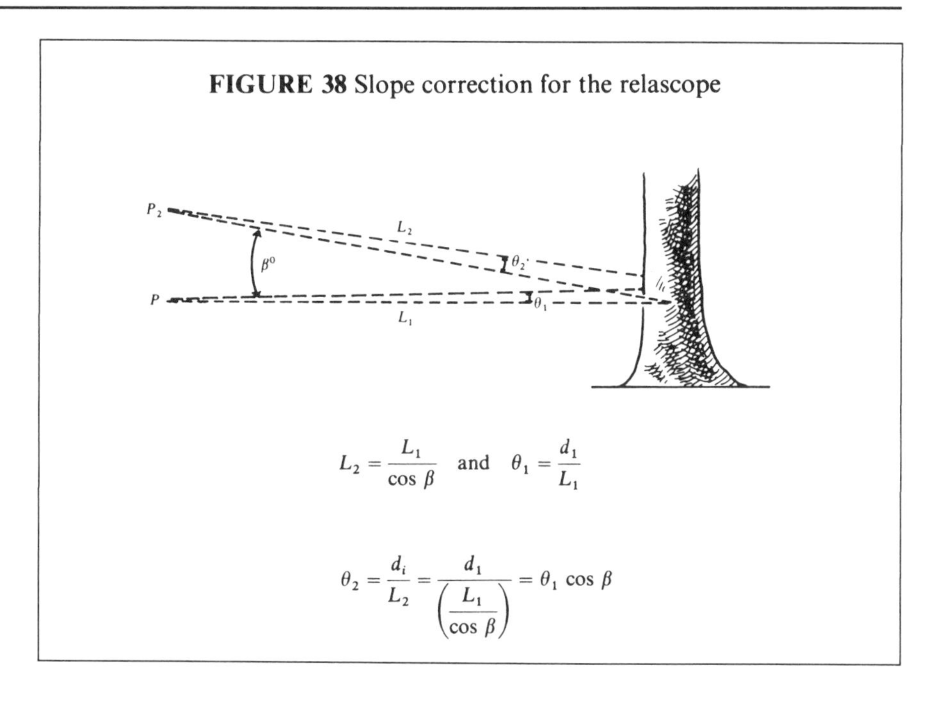

FIGURE 38 Slope correction for the relascope

$$L_2 = \frac{L_1}{\cos\beta} \quad \text{and} \quad \theta_1 = \frac{d_1}{L_1}$$

$$\theta_2 = \frac{d_i}{L_2} = \frac{d_1}{\left(\frac{L_1}{\cos\beta}\right)} = \theta_1 \cos\beta$$

This calculation is normally done for say 5 cm diameter classes; then the total number of stems per hectare N is calculated by summing the numbers per hectare N_j in each class.

$$N = \sum N_j$$

Usually the g_j used for each class is that corresponding to the mean diameter of the class, even though this introduces a small bias and numbers are slightly overestimated.

The formula

$$N_j = \frac{n_j F}{g_j} \text{ is equivalent to } N_j = \frac{n_j}{p_j}$$

Where p_j is the probability of counting a tree of size g_j per hectare, i.e. $p_j = g_j/F$, i.e. probability is proportional to the basal area of the tree and inversely proportional to the relascope factor.

3.1.4.2.2 Characteristics of different types of relascope

There are three main types of relascope all defining a fixed angle or angles of sight (see Figure 39):

- simple relascope incorporating a fixed distance from the eye to a gauge of fixed width
- glass wedge or circular prisms
- mirror relascope, e.g. the Spiegel relaskop

The simplest relascope consists of a metal plate with a notch – say 1 cm in width – attached to the end of a chain 50 cm long. The fixed angle is 1/50 radians and the relascope factor F is

$$F = \frac{\theta^2 10^4}{4}$$

$$= \frac{(0.0004)(10^4)}{4}$$

$$= 1$$

Common relascope factors and fixed angles are

Relascope factor	angle, radians
1	0.02
2	0.0282
4	0.04

A wedge prism is a convenient form of relascope. The prism is held over the plot centre and the image of the bole seen through the prism and over its top, appears as in Figure 39b. A tree subtends an angle greater than the reference angle when its displacement as seen through the prism is less than its diameter. The relascope factor of a wedge prism depends upon the prism angle and the refractive index of the material (Beers & Miller, 1964). The prism must be held so that the displacement is minimal.

The characteristics of the three types of relascope

	Simple gauge	Glass wedge	Spiegel
Available from	Anywhere	USA and Europe	Austria
Relative cost	Very cheap	Moderately expensive	Expensive
Ease of use	Simple	Simple	Needs training
Slope correction	Applied after collecting data	Can be incorporated, but slope has to be measured	Incorporated automatically
Other uses	—	—	Incorporates a hypsometer, several reference angles, a range finder and is designed to facilitate the measurement of tree diameters up the bole
Miscellaneous	—	—	Has screw fitting for a tripod; the wide angle relaskop is very useful in tropical high forest

FIGURE 39 Types of relascope

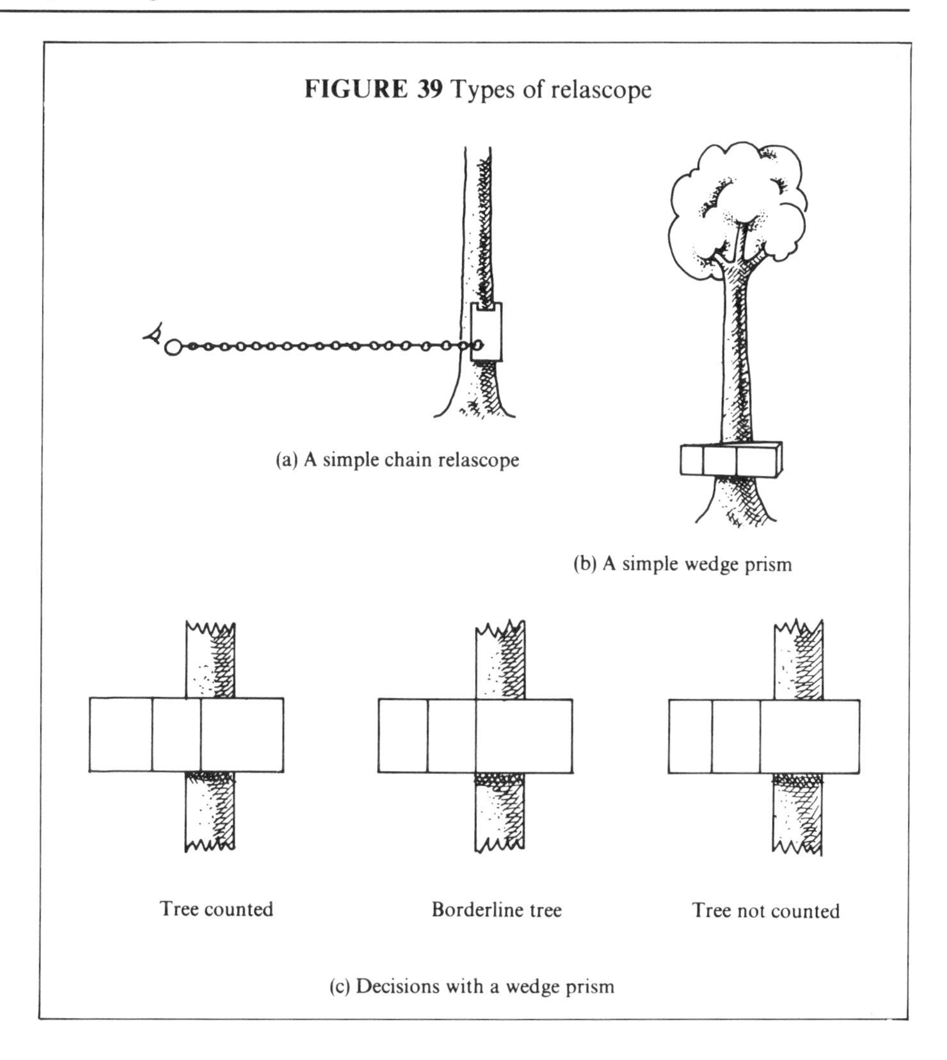

(a) A simple chain relascope

(b) A simple wedge prism

(c) Decisions with a wedge prism

The Spiegel relaskop is now made in three forms

- the ordinary Spiegel relaskop
- the wide angle relaskop
- the tele-relaskop

These forms are described by Findlayson (1969) and the originator Dr Walter Bitterlich (undated). The ordinary relaskop incorporates

- a hypsometer with scales for three fixed distances
- a range finder to measure these three fixed distances
- three major fixed angles of 0.02, 0.0282 and 0.04 radians; subdivisions provide angles of 0.005, 0.010 and 0.015 radians as well

The scales are illustrated in Figure 39d. The great advantage of the Spiegel relaskop is that there is a built-in slope correction factor. This is achieved by varying the width of the scales in view as the instrument is tilted away from the horizontal.

FIGURE 39(d) The scales of the Spiegel relaskop calibrated in absolute units

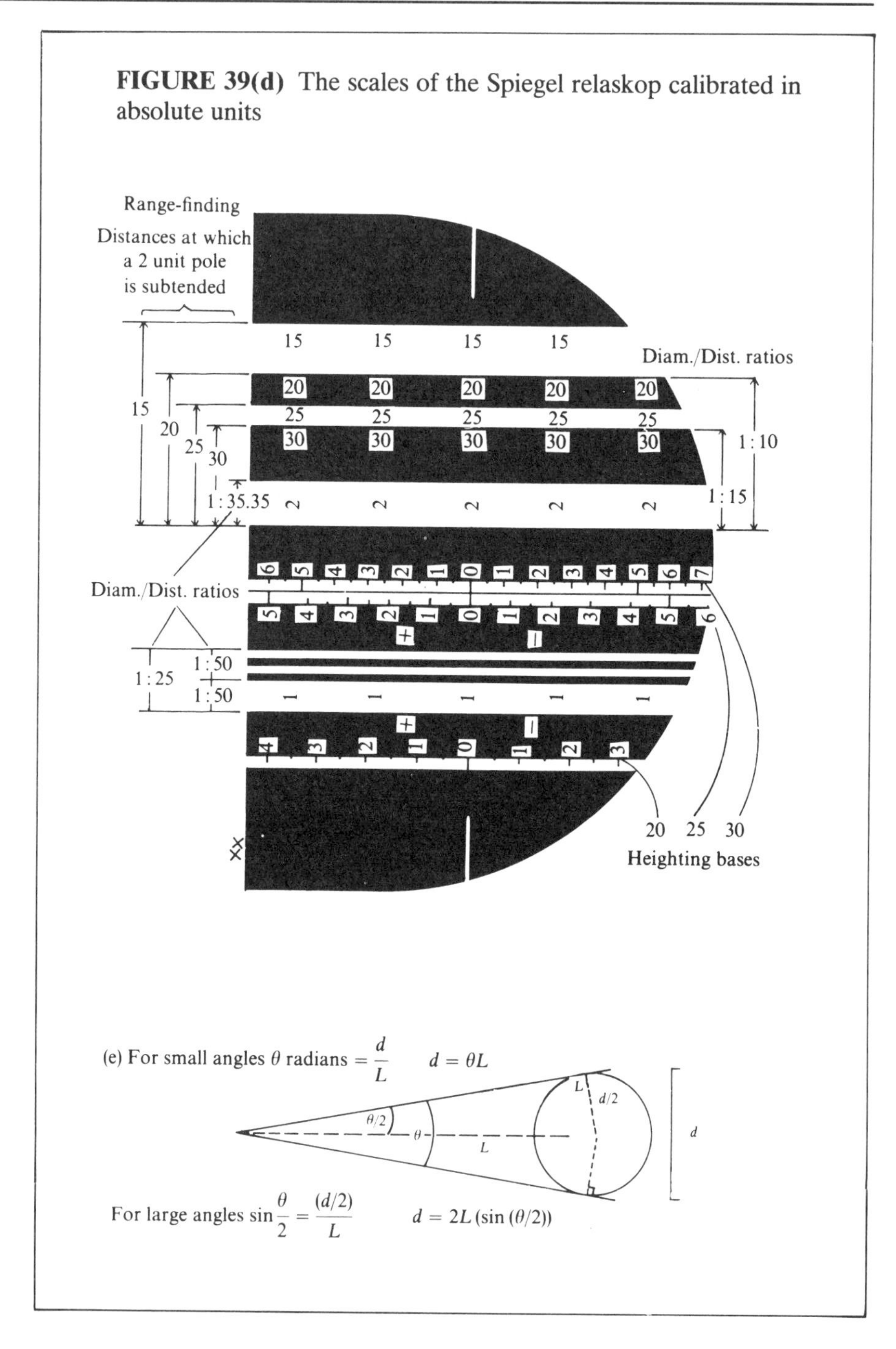

The wide angle relaskop is specially adapted to measure diameter up the bole of a tree. This relaskop uses, as its name implies, wide angles, so the approximation that was referred to in Section 2.2.4.3 is inapplicable (see Figure 39e). This relaskop incorporates a large number of reference angles and height scales giving direct height readings for distances from 4 to 20 m in multiples of

2 m. This instrument is invaluable for measuring diameters in tropical high forest where, because the boles are large and visibility is restricted, wide angles are essential. Nevertheless, accuracy diminishes as the angles of sight increase and they should not be used with extremely steep angles of sight.

3.1.4.2.3 Using the relascope in the field

A representative sample of a stand must be examined, so that relascope counts to estimate the basal area per hectare must be done at more than one point. The question of the number of samples will be discussed in Chapter 4, but a crude rule of thumb suggests that 15 samples per stand should be used. The points may be systematically or randomly sited.

At each sampling point a complete circular sweep is carried out and the number of trees subtending an angle greater than that of the relascope is recorded. Borderline trees are checked by calculating θ from the tree diameter and distance and where necessary exact border trees are counted as ½. The following precautions must be taken to avoid personal bias:

- the profile of every tree must be viewed at breast height; an assistant should outline the profile at the correct height with a white sheet of paper
- leaning trees should be viewed with the gauge sighting the diameter at right angles to the central axis of the bole
- the relascope must be held over the centre point of the sweep except when sighting hidden trees – see below
- the line of sight to breast height must be horizontal or a slope correction employed
- care must be taken to view trees hidden or partially hidden behind a nearer tree. If necessary the relascope may be moved to one side provided that the distance to the tree is not altered
- care must be taken to view individual silhouettes and to separate the overlap of neighbours
- the count must be made without error; starting points must be marked and counted trees marked for tallying. It is easy to forget the count
- borderline trees must be checked mathematically
- a tally factor to give a count of between 7 and 12 trees is often recommended. Some books suggest much higher counts, I think erroneously. Inexperienced observers obtain more consistent results with even higher relascope factors that count less than five trees per sweep
- inexperienced observers should use a tripod or rest for the relascope

Basal area in plantations with a closed canopy normally varies from about 10 to 60 m^2 ha^{-1}, the exact figure varying with species, site, age and intensity of thinning. Consequently relascope factors of between 2 and 4 are commonly used in plantations – a factor of 4 perhaps being the more common – although one of 8 is sometimes needed.

3.1.4.3 *Vertical Point Sampling with a Relascope*

An adaption of horizontal point sampling was developed by several authors (see Strand, 1954, 1958; Grosenbaugh, 1964; Kitamura, 1964) in which the probability of the selection of the trees was proportional to their height squared. As with a relascope, a gauge was held vertically over a point and a 360° sweep made, but viewing the tree heights around the point rather than their boles. In this case the gauge angle was large – about 35°–45° – but was not fixed. *Rather the gauge showed a fixed ratio of a vertical height to a horizontal distance.* The gauge is held at a fixed horizontal distance from the eye in order to determine whether the ratio of the tree height to the horizontal distance to the tree base was equal to or greater than the ratio of the gauge. As with horizontal point sampling, ratios of the tree parameters to the area of the circle defined by the particular ratio gauge and tree height may be calculated. A simple gauge is illustrated in Figure 39(f). In vertical point sampling great care has to be taken to ensure that the arm EF of the gauge is kept horizontal and the gauge AB kept vertical. If available, the most practical instrument for vertical point sampling is the Spiegel relaskop, using the % scale. A suitable % (= ratio) is chosen, e.g. 60%. Then, if the sum of the % readings to the top and bottom of the trees, with appropriate sign, is equal to or greater than the chosen one – the tree is 'in'.

Derivation of formulae

As in Figure 39(f):

$$h_1/R = \tan\alpha$$
$$h_2/R = \tan\beta$$
$$H = R(\tan\alpha + \tan\beta)$$

and generalizing for situations when the eye is at, above or below the tree base:

$$H/R = q = (\tan\alpha \pm \tan\beta)$$

Then the area (a_i) of the circle defined tree of total height h_i

$$a_i = \pi(h_i/q)^2$$

and the basal area per hectare represented by this tree

$$\begin{aligned}(G_i) &= 10^4 \cdot g_i/a_i \\ &= (10^4 \cdot \pi \cdot d_i^2/4)/(\pi \cdot h_i^2/q^2) \\ &= ((10^4/4) \cdot q^2) \cdot (d_i^2/h_i^2)\end{aligned}$$

and summing for all trees in respect to their individual areas

$$G = ((10^4/4) \cdot q^2) \cdot \sum(d_i^2/h_i^2) \text{ m}^2 \text{ ha}^{-1}$$

Similar formulae can be deduced to calculate the numbers of trees per hectare and, given suitable single or two parameter volume tables, volume per hectare. Husch *et al.* (1982) deal with these formulae in detail; the sampling aspects of these techniques will be dealt with in later chapters.

FIGURE 39(f) Vertical point sampling

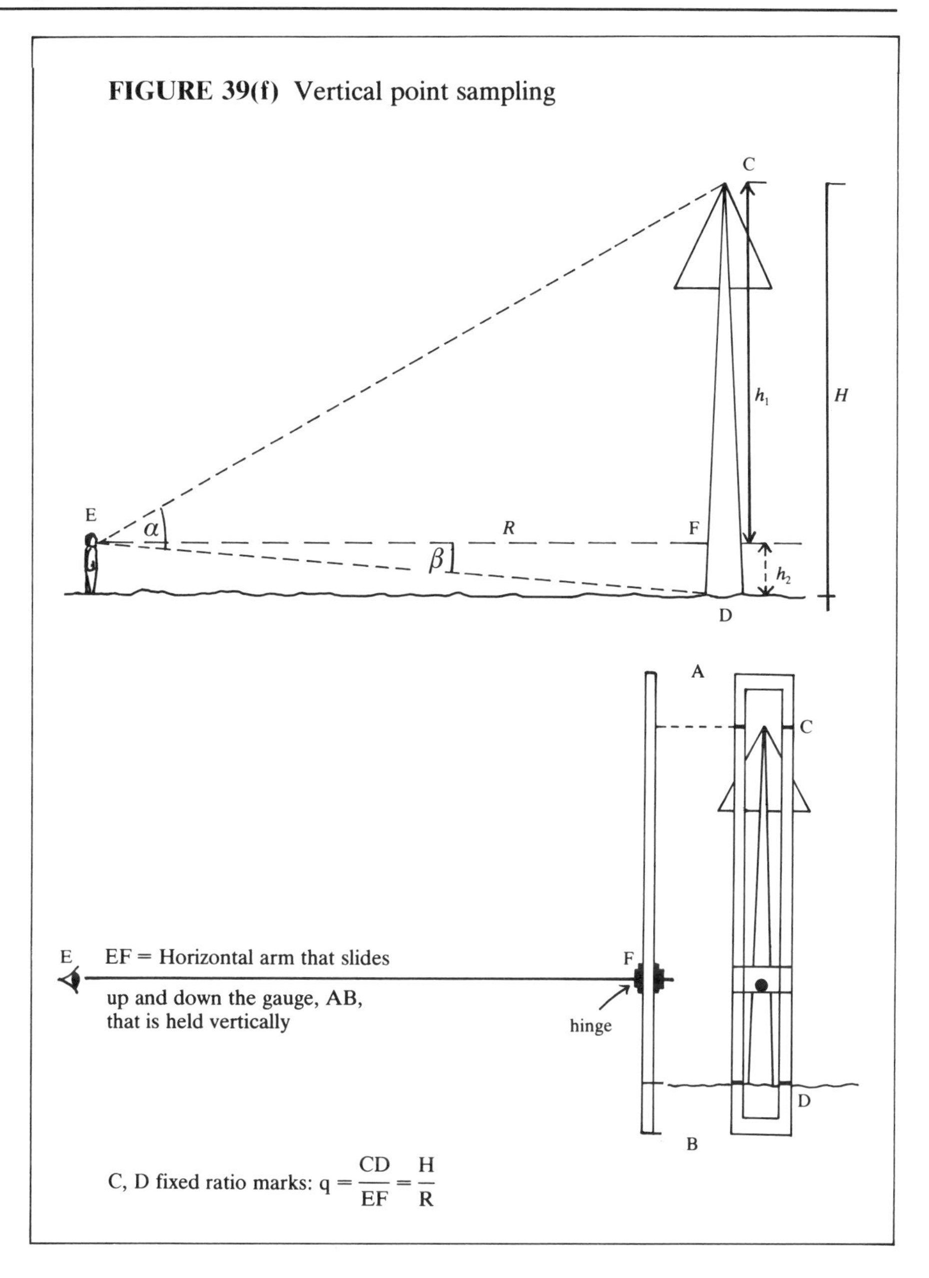

Obviously the major disadvantage of vertical over horizontal height sampling is the need to measure diameter and height of the trees having a height to distance ratio equal to or greater than the reference ratio. Unfortunately visibility of tree tops and bases often makes the technique impracticable. Probability of selection is seen to be proportional to tree height squared, h_i^2.

If it is feasible to use a fixed ratio of 1, then the height of the tree above the observer equals the distance from the tree and so this may be used in place of height. This may be useful in some instances where sophisticated instruments are lacking. Then

$$G = (10^4/\pi) \cdot \sum(g_i/h_i^2) \text{ and}$$
$$V = (10^4/\pi) \cdot \sum(v_i/h_i^2)$$

Nevertheless some technique for measuring or estimating the volumes of the selected trees is necessary, and the problem of visibility still has to be overcome. In some instances, *importance sampling* (see Section 5.2.4.3.4) may be used so that volume can be estimated from a single diameter (that is likely to be within reach from ground level) and the height that has been measured already.

3.1.5 Height

Four different parameters of height are commonly used to describe uniform plantations.

- mean height, unweighted
- mean height weighted by basal area, known as Lorey's mean height
- dominant height
- top height

N.B. Mean and average are synonyms.

All four parameters are closely inter-correlated and can be predicted one from the other. Normally in plantations, unless otherwise defined, height refers to total height; sometimes height to a stated diameter is specified.

3.1.5.1 Mean Height

This is the arithmetic average, i.e.

$$\bar{h} = \sum^{N} h_i/N$$

which is commonly estimated by

$$\bar{h}' = \sum^{n} h_i/n$$

Where:
$\bar{h}$ = mean height
$\bar{h}'$ = estimated mean height
h_i = height of tree i
N = number of trees in the population
n = number of trees used to estimate $\bar{h}'$

The sample of n trees may be either randomly or systematically selected but should be representative so that $\bar{h}'$ is an unbiased estimate of $\bar{h}$. This parameter is more often replaced by Lorey's mean height – see below.

3.1.5.2 *Weighted Average or Lorey's Mean Height*

This is the average height weighted by basal area so that trees with a large basal area – and therefore of a large volume – contribute more to the mean than trees of a small basal area. This is the more common expression of the average height of a stand of trees.

$$\bar{h}_L = \frac{\sum^N g_i h_i}{\sum^N g_i}$$

which is commonly estimated by

$$\bar{h}'_L = \frac{\sum^n g_i h_i}{\sum^n g_i}$$

Where:
$\bar{h}_L$ = Lorey's mean height
$\bar{h}'_L$ = estimate of Lorey's mean height
N = number of trees in the population
n = number of trees used to estimate $\bar{h}_L$
g_i = basal area of tree i
h_i = total height of tree i

A sample is automatically weighted by basal area if it is selected by a relascope; similarly the average height of the trees selected in a vertical point sample represents Lorey's mean height. However, empirical tests have shown that such small samples tend to be biased and overestimate $\bar{h}_L$ as the smaller trees with low probabilities tend to be under-represented (see Section 3.1.4.1). Lorey's mean height is useful as the following relationships may be deduced.

As $V = G h_L \bar{f}_w$

$$\bar{h}_L = \frac{V}{G\bar{f}_w} \neq \frac{\sum^N h_i}{N}$$

$$\bar{f}_w = \frac{V}{G h_L} = \frac{\sum^N g_i h_i f_i}{\sum^N g_i \left(\frac{\sum^N g_i h_i}{\sum^N g_i} \right)}$$

$$= \frac{\sum^N g_i h_i f_i}{\sum^N g_i h_i}$$

Where:
V = stand volume, $m^3\ ha^{-1}$
$\bar{v}$ = mean volume, m^3
G = stand basal area, $m^2\ ha^{-1}$
$\bar{f}_w$ = stand mean weighted form factor = $V/\sum^{N} g_i h_i$
N = number of stems per hectare.
Other symbols as above.

Unfortunately mean height especially and also Lorey's mean height of a stand may be altered by thinning if the mean height of the thinning and of the remaining crop are not identical – for example, removing the wolf trees from a crop or carrying out a heavy crown thinning may depress the mean height of the crop remaining. Conversely the removal of a heavy but low thinning may raise the mean height considerably. Lorey's mean height is most changed by crown thinnings removing trees of a large basal area, but is less affected by low thinnings.

3.1.5.3 *Top and Dominant Height*

In an attempt to find a crop parameter less affected by thinning, measurers have defined mean crop heights using only the largest trees in the crop. 'Largest' has been defined

- either in terms of the tallest trees, i.e. dominant height or the average total height of the 100 tallest trees per hectare. These are not always easy to identify from the ground
- or in terms of the fattest trees, i.e. top height or the average total height of the 100 trees per hectare with the largest diameters at breast height

Unfortunately these definitions of dominant and top height are not universal; recently the two terms have been accepted as synonyms. This text will use dominant height throughout and accept that in practice it will be defined using the 100 trees per hectare with the largest dbh, because these are easier to identify in the field.

Dominant height is commonly used as a determinant of site quality because it is relatively unaffected by thinning.

3.1.5.4 *Estimating Mean Heights*

An unweighted mean height is commonly derived from a systematic, random, or clustered sample.

A weighted mean height is commonly derived from a sample of trees selected using a relascope; normally several sampling points are used. They may be randomly or systematically sited.

Estimates of either dominant or top height are derived using small sample plots sited systematically or randomly in a stand. At each plot the number of trees measured for height is proportional to the plot area and is in the ratio of 100 trees per hectare; e.g. if the sampling units are 0.01 ha then one tree per unit is measured for height – either the tallest or the fattest. Similarly if the sampling units are 0.05 ha then five trees are measured in each unit. Rennolls (1978) has suggested that the definition of dominant height should specify that it is estimated from the average of the heights of the fattest tree in sampling units of 0.01 ha.

3.1.5.5 Height Distributions in Plantations

Heights usually have a nearly Normal or slightly skewed distribution. The exact distribution varies with species, site and thinning history. Fertile sites, shade-tolerant species and unthinned crops tend to have higher coefficients of variation than less fertile sites, more light-demanding species or subjectively thinned crops.

3.1.6 Form

The coefficient of variation of the form of trees in an even-aged crop is high, although the absolute variation tends to be independent of volume for trees over 0.01 m^3. The coefficient of variation of form is much higher than of basal area; consequently more samples have to be taken to estimate the average form factor than to estimate the average basal area with equal precision. (These considerations count against the use of form as the dependent variable in volume estimation.)

As shown in Section 3.1.5.2, the average form factor to be estimated is that weighted by the product of basal area and height, i.e.

$$\bar{f}_w = \frac{\bar{v}}{\bar{g}\,\bar{h}_L} = \frac{\sum^{N} v_i}{\sum^{N} g_i h_i}$$

Where:

$\bar{f}_w$ = average weighted form factor
$\bar{v}$ = average volume per tree in the population
$\bar{g}$ = average basal area
$\bar{h}_L$ = Lorey's mean height
v_i, g_i, h_i = volume, basal area and height of the ith tree

In young crops up to the age when height is 8 m or thereabouts, $\bar{f}_w$ is greatly affected by the relative height of breast height to total height. $\bar{f}_w$ also varies with site, genotype and tending regimes – especially thinning, pruning and fertilization.

3.1.7 Crown Size and Canopy Closure

These two stand parameters together with stand average height are used as independent variables to predict stand volume from measurements taken on aerial photographs. The high correlation between crown and bole diameters and crown area and bole basal area makes crown area a useful predictive variable for stand volume. Normally a somewhat subjective ranking of average crown size compared with a scale made from circles of known radius is used to estimate mean crown area, although measurements on individual trees are feasible on large scale photography and on scattered trees, especially when shadows are contrasted with snow. In stands with complex structures such as uneven-aged forest, unexploited miombo woodland and in tropical evergreen or mixed semi-deciduous forest, and with photographs of less than 1:20,000 scale, individual tree measurements are rarely feasible. Canopy closure, or its synonym 'crown cover' or the proportion of a unit area of ground covered by a vertical projection of tree crowns is also used as a variable to predict stand volume from aerial photographs. On the ground canopy closure is measured using either a spherical densiometer or a moosehorn; both instruments are liable to considerable variation between observers and are crude counts that are correlated with the interception of light by the canopy elements – branches and foliage. Instruments measuring the light intensity at or near ground level are much affected by cloud cover, etc. at the moment of measurement (Vales & Bunnell, 1988).

For pure crops with crowns of much the same size, a constant ratio (K/d) of crown diameter (K) to the bole diameter (d) that is independent of tree age and crown size implies a maximum basal area per hectare for complete crown cover. Further, if the crowns are conceived as perfect non-overlapping circles all of equal diameter, then at square spacing the theoretical maximum canopy closure is 78.5% or at triangular spacing is over 90%. Such relationships provide a crude means of checking growth models to ensure that the inter-relationships of basal area accumulation and numbers of stems per hectare remain reasonable.

EXAMPLE 28 Estimating maximum basal area per hectare for known spacing and K/d ratios

Where: N = number of stems per hectare at full stocking
K = crown diameter, m
d = stem diameter, m

$$z = \frac{K}{d} \text{ the crown/bole diameter ratio}$$

The maximum number of stems per hectare N, assuming square spacing, for trees of a given diameter is given by:

$$N = \frac{10^4}{K^2}$$

and
$$G_{max} = \frac{\pi N d^2}{4}$$
$$= \frac{\pi 10^4 d^2}{4K^2}$$
$$= \frac{\pi 10^4}{4z^2}$$

For a crown/bole diameter ratio of 10, crown diameter and bole diameter in m, the maximum feasible basal area per hectare, G_{max}, with square spacing is

$$G_{max} = \frac{(0.7854)10^4}{10^2}$$
$$= 78.5 \text{ m}^2 \text{ ha}^{-1}$$

or for a K/d ratio of 15 at square spacing

$$G_{max} = \frac{(0.7854)10^4}{15^2}$$
$$= 34.9 \text{ m}^2 \text{ ha}^{-1}$$

3.1.8 Volume per Hectare

Normally volume per hectare is estimated by means of small sampling units by

- direct measurement of the volume by felling or measuring the volume of each standing tree
- indirect estimates of tree volumes through single tree volume or taper/form tables
- indirect estimates of volume per hectare using a stand volume table

The measurements or estimates may be made on all trees in each sampling unit or may be made on a sub-sample. The sampling units may be plots or points.

3.1.8.1 Direct Measurement of the Volume of Small Sample Plots by Felling or Measuring Standing Trees

The volume per hectare can be estimated using the formula

$$V = \frac{\sum^{n}\left(\sum^{m_i}\right) v_{ij}}{na}$$

Where:
V = average volume per hectare, m², estimated from n samples each of a hectares
v_{ij} = volume of an individual tree measured on the ith plot after felling or measured standing
m_i = total number of trees in the ith plot, $1 \ldots i \ldots n$
n = number of plots
See Example 29.

The volume of standing trees can be measured directly either by climbing, using a Spiegel relaskop to measure log lengths and mid-diameters or by height accumulation (see Section 2.7.2.2 and 2.7.2.4).

Alternatively tree volume may be estimated using single tree volume tables.

If sub-sampling for volume is practised there are three common methods used to calculate volume per hectare

- the mean tree method
- in proportion to basal area or the mean form height method
- the regression of volume on basal area method

Care must be taken in the application of the following methods to ensure that the method of calculation matches the method of sample tree selection which may be with equal probability or with probability proportional to some measure or estimate of tree size.

EXAMPLE 29 Calculation of volume per hectare (1)

In an inventory of a stand of *Pinus patula* the following data were collected.

$n = 5$

$a = 0.005$ ha

Trees	1	2	3	4	5	Total
Plots			Volumes (m^3/tree)			
1	0.42	0.36	0.39	0.27	–	1.44
2	0.38	0.37	0.41	0.40	0.41	1.97
3	0.29	0.36	0.31	0.34	–	1.30
4	0.41	0.36	0.34	0.33	–	1.44
5	0.30	0.40	0.39	0.27	–	1.36

$$\sum^{n}\sum^{m} v_{ij} = 7.51 \text{ m}^3$$

$$\bar{V} = \frac{\sum^{n}\sum^{m} v_{ij}}{na}$$

$$= \frac{7.51}{(5)(0.005)}$$

$$= 300 \text{ m}^3 \text{ ha}^{-1}$$

In this example all the trees in each sample plot were measured standing with a Spiegel relaskop.

3.1.8.1.1 Mean tree method

This method uses the average volume of the sub-sample trees in each plot and from this and the number of trees in each plot, the volume of each plot and volume per hectare are calculated:

$$\bar{v}_i = \frac{\sum^{s_i} v_{ik}}{s_i} \quad \text{m}^3 \text{ per tree}$$

$$v_i = m_i \bar{v}_i \quad \text{m}^3 \text{ in plot } i$$

$$V = \frac{\sum^{n} v_i}{na} \quad \text{m}^3 \text{ ha}^{-1}$$

Where:

V, m, n and a are as defined in Section 3.1.8.1

v_{ik} = volume of the kth individual tree in sampling unit i used as a sub-sample and measured for volume

v_i = estimated total volume of the m_i trees in the ith sampling unit

s_i = number of trees in the sub-sample measured for volume in plot i

The accuracy of this method depends upon the sample trees reflecting the true plot mean volume per tree; this may be facilitated by using sample trees having the mean basal area. Hence their choice may be governed by Weiss's 40% rule (see Section 3.1.3.2.1).

EXAMPLE 30 Calculation of volume per hectare (2)

In an inventory of a stand of *Cupressus lusitanica* the following data were collected.

$n = 5$
$a = 0.01$ ha

i	m_i	s_i	v_{ik}	$\sum^{s_i} v_{ik}$	$\bar{v}_i$	$v_i = \bar{v}_i m_i$ or	$v_i = \bar{v} m_i$
1	10	4	0.14 0.12 0.13 0.09	0.48	0.120	1.20	1.39
2	12	4	0.13 0.12 0.14 0.13	0.52	0.130	1.56	1.67
3	9	3	0.11 0.12 0.20	0.43	0.143	1.29	1.25
4	11	4	0.10 0.13 0.13 0.09	0.45	0.113	1.24	1.53
5	12	2	0.28 0.20	0.48	0.240	2.88	1.67
	54	17		2.36		8.17	7.51

$$\bar{v} = 2.36/17 = 0.139$$

$$V = \frac{\sum^{n} v_i}{na} = \frac{8.17}{(5)(0.01)}$$

$$= 163 \text{ m}^3 \text{ ha}^{-1} \text{ (N.B. Note the high weight of plot 5)}$$

OR using a pooled mean volume

$$V = \frac{7.51}{(5)(0.01)} = 150 \text{ m}^3 \text{ ha}^{-1}$$

This is an inappropriate method unless s_i is sufficiently large to afford a representative sample and a precise estimate of the tree of mean volume in a plot. A sub-sample of at least 20 trees per plot is normally necessary to provide such a result. Alternatively the sub-sample trees in all plots can be pooled to provide a pooled tree of mean volume.

$$\bar{\bar{v}}_{\text{pooled}} = \frac{\sum^{n}\sum^{s_i} v_{ik}}{\sum^{n} s_i} \quad \text{m}^3 \text{ per tree}$$

$$v_i = m_i\bar{\bar{v}} \quad \text{m}^3 \text{ in plot } i$$

$$V = \frac{\sum^{n} v_i}{na} \quad \text{m}^3 \text{ ha}^{-1}$$

Where symbols are as above. See Example 30.

However, this may result in a biased estimate if different plots provide different numbers of trees in the sub-sample and contain different sizes of tree.

3.1.8.1.2 Mean form height method

The diameter of each tree in the plot is measured. A sub-sample of trees is selected using either a systematic procedure, or using a relascope so that the probability of selection for the sub-sample is proportional to the tree's basal area. Both the diameter and the volume are measured on each tree in the sub-sample. The mean form height is estimated from this sub-sample as the ratio of their total volume to their total basal area. The calculation may be done in two ways:

1. Using a mean form height for each plot:

$$\overline{fh}_i = \frac{\sum^{s_i} v_{ik}}{\sum^{s_i} g_{ik}}$$

$$v_i = \overline{fh}_i \sum^{m_i} g_{ij} \quad \text{m}^3 \text{ per plot}$$

and

$$V = \frac{\sum^{n} v_i}{na} \quad \text{m}^3 \text{ ha}^{-1}$$

2. With a mean form height pooled over all plots:

$$\overline{\overline{fh}} = \frac{\sum^{n}\sum^{s_i} \frac{v_{ik}}{g_{ik}}}{\sum^{n} s_i}$$

Then

$$v_i = \overline{\overline{fh}} \sum^{m_i} g_{ij} \text{ m}^3 \text{ per plot and } V = \frac{\sum^{n} v_i}{na} \text{ m}^3 \text{ ha}^{-1}$$

Where:
$\overline{fh}_i$ = the mean form height of the s_i sample trees in plot i
$\overline{\overline{fh}}_i$ = the mean form height calculated from $\sum^{n} s_i$ sample trees
s_i = number of trees in the sub-sample of plot i measured for volume
n = number of plots
v_{ik} = volume of the kth sub-sample tree in plot i measured for volume, m^3
g_{ik} = basal area of the kth sub-sample tree in plot i measured for volume, m^2
m_i = number of trees in sample plot i
g_{ij} = basal area of the jth tree in the ith plot, m^2
V = estimate of volume per hectare, m^3
a = area of sampling unit or plot

As basal area is highly correlated with volume, the use of basal area rather than numbers of trees as the multiplying factor to convert the sub-sample volume to the plot volume is expected to provide a more precise estimate, especially where the probability of selection of the sub-sample trees is proportional to a tree's basal area. Hence the mean form factor method with sub-sample selection by relascope is recommended in preference to the mean tree method. The UK Forestry Commission Tariff system (see Section 2.2.4.8) is an adaptation of this method.

EXAMPLE 31 Calculation of volume per hectare (3)

In an inventory of a stand of *Pinus patula* the following data were collected.

$n = 5$
$a = 0.01$ ha

i	m_i	$\sum^{m_i} g_{ij}$ m²	s_i	d_{ik} cm	g_{ik} m²	v_{ik} m³	$\sum^{s_i} g_{ik}$ m²	$\sum^{s_i} v_{ik}$ m³	$\overline{fh}_i$
1	10	0.124	4	13.4	0.014	0.14	0.049	0.48	9.80
				11.8	0.011	0.12			
				13.4	0.014	0.13			
				11.3	0.010	0.09			
2	12	0.132	4	12.9	0.013	0.13	0.053	0.52	9.81
				12.4	0.012	0.12			
				13.8	0.015	0.14			
				12.9	0.013	0.13			
3	9	0.119	3	11.3	0.010	0.11	0.044	0.43	9.77
				12.9	0.013	0.12			
				16.4	0.021	0.20			
4	11	0.100	4	10.7	0.009	0.10	0.044	0.45	10.23
				11.8	0.011	0.13			
				13.4	0.014	0.13			
				11.3	0.010	0.09			

Example 31 continued

5	12	0.140	2	19.9	0.031	0.28	0.048	0.48	10.00
				14.7	0.017	0.20			
Totals	54	0.615	17		0.238	2.36			

$$\overline{\overline{fh}} = \frac{\sum^{n}\sum^{s_i} v_{ik}}{\sum^{n}\sum^{s_i} g_{ik}} = \frac{2.36}{0.238} = 9.92$$

i	$\sum^{m_i} g_{ij}$	$\overline{fh}_i$	V_i
1	0.124	9.80	1.22
2	0.132	9.81	1.29
3	0.119	9.77	1.16
4	0.100	10.23	1.02
5	0.140	10.00	1.40
	0.615		6.09 m³ in 0.05 ha

$$V = \frac{6.09}{(5)(0.01)} = 121.8 \text{ m}^3 \text{ ha}^{-1}$$

OR using a pooled mean form height of 9.92

$$V = \frac{(0.615)(9.92)}{(5)(0.01)} = 122.0 \text{ m}^3 \text{ ha}^{-1}$$

3.1.8.1.3 Regression of volume on basal area method (see Section 5.1.1.8)

The sub-sample trees from all the sampling units are pooled and a simple linear regression of volume on basal area is calculated. The regression equation is then applied to all the trees in each sampling unit separately.

$$v_{ik} = a + b(g_{ik}) \qquad \text{m}^3 \text{ per tree}$$

$$v_i = m_i a + b\sum^{m_i} g_{ij} \qquad \text{m}^3 \text{ per plot}$$

$$V = \left(\sum^{n} m_i a + b\sum^{n}\sum^{m_i} g_{ij}\right) / nr \qquad \text{m}^3 \text{ ha}^{-1}$$

Where:
a, b = regression constants
r = area of sampling unit
$v_{ik}, g_{ij}, g_{ik}, m_i$ and n as before

This is normally the preferred method, especially where computers are available and used to calculate results of inventories.

EXAMPLE 32 Calculation of volume per hectare (4)

Using the same data as in the previous example, the volume on basal area regression is calculated and volume per hectare derived using the regression equation:

$n = 17$
$a = 0.01$ ha

v_{ik}	g_{ik}	$(g_{ik})^2 \cdot 10^3$	$(v_{ik}g_{ik}) \cdot 10^2$
0.14	0.014	0.196	0.196
0.12	0.011	0.121	0.132
0.13	0.014	0.196	0.182
0.09	0.010	0.100	0.090
0.13	0.013	0.169	0.169
0.12	0.012	0.144	0.144
0.14	0.015	0.225	0.210
0.13	0.013	0.169	0.169
0.11	0.010	0.100	0.110
0.12	0.013	0.169	0.156
0.20	0.021	0.441	0.420
0.10	0.009	0.081	0.090
0.13	0.011	0.121	0.143
0.13	0.014	0.196	0.182
0.09	0.010	0.100	0.090
0.28	0.031	0.961	0.868
0.20	0.017	0.289	0.340
2.36	0.238	3.778	3.691

$$b = \frac{\sum^{n} vg - \frac{\sum^{n} v \sum^{n} g}{n}}{\sum^{n} g^2 - \frac{\left(\sum^{n} g\right)^2}{n}}$$

$$= \frac{\left(3.691 - \frac{(2.36)(0.238)(10)^2}{17}\right)}{\left(3.778 - \frac{(0.238)^2(10)^3}{17}\right)} = \frac{(0.387)(10)}{0.446} = 8.677$$

$$a = \bar{v} - b\bar{g} = 0.139 - 0.121 = 0.018$$

$$v = 0.018 + (8.677)(g) \text{ m}^3 \text{ per tree}$$

$$\sum^{n} v_i = \left(\sum^{n} m_i\right) a + b \sum^{n} \sum^{m_i} g_{ij}$$

$$= 54(0.018) + (8.677)(0.615) = 6.308 \text{ m}^3 \text{ in } 0.05 \text{ ha}$$

$V = 6.308/((5)(0.01)) = 126$ m^3 ha^{-1} (see Section 5.1.1.8)

3.1.8.2 *Volume Estimation Using Point Sampling*

In miombo woodland or other forest types where boles are short and log length is obviously determined by forking, volume per hectare may be estimated directly using a Spiegel relaskop. At random points in a stand a sweep is carried out sighting the Spiegel relaskop at the mid-lengths of each saw log of each tree viewed. Where a tree has more than one log, then sights are made at the mid-length of each. If the mid-point subtends an angle greater than the reference angle of the relaskop, then the length of the saw log (L_i) is measured and accumulated. Then the volume per hectare is estimated by

$$V = F\left(\sum^{n} L_i\right) \quad \text{m}^3\ \text{ha}^{-1}$$

Where:
F = relascope factor
n = number of saw logs whose length has been measured

Inaccuracies will occur if the boles are not vertical. Checking borderline cases is also difficult (Temu, 1979).

Each log whose mid-length profile subtends an angle equal to or greater than the reference angle of the relaskop represents F square metres of cross-sectional area per hectare, which when multiplied by the length of the log gives an estimate (FL_i) m^3 ha^{-1}. Hence

$$V = F\left(\sum^{n} L_i\right) \quad \text{m}^3\ \text{ha}^{-1}\text{, as above}$$

FOR THE ADVANCED STUDENT

Where a simple mathematical expression of volume in terms of single tree parameters is available, then a formula may be deduced to convert tree counts and measurements directly to volume per hectare. For example:

Where $v = a + b(d^2h)$

then (see Section 3.1.4.2)

$$\frac{\text{volume of } n \text{ trees}}{\text{area of circular sweep}} = \sum^{n}\left(\frac{a + b(d_i^2h_i)}{\pi L_i^2}\right) \quad \text{m}^3$$

$$= \sum^{n} \frac{\theta^2 a}{\pi d_i^2} + \frac{\theta^2 bh_i}{\pi} \quad \text{m}^3$$

$$\text{volume per hectare} = \sum^{n} \frac{\theta^2 10^4 a}{d_i^2 \pi} + \theta^2 \frac{10^4 bh_i}{\pi} \quad \text{m}^3\ \text{ha}^{-1}$$

but $\theta^2 10^4 = 4F$

$$\text{volume per hectare} = \frac{4Fa}{\pi}\sum^{n}\left(\frac{1}{d_i^2}\right) + \frac{4Fb}{\pi}\sum^{n} h_i \quad \text{m}^3\ \text{ha}^{-1}$$

If
$$\frac{4Fa}{\pi} = k_1 \text{ and } \frac{4Fb}{\pi} = k_2$$

$$V = k_1 \sum^{n} \left(\frac{1}{d_i^2}\right) + k_2 \sum^{n} h_i \qquad \text{m}^3\ \text{ha}^{-1}$$

This formula implies that one must measure the diameter and height of each tree tallied with the relascope.

Where vertical point sampling is used similar expressions can be derived to convert the diameters and heights of trees selected by the angle gauge to volume per hectare:

$$V = (10^4\theta^2) \cdot \sum \frac{v_i}{h_i^2} \quad \text{symbols as above and in Section 3.1.4.8.3}$$

A modification of vertical point sampling for volume is critical height sampling (Strand, 1954, 1958; Kitamura, 1964; Bitterlich, 1976; McTague and Bailey, 1985; Deusen, 1987; Lynch, 1990). In this technique of point sampling a Spiegel relaskop is used to determine the height of trees within a sweep at which their bole exactly subtends the same angle as the relaskop. Then

$$V = F \cdot \sum h_{ic}$$

Where:
V = volume, $\text{m}^3\ \text{ha}^{-1}$
F = relaskop factor
h_{ic} = critical height of the ith tree
The formula makes assumptions about the shape of the tree bole and necessitates good visibility from the sampling point up the bole of each tree to its critical height. In the case of trees near the sampling point this may involve steep angles of sight and is frequently impracticable. Bitterlich generalized the tree taper thus

$$\frac{b_{ci}}{b_i} = \frac{l_{ci}}{l_i}^{\lambda}$$

Where:
b_{ci} = cross-sectional area at critical height in relative units, e.g. units of Bitterlich's relaskop
b_i = cross-sectional area at bh in relative units
l_{ci} = length from the tip to the critical height
l_i = length from tip to bh
λ = a power term equalling 1 for a parboloid, 2 for a cone, 3 for a neiloid

and designed a simple hand held calculator program to solve for λ from a few measurements of taper taken on sample trees in the field. Then using this value for λ, he solved the taper equation for the critical height of each sample tree knowing its

- height in metres
- d_{bh}/d_c in relative units, e.g. in the relascope units of the telerelaskop

The estimated critical heights were then used in the formula

$$V = F \cdot \sum h_{ic}$$

The total volume per hectare estimated by critical height sampling is slightly biased as the formulae imply the shape of a cylinder below bh; in theory this could be avoided by counting in trees that subtend an angle equal to or greater than the reference angle below bh. If a taper model is assumed, then the volume estimated can be considered analagous to the silve or other type of tariff volume.

3.1.8.3 Stand Volume Tables

Instead of felling sample trees or measuring them standing or using a single tree volume table, volumes per hectare of even-aged crops of single species may be predicted directly using a stand volume table. The commonest stand volume table is derived from a simple linear regression of volume per hectare, V, on the combined variable – basal area per hectare multiplied by some measure of height representative of the crop; often dominant height is used because it is convenient, being objectively defined as the height of the 100 fattest or tallest trees per hectare (see Section 3.1.5.3).

$$v = a + b(Gh_{\text{top}})$$

Stand volume tables normally have confidence limits at $p = 0.05$ of about 5–10% of the mean stand volume using 20 samples per stand. They are simply constructed either by felling representative samples in a defined population of stands or, but less satisfactorily, by measuring the standing volumes of all the trees in small plots, or predicting their volumes using a single tree volume table. In this latter case the error of prediction of the average stand volume must be derived from the sum of the three sources of error, i.e. the residual variance in the single tree volume table, the residual variance in the stand volume table, and the variance of the sampling units themselves. If trees are measured standing or samples are felled then only the last two sources of variation are concerned.

3.1.8.4 Other Methods of Estimating Stand Volume

In more uneven-aged stands the use of the simple methods outlined above may be inappropriate because the variation in form height is excessive. Then the *height curve method* or sampling by diameter classes may be more efficient.

The *height curve method* uses a two parameter volume table with diameter and height as the predictor variables. A representative sample of trees is measured to establish a relationship between height and diameter. Then

- a representative sample is measured to establish the frequency of trees by diameter classes. This may be done using a systematic sample; at least 200 trees are likely to be required (Hamilton, 1975).
- the volume of the mean tree in each diameter class is estimated from the two parameter volume table, using height obtained from the height on diameter model

- the volume of the stand is estimated as the sum of the volume in each diameter class, i.e.

$$TV = \sum^{m} N_i \bar{v}_i \quad m^3$$

Where:
TV = total volume
N_i = number of stems in the ith diameter class
v_i = mean volume per tree in the ith diameter class
m = number of diameter classes

Sampling by diameter classes determines total volume in a similar way:

- a representative sample is measured to establish the frequency of trees by diameter classes
- in each diameter class a representative sample of trees is measured for volume, or their volume is estimated using a volume table
- the mean volume for each diameter class is calculated
- the volume of the stand is estimated as the sum of the volume in each diameter class, as above

$$TV = \sum^{m} N_i \bar{v}_i \quad m^3 \text{ (symbols as above)}$$

In fact this is a stratified sampling design using the diameter classes as strata. Consequently the normal rules for allocating sampling units to the strata in order to obtain the most precise estimate within a given cost apply (see Section 5.4.2.1.2). If random sampling is employed the error is a function of the number of trees and the variance in each diameter class. Most sampling units should be allocated to the diameter classes with the highest volume and variance; especially the larger diameter classes should be adequately represented as their variance and contribution to the total volume may be expected to be high.

FOR THE ADVANCED STUDENT

3.1.8.5 Volume Distributions

The standing volume in an even-aged crop of one species can be described by the estimated average volume per hectare, its standard deviation and the mean volume per tree. However, these three parameters

- omit a measure of the volumes per hectare using different top diameters
- fail to define the distribution of volumes per tree around the mean volume; volumes per tree are usually skewed with the mean volume higher than the mode, i.e. negatively skewed

3.1.8.5.1 Volumes per hectare to different top diameters

The commonest way to predict stand volume to different top diameters is from functions of total volume

$$V_x = f(x, V_t)$$

Where:
V_x = volume to a top diameter over bark of x cm, $m^3 ha^{-1}$
V_t = total volume over bark, $m^3 ha^{-1}$

This function is not linear. Care has to be taken if, owing to sampling errors in the model, the prediction of V_x is greater than V_t; Williamson (1976) recommended using a logarithmic function; alternatively the volumes can be expressed as ratios of total volume and the models constrained so that a volume to a greater diameter must be less than the volume to a lesser diameter. In a crop in which all trees are identical the distributions of volume in the tree and in the stand are identical, but as variance and skew increases, the distributions in the tree and stand diverge. Consequently, for more precise predictions of volume distributions to different top diameters, not only total volume and mean volume per tree, but the probability function of the volume per tree distribution is needed.

3.1.8.5.2 Volumes per hectare by categories of tree size

Stand volume within limits defined by breast height diameters or tree volumes can be calculated if the stem frequency distribution is known. Two stands with identical volume per hectare and average volume per tree will have very different volumes for trees above a stated minimum diameter or volume if the frequency distributions are different – perhaps due to different thinning histories.

Stand volumes may be obtained by summing the volumes per diameter or volume class within the desired limits. More elegant and general solutions may be found if the probability distribution of the stem volume frequencies can be predicted and volumes obtained by integration. The most useful but most sophisticated models of stand volume to different top diameters or log specifications are provided by probability functions of the diameter or volume per tree distribution, coupled with taper functions to provide volume per tree by integration.

3.1.9 Biomass per Hectare

Biomass stand tables are usually calculated by destructive sampling to ascertain the green weight component and moisture content of the different parts of the tree. Biomass per hectare is then synthesized by summing the biomass of each species and diameter class (see Section 2.8).

3.1.10 Growth

Growth in even-aged stands of one species can be measured in terms of

- increase in mean diameter per tree
- increase in mean basal area per tree
- increase in mean (or dominant or top) height
- increase in mean form factor
- increase in mean volume per tree
- increase in basal area per hectare
- increase in volume per hectare
- change in stem numbers per hectare

Growth is best measured by repeated measurements in permanent or temporary sample plots with individually identified trees. This topic along with that of growth prediction will be dealt with in Chapters 5 and 7.

3.2 Descriptions of Even-aged Crops of More Than One Species

Even-aged crops of more than one species are normally described in quantitative terms as if they were pure, provided that one species forms a dominant part – usually over 80% – of the crop by volume. If, however, individual species form 20% or more of a crop by volume, basal area or stem numbers, then each is described separately as in the preceding sections. Such mixtures increase the costs of measuring standing crops but otherwise cause little difficulty. In contrast, the prediction of growth is far more complex and usually the methods for mixed uneven-aged crops have to be employed; such predictions are far less precise than those for pure crops because knowledge of the interactions between species and sizes is fragmentary.

3.3 Parameters Describing Uneven-aged Crops of a Single or, More Commonly, Several Species

Because of the variation in the distribution of age, size and species in this type of forest, parameters such as average age, average stem number per hectare, average stem size or average total volume per hectare and their standard deviations are ineffective measures for representing the population. In their place usually more complex data have to be used based on non-normal distributions – usually for each participant species separately. The most common are

- diameter or basal area frequencies
- volume frequencies

3.3.1 The Diameter and Basal Area Frequency Distribution

FOR THE ADVANCED STUDENT

3.3.1.1 The de Liocourt or Negative Exponential Model

Many mixed stands with a continuous series of age classes and continuous recruitment by natural regeneration illustrate a diameter distribution in which each diameter class has fewer stems than the adjoining, smaller diameter class; also the ratio of the number of stems in a class to the number in the adjoining and smaller class is constant. The shape of this distribution is shown in Figure 40. The mathematical model for this type of distribution is

$$Y = ke^{-aX} \text{ where } k \text{ and } a \text{ are constants}$$

EXAMPLE 33 A negative exponential diameter distribution

$$Y = 10^4 e^{-0.07X}$$

e = 2.7183, Y = stems per hectare, X = dbh class, cm

When $X = 10$, $Y = 10^4(2.7183)^{(-0.07)(10)} = 4966$
20, $Y = 10^4(2.7183)^{(-0.07)(20)} = 2466$
30, $Y = 10^4(2.7183)^{(-0.07)(30)} = 1225$
40, $Y = 10^4(2.7183)^{(-0.07)(40)} = 608$
50, $Y = 10^4(2.7183)^{(-0.07)(50)} = 302$
60, $Y = 10^4(2.7183)^{(-0.07)(60)} = 150$

The constants k and a in the model vary with the species and site. The constant k reflects the stocking of very small seedlings, so that models of prolific seeders have large k values; the constant a governs the relative frequencies of successive diameter classes – a comparatively large value for a being associated with high mortality between classes and low stockings of large trees. For example, the model of most pioneers that are prolific seeders like most *Betula* spp., *Trema* spp., *Macaranga kilamandscharica,* etc. will be expected to have a large k and a rapid fall in numbers in each succeeding size class (large diminution quotient) which implies a relatively large value for a. The model for a wide crowned, light-demanding species with a heavy, large, bird and animal disseminated fruit such as *Maesopsis eminii* would be expected to have a smaller k but a large a, whereas the model portraying a shade-tolerant 'climax-type' species such as *Fagus sylvatica, Cynometra alexandri, Strychnos mitis, Rapanea rhododendroides* or *Uapaca guineensis* would be expected to have a large k and a relatively small a. Better sites that support large stockings than poor sites would be expected to have relatively smaller values for a (and possibly larger values for k) than poor sites.

Manipulation and calculation of this negative exponential model may be facilitated by transformation to logarithms and plotting on semi-logarithmic graph paper. Then the curvi-linear relationship illustrated in Figure 41(a) is transformed into the linear relationship shown in Figure 41(b).

On taking logarithms of both sides,

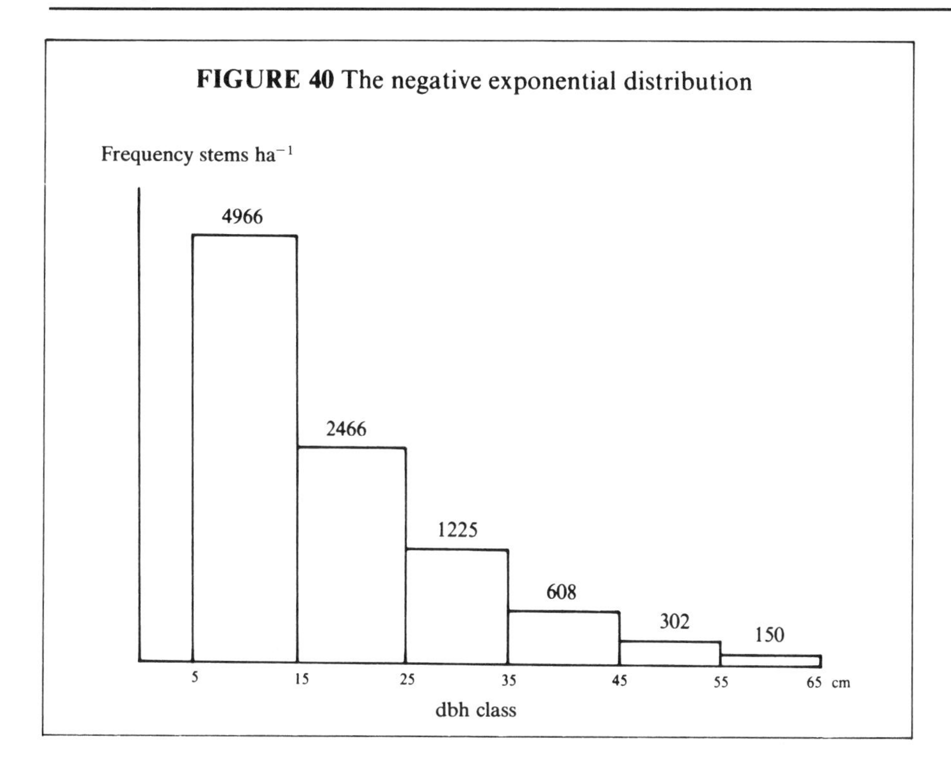

FIGURE 40 The negative exponential distribution

$Y = ke^{-aX}$ is transformed into the linear equation

$$\log Y = \log k - aX(\log e)$$

but k, a and e are constants and only Y and X are variables, so this equation may be rearranged as

$$\log Y = \log k - (a \log e)X$$

which is a simple linear equation where $\log k$ is the regression constant and (a log e) the regression coefficient.

N.B. Both Y and k (numbers per hectare) are in the logarithmic scale but X, the diameter, is untransformed. The constants k and a may be calculated from measurements made in the forest (Moser, 1976).

EXAMPLE 34 Calculation of k and a from data obtained from a forestry inventory

The number of stems per hectare in the 60 cm diam class = 100

The number of stems per hectare in the 59 cm diam class = 107

The diminution quotient $q = 107/100 = 1.07$

but $q = \dfrac{N_{59}}{N_{60}} = \dfrac{ke^{-a59}}{ke^{-a60}} = e^{a} = 1.07$

so $\log q = a \log e$ and $a = \dfrac{\log q}{\log e}$

Example 34 continued

Therefore $Y = kq^{-X}$
and log $Y = \log k - (\log q)X$
and log $k = \log Y + (\log q)X$

$$\log k = 2 + (0.0294)60 = 2 + 1.7640 = 3.7640$$
$$k = 5808$$
$$q = 1.07 = e^a$$
$$a = 0.068$$

N.B. If the observations are made in 2 cm diameter classes then

$$\frac{N_{58}}{N_{60}} = \frac{N_{58}}{N_{59}} \frac{N_{59}}{N_{60}} = q^2$$

A diminution quotient q estimated from a single pair of data is liable to very large sampling errors. Usually it would be calculated through a regression using data collected for all size classes and the logarithm transformation

FIGURE 41 Transformation of the negative exponential distribution

$Y = ke^{-aX}$ $k = 10^4$ $a = 0.07$

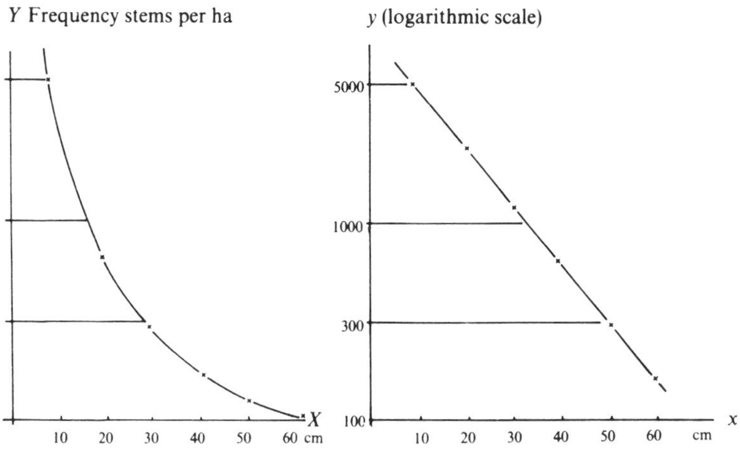

$Y = ke^{-aX}$
(a)

$\log Y = \log k - (a \log e)X$
(b)

The negative exponential model is frequently used as a standard of comparison for natural or mixed stands managed on a polycyclic cutting system with continuous regeneration under the shelter of the mature trees, and as an aid in designing cutting schedules in such stands (Figure 42).

Rollet (undated) compiled data from studies of tropical evergreen lowland forest from South America, Asia and Africa. He suggested that the negative exponential model may be of use for predicting numbers of stems of all species greater than 20 or 30 cm dbh, but is unreliable for diameters less than 20 cm.

He found that the frequency in the smaller sizes was greater than that predicted by the negative exponential and suggested that the smaller diameter size classes should be modelled separately using a truncated log normal distribution. He also studied the diameter frequency distribution of individual commonly occurring species and found tremendous variation in the form of the distributions varying from completely erratic, symmetrical to those similar to a negative exponential. Both Rollet (undated) and Dawkins (1958) have proposed pan-tropical stand tables, but both were based on averages from inventories rather than predictions derived from either an empirical or a deductive model. Table 1 compares Rollet's and Dawkins' stand tables.

FIGURE 42 The design of cutting schedules in mixed forest

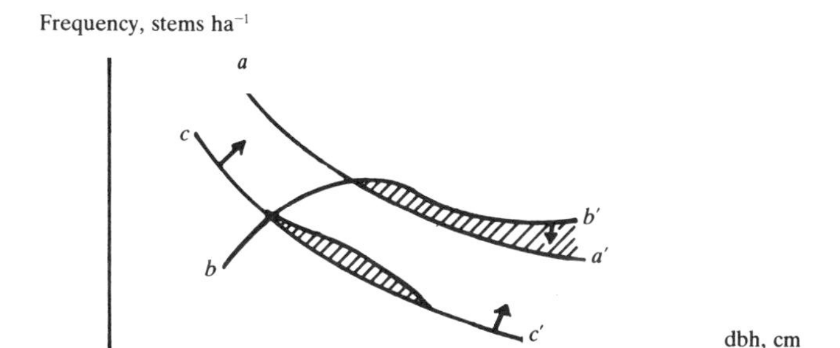

In Figure 42 curve $(a - a')$ represents an ideal stocking and is a negative exponential. Curves $(b - b')$ and $(c - c')$ represent actual stockings in two stands B and C to both of which the ideal stocking is thought to apply. With the aim of improving regeneration, cutting in stand B should concentrate on the excess stock indicated by the shading; whereas in Stand C the aim is to remove the excess stocking as shown by the shading in the diagram and, by cutting less than the growth, gradually increase the growing stock in all other size classes so that the stocking approaches the curve $(a - a')$ in all size classes.

3.3.1.2 *Basal Areas in Tropical Moist Evergreen Forest*

The stand or diameter frequency table provides the raw data for calculating the basal area by size classes. As with diameters, Rollet reports very considerable variation from stand to stand and country to country. Dilmy (1971) and numerous authors have reported that basal areas (and therefore volumes and biomass) are highest on well drained soils and very much lower on inundated sites and permanent swamps. For example, whereas the total basal area of well drained mixed semi-deciduous high forest in Uganda may be as high as 40 m^2 ha^{-1}, that of *Mitragyna stipulosa* swamp forest is less than 20 m^2 ha^{-1}. Table 1 also shows Dawkins' and Rollet's stand tables converted to m^2 ha^{-1} by 10 cm diameter classes, and indicates that although the distributions have a similar form, that of Rollet implies a very much lower basal area per hectare than that of Dawkins.

Table 1 Pan-tropical stand structures

Dbh class	20-29		30-39		40-49		50-50		60-69	
Authority	*N*	*G*	*N*	*G*	*N*	*G*	*N*	*G*	*N*	*G*
Rollet	80	4.99	41	4.99	21	4.26	12	3.47	6	2.60
Dawkins	101	6.33	42	5.15	20	4.00	11	3.36	7	3.03
Dbh class	**70-79**		**80-89**		**90-99**		**⩾ 100**		**Total**	
Authority	*N*	*G*	*N*	*G*	*N*	*G*	*N*	*G*	*N*	*G*
Rollet	3	1.92	2	1.48	1	1.08	2	1.98	168	26.77
Dawkins	5	2.78	4	2.53	2	2.23	5	7.19	197	36.50

N = stems per hectare (rounded to a whole number)
G = basal area per hectare, m^2 ha^{-1}

4 FOREST INVENTORY

4.1 PLANNING AND EXECUTING A FOREST INVENTORY

An inventory of a forest area can provide information for many different purposes; it may be part of:

- a natural resource survey with the aim of allocating land to different uses, i.e. land planning
- a national project to assess the potential for forest and wood-based industry development
- a wood-based industry feasibility study
- a management plan for a forest, *either*
  - providing information as a basis for long term planning, *or*
  - providing information for short term programmes such as a schedule of compartments or crops to be felled or thinned each year

Natural resource surveys require information on the location, extent and type of forest and wood resources, but need only relatively imprecise information on quantities, that is m^3, and only a crude classification by quality, that is, perhaps, forest types or species groups with similar wood properties rather than individual species. Most of this information can be obtained from up-to-date maps and aerial photographs, supported by a small amount of field work.

National projects to assess the potential for wood using industry and wood-based industry feasibility studies require more precise information on the location, quality and quantity of raw wood material. The area from which wood can be supplied to an industry, either on a sustained basis or for a limited period, may be called its 'wood catchment area'. For example, on a sustained yield basis a 200,000 tonne a year pulp mill will need a larger catchment than a 60,000 tonne a year particle board mill; or a plywood mill salvaging large over-mature trees and requiring 10,000 tonnes of veneer logs a year will need a larger catchment area if production is planned to last 15 years than if it is to be closed after only seven years. If the location and size of the industrial complex being planned is uncertain, then a comprehensive feasibility study will require information about the resource in each of the locations under consideration, and for different sized catchment areas at each. Careful planning and choice of inventory design and efficient inventory planning are needed to minimize costs and obtain satisfactory results.

Forest management planning may concern either or both of two time horizons:

- a long term approach, often as long as the rotation period of the crops
- a shorter term, often up to 5 or 10 years, for which relatively detailed programmes will be designed

Two sets of information may be needed – general information on area, species and growth rates for the whole forest, and more detailed information on those stands that may be partially or wholly harvested in the planning period, and their predicted outturn of forest produce.

Whatever the purpose and scale, forest inventory provides only part of the information required by the planner and manager, that is the part concerning the growing stock of trees. The remainder, for example information on markets, prices, labour costs, etc. must be obtained from other sources and surveys. Very often a forest inventory is only part of a whole system of information collection, and must be seen and designed with the overall purpose in mind.

Inventories for forest management planning are usually one of two types:

- a single inventory to provide information on the current growing stock and rates of growth
- a recurrent inventory to monitor growth rates and other changes in the forest. The design of this type must allow comparisons of the results from successive measurements made at relatively short intervals – say 5–10 years

Inventories for planning harvesting: the capital investment needed for modern harvesting operations and road construction is now very large. Decisions on the method of felling extraction and haulage and on the spacing and alignment of roads are now more critical in efficient management of the forest and wood-based industry than ever before. Effective decision making usually needs particular and accurate information so special surveys are frequently required; for example in tropical evergreen lowland forest harvest planning may entail identifying, locating and measuring all marketable trees within the current year's felling area in order to plan road making, extraction, log loading points, transport, processing and marketing.

These very different objects in the different categories of resource survey and forest inventory necessitate different methods, but all have in common the need for efficient inventory planning if the information required is to be obtained accurately and efficiently.

There are also two schools of thought about the strategy of inventory planning. In one, as advocated by Frayer (1974), the uncertainty concerning the information actually required by future forest managers and planners dictates that field data collection should err on the side of comprehensiveness. Frayer argues that the major element in the cost of field data collection is in the planning, organization and time spent on reaching the sampling unit and that, once there, the marginal cost of collecting additional information is low. The argument is developed to advocate the most comprehensive form of data collection, even without plans for its immediate use. For example, this school would advocate enumerating all species and size of woody plants (though, of course, not necessarily all at the same sampling intensity) in tropical evergreen forest, even though most were unmarketable. In contrast, Philip (1976) stresses that the useful life of information collected in the course of a forest inventory is relatively short, and emphasizes the speed needed in the execution of an inventory if the staff are to maintain their enthusiasm and accuracy. Philip concludes that only data whose use and analysis have been planned should be collected. He assumes that should additional information be required, then it may be collected efficiently using regression estimates linked to information already available. For example, if a survey is done omitting species *X*, but an associated species *Y* has been included, then supplementary information about *X* may be estimated efficiently using a new sub-sample to define the relationship between *X* and *Y*. Normally there is no need to re-enumerate the whole forest.

In tropical countries, especially in tropical moist evergreen forest with its complex flora, and where there is a shortage of workers trained in inventory field techniques, limited inventory objects and limited data collection regimes are more appropriate than more comprehensive schemes of data collection. None the less in some long term surveys monitoring changes – especially changes whose exact nature cannot be clearly foreseen, a broader approach may be desirable (Dawkins, 1978). Also, inventories may be planned as a complete exercise, or as part of a programme that will continue to provide updated information to managers over a long period, i.e. continuous or recurrent inventory.

The professional staff involved in planning and executing a forest inventory may need training to acquire extra skills in:

- cartography, aerial photographic interpretation and photogrammetry (Spurr, 1960; Howard, 1991; Avery & Berlin, 1992)
- statistical principles and analyses, and data capture and processing techniques
- tree and crop mensuration techniques
- planning, budgeting and financial forecasting
- man management
- costing, operation control, and experience in the logistics of maintaining and controlling inventory crews in the field
- mechanical engineering to maintain vehicles, equipment, etc.

The information derived from an inventory may be used over a period of several years and the penalty resulting from errors in that information may be high. Careful planning is needed to ensure that the objects of the inventory are achieved as efficiently as possible. Failure in either planning or execution can result in:

- inaccurate estimates
- incomplete information
- excessive cost

The methods employed in a particular inventory, a record of the experience gained and the results must be recorded in an inventory report. It is very important to give the reasons behind the decisions and, especially, for any unusual procedures. Then once the inventory has been completed the efficacy of the plan can be assessed and the correctness of assumptions checked. Such reports are then sources of information in planning later inventory operations in the same or similar areas. The following format and headings for such an inventory report will help the inexperienced inventory planner to foresee and avoid some of the difficulties that otherwise might lower the effectiveness and efficiency of the work. However, the stages in planning are not distinct and cannot be ordered; for example a realistic estimate of the cost of providing the information needed to satisfy the objects at first declared may force the planners to re-define the objects.

One useful approach common to many planning situations is summarized in Figure 43.

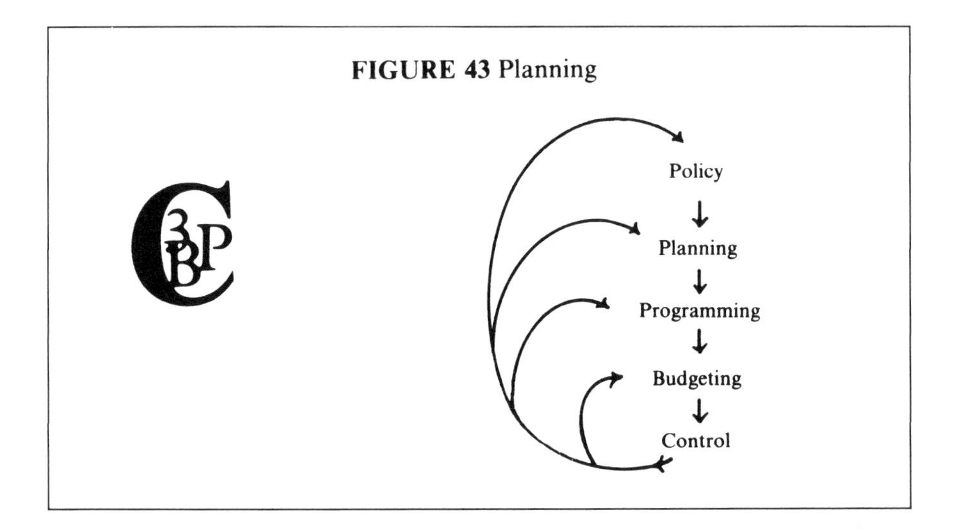

FIGURE 43 Planning

The 3PBC mnemonic reminds the planner of the five important heads. *The links between the letters* imply that the planning process is continuous or dynamic and the enclosure of the 3PB within the C stresses the importance of the controls.

The main purpose of the head *policy* is to ensure that the objects of the inventory are both defined and agreed among all concerned. *Planning* involves the choice of methods to be employed. *Programming* details the procedures and their order and timetable, whereas the provision of staff, materials, money and management required to complete the programmes are included in the *budget*. *Controls* compare actualities and predictions to provide a continuous monitoring system to ensure that wrong assumptions are identified and their implications to the programme recognized as soon as possible and incorporated into revised plans, programmes and budgets. Vanclay (1992) summarizes many considerations in inventory planning in his article entitled 'Before you begin your inventory'.

Further reading

ON FOREST INVENTORIES

Goulding, C.J. & Lawrence, M.E. (1992) *Inventory Practice for Managed Forests.* FRI Bulletin No. 171, Rotorua, New Zealand.

Lanly, J.P. (1973) *Manual of Forest Inventory, With Special Reference to Mixed Tropical Forests.* FAO, Rome.

4.1.1 The Inventory Report – Part I

The aim of this part of the report is to ensure that the inventory produces the desired information and results efficiently. The following outline provides a check list of topics to be considered and decisions to be taken. The items in the outline are:

- objects
- sources of information
- field measurements and ready reckoners
- statistical considerations
- staff management
- training field teams
- field staff
- field instructions
- field checks and controls
- management of the inventory teams in the field
- calculations, analyses and data capture
- programmes, budgets and logistics

4.1.1.1 Objects

The objects of the inventory must be defined unambiguously, written down and agreed by all concerned. They must be feasible – that is framed within any constraints such as the amount of money available – and may have to be revised if new demands are made on the inventory or new information is collected during the preliminary planning stages. Often those initiating the inventory – particularly where it is part of a wider natural resources survey, or land planning exercise or an industrial feasibility study – are unfamiliar with the techniques, difficulties and costs of forest inventories; in such cases the objects may have to be defined by the forest inventory planner himself. Nevertheless the complete understanding of and agreement to the objects must be sought from those who will use the information collected.

In addition to the general objects, the geographical limits to the area to be surveyed and the definitions of the species and tree sizes to be included must be stated. The limits of the resources available – such as money, time, supervisory and technological skills, etc. – must be given and, if necessary, the maximum tolerable level of sampling errors must be set. It is feasible that the constraint on the money available will conflict with this maximum sampling error; then, either more money must be provided or a relaxation in the maximum sampling error permitted must be agreed.

Clarification of the objects is aided by the design and presentation of an outline of the final inventory report including the form and tabulation of the results. For example, in an inventory of tropical high forest the format of the table showing the stand structure by species, species groups and diameter classes should be presented so that no misunderstandings about the species to be included, their groupings, and the limits to the size classes are possible.

4.1.1.2 *Sources of Information*

Assembly and digestion of the information already available about an area are essential. Common sources are

- maps – topographical, geological, soil and vegetation
- aerial photographs and other imagery such as SLA (Side Looking Airborne) radar or satellite imagery
- reports from previous inventories
- forest records

If these sources are inadequate then pilot surveys to collect data on species and size frequencies and variances will be essential. Nowadays extensive forest inventories of natural vegetation or intensive inventories of plantations are rarely done without recent aerial photographs or other imagery and up-to-date maps – often made from aerial photographs. If these are not available then their procurement must be included in the planning procedures and budgets.

4.1.1.3 *Field Measurements and Ready Reckoners*

The choice and definition of the population and parameters to be included in the inventory has to be made. Also the precise field techniques and procedures must be decided and written into field instructions, for example:

- the point of measurement on buttressed trees
- the definition of the point of measurement for height, e.g. total height, or timber height, or bole length
- the treatment of forked or defective stems

These decisions must be derived from the nature of the final output desired from the inventory, for example:

- if diameter distributions are required, then total basal area estimates from relascope sweeps alone will not be adequate
- if volumes per tree or per hectare are required, then the means of estimating volume and the field measurements needed will depend upon the types of ready reckoner such as volume tables or volume equations, available. If such ready reckoners are not available and are needed, then the inventory plan must include the collection of data, their analysis and the calculations needed to prepare them. Alternatively detailed and precise

instructions must be given to the field staff to measure volume directly – by height accumulation or other techniques

4.1.1.4 *Statistical Considerations* (see Chapter 5 for more detailed considerations)

Decisions on sampling concern

- the design or pattern of sampling
- the nature of the sampling units used
- the statistical analyses and the calculation of estimates and, where required, their precision

Often the assistance of a forest statistician or biometrician is needed. The design also concerns

- subdivision of the inventory area into more homogeneous parts
- the choice between using subjective and objective selection of the sampling units
- the choice between systematic and random sampling designs if objectivity is required

Considerations of the nature of the sampling units include:

- the choice of sampling unit – plot, point, transect, etc.
- the choice of shape of the sampling unit – circular, square, rectangular, etc.
- the choice of the size of the sampling unit

These and similar considerations affect enormously the cost effectiveness of the inventory. To make rational choices requires a great deal of detailed information – both statistical and logistic; much of this information may have to be collected in preliminary pilot trials, though some may be derived from accounts of inventories in similar conditions elsewhere.

4.1.1.5 *Staff Management*

Forest inventory is usually onerous, laborious and somewhat boring and is frequently done in remote areas and in harsh physical conditions with additional hazards from heat, rain, noxious insects, stinging plants, steep terrain, etc. The information collected will only be as good as those who do the field work. Therefore the planner and executive supervisor of the inventory must pay great attention to the morale, well-being and training of the field staff.

One aspect of staff management that must be decided is whether to select and train special teams to be employed on inventory field work for protracted periods and far from their home areas, or whether to train persons to do this work for relatively short periods only, recruited locally near to the inventory area. Usually the second alternative is preferable as the field instructions and training needed for field teams should be made as simple as possible. Sufficient teams should be engaged and trained so that the field work is completed over a relatively short period – 3–6 months – during which time it is feasible to maintain the morale of the teams. However, specialists will also be needed.

The importance of effective personnel management cannot be overstressed. Common complaints of field staff are:

- lack of effective transport
- leaking tents
- irregular payment of salaries
- inadequate protective clothing
- irregular deliveries of food
- inadequate first aid and medical facilities
- inadequate equipment and facilities for its repair and maintenance

In mixed forest such as tropical high forest and miombo woodland, one important constraint is the number of persons available who are capable of identifying species and using somewhat complex optical instruments such as the Spiegel relaskop. Planners must assess such constraints realistically.

Each section or field team contributing to an inventory should be informed of the general outline and programme as well as their own contribution; then, should they be unable to attain their targets, they can contribute to or suggest a feasible revised programme to the planner.

4.1.1.5.1 Training field teams

The skills needed by the staff involved in a forest inventory vary with their responsibilities and the techniques being employed.

Field supervisors commonly need training to acquire extra skills in:

- map reading, surveying and simple appreciation of aerial photographs or other types of imagery
- vehicle, tool and equipment maintenance
- tree measuring and data recording techniques
- botanical identification, specimen collection and preservation techniques

Field workers commonly need training to acquire extra skills in:

- simple ground surveying techniques – such as the correct handling of a prismatic compass and chain or steel band
- tree measuring techniques
- botanical identification

4.1.1.5.2 Field instructions

As, usually, data are collected by a number of persons and field teams working in different areas, precise written instructions are essential to ensure that all use the same procedures so that data from different sources are comparable. Examples of situations that occur commonly in forest inventories and must be foreseen in field instructions are:

- method of sampling unit definition and location – especially action to be initiated when the surveyed location is inaccessible or outside the popu-

lation definition: for example, if the plot is situated in a swamp, or on a road or area of cultivation
- method of boundary survey, especially method of slope correction if the plot is in steep terrain
- method of boundary demarcation and treatment of trees on the boundary
- definition of points of measurement – especially on trees with buttresses, forks or other defects
- procedures for measurement on abnormal stems
- procedures when the tree cannot be identified, etc.

Field instructions are often printed on weather resistant cards or plastic, and may be accompanied by ready reckoners such as:

- slope corrections
- critical distances for trees of different diameters when using an angle gauge
- field keys to assist in botanical identification, etc.

The instructions should also prescribe the precautions and checks to be carried out in the field to establish and correct human errors, and later to protect the data from loss. Data security is a major consideration for inventory planners and supervisors.

4.1.1.5.3 Field checks and controls

Checks of the field records and procedures are essential in order to assess and minimize the frequency and extent of human errors.

Checks mostly consist of repeating the measurements in a number of identifiable sampling units, or parts of units, and comparing the two sets of measurements and records. The frequency of such field checks should be greatest at the start of the inventory and be reduced thereafter as long as the comparison reveals that the standard of work is acceptable. The check reveals the efficacy of the training.

Normally checks should be done by or in the presence of the field team that was responsible for the first measurement – but with an independent supervisor. The inventory and field instructions must detail the checking procedures and frequency and, most important, the limits to the number and types of errors that render the records unacceptable. For example, gross errors must not exceed an average of 1 in 100 or occur more frequently than 2 in 50 observations; totals of additive data for a single sampling unit must not deviate from the check by more than 1%. A gross error may be a species misidentification, omission, addition or a single reading difference greater than, say, 5% of the check. The definition of an unacceptable record is not easy; different measures may require different definitions. For example, the diameter or girth of an irregularly shaped or sharply buttressed bole may be far less precise than that of a plantation grown eucalypt; similarly height measurements are less precise than girth measurements. Checks must cover:

- plot location and identity

- plot demarcation
- tree measurements and identification
- records, etc.

Should a check reveal errors outside the acceptable limits, then the team responsible must be given additional training and their work done since the previous acceptable check must be re-done. No inventory result can be better than the field work from which it was derived, so every effort must be made to collect data without error.

During the period of training, the teams must be taught to check themselves continually and to accept that independent checks are part of the inventory routine. The permanent field supervisor with each team must institute his own random checks many times each day. Checking is done to assess and minimize errors and not just to find fault. A 10% independent check should be the goal of the inventory supervisor.

4.1.1.6 The Management of the Inventory Teams in the Field

However good the training and intensive the checking, the field parties will not work well and accurately unless the effectiveness of the rest of the organization matches their own. Human nature is such that high morale among those in the field can only be maintained if they believe that the inventory organizers know of and to some extent share in their difficulties and hardships, and have tried to alleviate them as much as possible. In fact, one additional and very important role of the check done in the presence of an independent supervisor is to provide the opportunity for those responsible for planning and supervision to visit the forest and to be seen working alongside the field teams. At the same time as checking, the supervisors can

- see the forest for themselves
- familiarize themselves with the working conditions at all times and in all parts of the forest
- assess the efficacy of the back up service for the field teams, e.g. provision of food, transport, pay, equipment, etc.
- become familiar to and know the field teams

4.1.1.7 Calculations, Analyses and Data Capture

Before the field instructions can be prepared and before training the field teams can begin, the methods of calculation, analysis and data capture must be finalized. Detailed flow charts showing each step should be prepared and, where applicable, appropriate computer programs and data input instructions must be designed and tested. Then, but not before, the technique for data capture can be chosen and the design of the printed forms or other aid to data capture can be finalized. The original field record should be in a form that can be readily re-read, so that the comparison of the original data with that collected in the checking procedure may be compared immediately. Then

- the impact of the check is strong and affects the field teams beneficially
- the result of the check is known immediately
- retaining or repetition of the work can be put in hand without delay
- the field staff can themselves carry out checks additional to those done by the supervisory staff

The use of modern high speed computers for calculation and analysis means that some data items have to be coded; for example, species are given identity numbers. If this encoding is done in the field then the original record should show both the raw uncoded and the coded input so that the coding step itself can be checked.

The security of data against loss is vital but also copying must be avoided if errors are to be minimized. If for any reason field data have to be copied manually then a rigorous and independent check of the accuracy of copying must be done and certified to have been done. If field data are to be kept safe and in good order, field teams must be given adequate resources to ensure this. Good quality paper, water-proof holders for forms in current use and more permanent means for their safe, dry storage to protect and maintain them in correct order are essential. The field supervisors must be trained to ensure correct recording in the field, to check daily for omissions or ambiguity in field records and to store records safely. Safe lines of despatch to the inventory office for calculation, analysis and storage are equally essential.

4.1.1.8 Programmes, Budgets and Logistics

4.1.1.8.1 Schedules of progress and controls

Once the main decisions on the objects of and methods for an inventory have been taken, then follows the task of compiling detailed programmes for:

- preliminary organization
- acquisition of information, aids such as maps and aerial photographs, and equipment
- training
- field work
- calculation, analysis and presentation of the results

These programmes then become the bases of continuous control of the progress of the inventory so that the supervisor can compare the actual progress with that expected in the inventory plan. Deviations from the expected will be highlighted in periodic regular reviews and, either additional resources applied to make up lost time, or timetables and targets amended. A visual control chart is illustrated in Figure 44.

4.1.1.8.2 Budgets and control of resources

Inventory uses skilled supervisors, field staff, vehicles, equipment and money, all of which have other uses within the forest enterprise. Hence their allocation to the inventory affects the progress of other works; their use on the inventory must be as effective as possible and must be rigorously controlled to ensure that the allocation is adequate, is used efficiently and is not exceeded.

FIGURE 44 A control of progress in a simple forest inventory

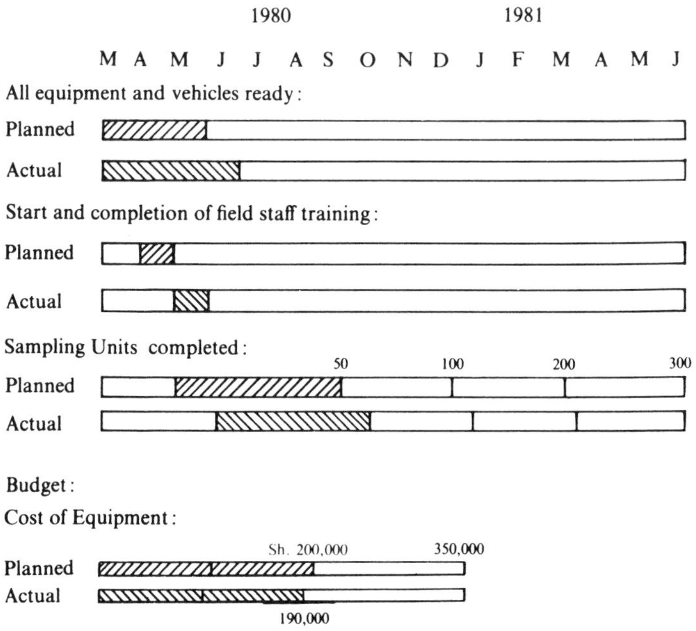

1980 1981

M A M J J A S O N D J F M A M J

All equipment and vehicles ready:

Planned

Actual

Start and completion of field staff training:

Planned

Actual

Sampling Units completed:

50 100 200 300

Planned

Actual

Budget:

Cost of Equipment:

Sh. 200,000 350,000

Planned

Actual

190,000

Budgeting commonly refers only to money, but this is too narrow a view. If untimely wet weather interferes with and delays the field work, not only will more money be required to complete the planned amount of data collection, but also persons, vehicles and equipment will be tied up for a longer period and not be available for use elsewhere. Consequently budgets for all these resources must be prepared and maintained through control comparing the planned and actual usages. These then reveal deviations from the plans and allow the initiation of revisions.

4.1.1.8.3 Logistics of staff, transport, equipment, food and other resources

Not only must the plan predict the use of resources but it must also ensure that the resources are available at the right place at the right time. For this an effective organization must be built. This organization must be given realistic and feasible targets and the means to achieve them. Field teams must be supplied with pay, food and transport when and where needed. If they are kept without these supplies then the progress of the field work will falter; similarly, if the field work is delayed then the delivery schedules for supplies will have to be altered at short notice.

Inventories of extensive tracts entail a great deal of transport of personnel, supplies and equipment. The provision and maintenance of an efficient transport service is frequently a key and often difficult task; without such a service the planned progress of the field teams cannot be maintained. The importance of adequate provision, effective planning and tight control of the resources and transport to service the field teams cannot be over-stressed.

4.1.1.8.4 Revision of plans and programmes

The controls built into the budgets and schedules of progress for the inventory provide a continuous flow of information back to the planner, enabling him or her to revise them as necessary to minimize the effect of unforeseen events and changes in the timetable. This continual revision is a normal part of planning and the execution of a plan. For events not to occur as expected is normal and to be anticipated but this does not negate the need for plans or render planning ineffective. Without plans, decisions taken on the spur of the moment may have unforeseen and unwanted consequences that could have been avoided had other feasible courses of action been considered. Part I of the inventory report includes the record of this continuous planning process.

4.1.2 The Execution of the Field Work and its Control

Of equal importance to this background organization is the organization of the field work on whose quality and progress the results of the inventory depend. Throughout the duration of the field work, the field teams will be both collecting data and feeding information on the progress of the inventory and use of resources back to the planner; they will also receive revised programmes, schedules and controls. The repetition here of information already included in Sections 4.1.1.5 to 4.1.1.7 is intended to stress the importance of these considerations both in planning and in execution.

4.1.2.1 Team Management

The success of the inventory will depend largely on the skill of the planner and field supervisors in matching their demands to the resources available – especially the abilities of the persons in the field. Effective personnel management must have a high priority in the training of the field supervisors. The capabilities of the field teams will be greatly enhanced by

- effective training
- effective organization
- adequate and efficient equipment
- good examples from all staff

Close companionship between the planner and those responsible for the tree measurements is essential; then any hardships and failures in the organization supplying the field teams is seen to be known and in part shared by all.

Also inadequacies in the instructions, equipment, etc. can be rectified without delay and the comparability between the work of different teams working far apart can be ensured. The planners must try to work in the field alongside the field teams for at least 10% of the duration of the field work – especially in the early stage. Short flying visits to teams camping in remote areas are ineffective and to be discouraged. As a working rule the duration of a visit to a field team should be at least three days. Then the real daily routine can be experienced. The field staff have time to exchange views with the planner and the planner has the opportunity to be in the field with a team long enough to encounter difficulties that occur only occasionally.

4.1.2.2 Control of Accuracy

The field instructions must define the level of errors that render the field work unacceptable. Normal field procedures must incorporate provisions for random checks continuously. For example:

1. On the completion of a field record sheet the field supervisor must check that there are no omissions; the design of the field forms must facilitate this check.

2. The field supervisor must ensure that all operators follow exactly the field instructions and should themselves incorporate continuous random checking by calling for measurements to be repeated. Also the supervisor must ensure the safety of the field records after completion until they are forwarded for calculation of the results.

3. Usually independent checks are done in the presence of an independent supervisor but with the original field team. Discrepancies must be identified immediately by comparing the two records while in the field, and the standard of field work accepted or additional training and repeating of past field work implemented without delay.

4. The frequency of these checks should be greatest early in the period allocated to the field work to ensure that the field training has been effective.

5. A possible and frequently encountered source of large errors is in the demarcation of the sampling unit. If the unit is small, for example 0.04 ha, then any error in the plot area or in the data collected will be exaggerated twenty-five fold when the data are converted to a 'per hectare' basis. Whatever their size, sampling units must be accurately surveyed and carefully demarcated; field checks must include checking this survey and demarcation of sampling units. Small plots can, with advantage, be marked by a continuous string or tape so that the boundaries cannot be mistaken by the measuring crew. Very large plots of long transects must be subdivided both on the ground and in the field records so that checks may be made on portions of a unit, and the measurements unambiguously related to the original records. Objective routines must be devised to ensure that only a representative number or portions of trees growing on plot boundaries are recorded; the details of the routine to be followed in such cases must be set out in the field instructions.

4.1.2.3 *Control, Feedback and Revision of Programmes*

Continuously throughout the periods of preliminary planning, organization and field work, information should flow back to the planner, for example:

- on the progress of flying to obtain special aerial photographic cover
- on the progress in ordering and receiving equipment
- on the progress of recruitment and training of staff
- on the progress of the field work
- on the rate of expenditure, etc.

This flow of information must be compared with the predictions of the plan and the allocations in the budget and schedules of work. Where discrepancies appear, then immediate action should be taken to minimize their effects on the achievement of the objects of the inventory. The earlier the intimation of such upsets the easier it is to revise the plans and programmes to cope.

Normally such information from the field is received in routine progress reports; consequently these must be submitted without delay. Each field team must know its own part in the whole programme and be in a position to suggest any remedial action or programme revision necessary.

4.1.2.4 *Security of Data*

Only in exceptional circumstances should original field data be copied by hand. If, owing to rain, the original field sheet is damaged, then a copy must be made immediately. The copy must be checked by an independent person against the original and the two copies attached together. As, normally, there will only be one copy of the field data, its security is very important. Field forms must be numbered to provide an unambiguous identity and stored in weather-proof, hard-backed ring files. Careful attention to the correct identification on field forms and their security while still in the field must be stressed during the training period.

The data are at risk both while in the field and during transit to the computing centre. Data must be available in the field for the purpose of the random check of accuracy. Immediately after, the data should be transferred to the computing centre – usually in the care of a senior officer. Precautions against loss during transit through a road accident or other mishap must be considered carefully. All supervisors must ensure that the instructions covering the security of data are followed without omission.

4.1.3 The Inventory Report – Part II

This second part of the inventory report summarizes the results and also the experience gained – that is experience in planning, in field work, and in the records obtained from the inventory on such aspects as:

- costs
- rate of progress in field work

- variances of different parameters in different crops
- success of recommended practices, etc.

4.1.3.1 *Record of Events*

The record of events provides an account of the general experience gained, especially on the rates of progress and duration of different stages. The information is assembled from the progress control charts used in the inventory.

4.1.3.2 *Results*

The objects of the inventory were defined in Part I and suitable output tables were designed. These tables are now completed and augmented as necessary to display the results. Additional explanation may be provided. Considerable resources of skilled manpower, specialist services, material and equipment, time and money were deployed for the inventory. An analysis of the activities and the use of resources is made and summarized to show the total of the resources used, their contribution in the final results, and their deployment in time. These are vital pieces of information for future inventory planning. Difficulties must be highlighted, critical points and paths in the planning system should be identified, and the causes of the deviations from the original predictions should be highlighted.

Suitable sub-headings for the analyses of time and costs might be:

- planning
- pilot surveys
- equipment
- training
- field work:
  - — transport
  - — salaries
  - — allowances
  - — services
- computing and analysis
- publication

Sometimes a special section is allocated to a discussion and record of the statistical analyses performed on the data to provide the population estimates.

4.2 RECURRENT FOREST INVENTORY

In both even-aged plantations of one or a few species as well as in mixed uneven-aged crops, forest inventory may be repeated at regular intervals. The aims of such recurrent or continuous forest inventories may vary from forest to forest.

On the national scale recurrent forest inventory is done to ascertain the balance between 'gain' and 'drain' in order to plan and control the development of the growing stock and the dependent industries. Gain is the sum of both growth of the existing forest and of its extensions; drain is the sum of harvesting and forest destruction. This type of information is vital when the protection of the forest and both cutting and regeneration operations are in the hands of large numbers of individuals, co-operatives and other organizations.

In extensive plantations recurrent forest inventory is the tool of management at the forest level providing information on:

- the results of treatments applied – for example the actual spacing achieved, the stocking once the plantations have been established, the stocking before and after thinning, damage, etc.
- the growth of the crops in order to compare the field results with predictions made from yield tables or other form of growth model

Recurrent inventory is especially useful where growth prediction is based on inadequate data and also where large wood processing industries with high capital investment are dependent on the forest. Recurrent forest inventory also produces the raw data for more detailed growth modelling in the future and is part of a management system providing up-dated information on the growing stock.

In uneven-aged mixed forest some form of recurrent forest inventory is the only means of estimating change within the forest and predicting future growth. In this type of forest mortality and recruitment as well as removals must be estimated in order to predict changes in the growing stock. Such inventories may be done using all permanent plots, or sometimes a combination of temporary and permanent plots. The latter is more suitable for very extensive forests where all permanent plots would be prohibitively expensive.

Effective recurrent forest inventory depends on

- efficient field work and plot maintenance
- efficient maintenance of records, security and continuity

Before a recurrent forest inventory system is introduced, great care and thought in the selection of suitable procedures and techniques are essential; the system must be sustainable within the resources likely to be available immediately and in the future, and must be designed to provide

- accurate field records
- continuous and true representation of the population by the sample
- continuity of measurement procedures in order to provide consistent data
- security of the records
- accessibility of data, analyses and results in a form that is easy to comprehend

4.2.1 Field Work for Recurrent Inventory with Permanent Plots – in Simple Crops such as Plantations

The type of permanent sample plot depends on the degree of variation in the crops. In even-aged plantations of one species, and in mixed crops with a simple structure, small sample plots with summaries expressed as average stand parameters are adequate. The main points for consideration are:

1. *Location.* The plots must be representative of the crop and easily relocated. Their location must be surveyed and marked on maps. This is simple in intensively managed and well roaded forests – but more difficult and costly in extensively managed areas.

2. *Demarcation.* Circular sample plots are defined by a centre point, other shapes by their corners. Trenches some 1–2 m by 0.3 m by 0.3 m are effective in boundary and corner demarcation and last long. Trees in sample plots should not be easily distinguishable from trees outside or field staff may treat them differently – in extreme cases leaving them unthinned, for example. Obvious paint marks should be avoided. Sampling units may be identified using a relascope at a marked centre point rather than a plot. In Switzerland plot centres are marked by metal tubes buried below ground and later located with a metal detector. The plots themselves are located systematically and near each, usually within 20 m of the plot centre, a conspicuous paint mark is placed near ground level on a large tree; the plot record gives precise survey data from the paint mark to the plot centre (Schmid-Haas & Werner, 1970).

3. *Plot size and shape* (see also Section 5.2.3). In plantations plot size depends upon the age of the crop and its tree spacing. The variance of the estimated mean volume per plot using single tree volume tables includes a component that is a function of the numbers of trees in the plot. Normally at least 20 trees per plot are needed to obtain a level of precision in accord with that of the variance between plots. Ideally all plots should be the same size; however, this is not essential so plots in older crops may be larger than those in denser, younger ones. All trees should be labelled individually and their positions marked on a plot chart. Then at subsequent enumerations missing stems can be identified. Plot size may be altered at a re-measurement provided that data for the old plot size are recorded; then the plot may be enlarged and the extra trees recorded separately in order to provide a 'previous' record at the next assessment. The survey of plots must include corrections for slope. With circular plots either one can demarcate a circle, the horizontal plane form of which has the desired area or, on broken ground, the radii to borderline trees may have to be corrected individually to establish the boundary of a circle on the horizontal plane. The recent literature on the methods of dealing with ingrowth when using point sampling is dealt with in Section 5.5.

4. *Tree identity.* Each tree in the sample should be numbered and its position plotted to scale on a plot chart. Numbered aluminium tags fastened with an aluminium nail are effective. The point of measurement may be accurately located by reference to the nail, e.g. 20 cm below the nail which is located 1.5 m above ground level.

5. *Calculations.* Before the field measurement and data capture procedures can be defined, the form of the calculations, analysis and summary must be decided. Usually this will include the determination of any volume equations to be used. The data summary for each plot and population will usually include:

v_1 volume at previous inventory
v_2 volume at current inventory
v_m mortality in period between inventories
v_r recruitment in period between inventories
v_f removals in period between inventories
gross increment
net change
stand structure frequency tables by species, diameter or volume classes
stand characters by species – if even-aged N, V, G, $\bar{h}$, $\bar{v}$, $d\bar{g}$, etc.

6. *Measurements.* Often all trees are measured for dbh. Other additional measurements will be needed for estimating volume – often taken on a subsample of trees only, e.g. height, timber height, bole length, diameter above buttress, etc. If heights are measured, either two measurements must be taken on each tree from opposite sides or the point of observation recorded on the plot chart; otherwise re-measurements may be inconsistent.

7. *Checking.* The previous records should be available to the field teams in order to spot and recheck inconsistent records. This is more important than to have independent observations and resultant inconsistencies that can be resolved only by a second field visit.

4.2.2 In Crops with a Complex Structure

In Uganda in tropical overgreen and semi-deciduous forest, permanent sample plots of 1.0 ha were established. Location is by an access route from a permanent road demarcated by a line of direction trenches 2.0 m by 0.3 m by 0.3 m at some 20 m intervals. No attempt is made to measure all the trees within the sample plot; instead the plot is subdivided into 25 sub-plots, each 20 m by 20 m, and the four largest trees from a short list of some 30 species within each sub-plot (i.e. 100 per ha) are located, marked and measured. The list of species to be measured is deemed to include those that would form the future utilizable crop.

The intention is to collect information on the growth rates of each species by size classes, in order to use the information in stand prediction (see Section 3.3.2) for growth estimation.

Further reading

Adlard, P.G. (1990) *Procedures for Monitoring Tree Growth and Site Change: a Field Manual.* Tropical Forestry Paper No. 23, Commonwealth Forestry Institute, Oxford.

Synnott, T.J. (1979) *A Manual of Permanent Plot Procedures for Tropical Rainforests.* Tropical Forest Paper No. 14. Commonwealth Forestry Institute, Oxford.

Synnott, T.J. (1980) *Rainforest Silviculture: A Research Project Report.* Commonwealth Forest Institute Occasional Paper No. 10, Oxford.

5 STATISTICAL PRINCIPLES IN FOREST INVENTORY

5.1 DEFINITIONS

Any forest is a thing of infinite variety and the forest manager usually needs different information for different parts of a forest. For example, survival is often the most important parameter in very young plantations. In older plantations total standing volume may need to be estimated, whereas in mixed forest or woodland of different species and age, volume by classes defined in terms of groups of species and diameters may be needed. Consequently a forest is often considered as a collection of distinct parts rather than one homogeneous whole. Each part may be termed a 'population'. By definition a population is a collection of similar objects that the sampler wishes to describe by a simple parameter or set of parameters.

For example, a population may be a number of stands or parts of the forest which are sufficiently similar to be described by a single set of average data such as:

- species
- average number of stems per hectare
- average basal area per hectare
- average volume per hectare, etc.

Or a population might be a collection of logs at the roadside of one species but with somewhat variable top diameter and length that can be described by their

- average volume per log

The essence of a population is that for the purposes of the sampler it can be described adequately by the estimated average value or values of the parameters measured.

A population consisting of different areas, some of which have 40-year-old teak trees grown in plantations and others have 5-year-old coppice regrowth of *Eucalyptus camaldulensis,* to be described by

- average volume per stem
- average volume per hectare

is unlikely to yield information of use to the forest manager. These averages would not adequately represent the whole – the different parts of which are too diverse to be described satisfactorily by one simple set of parameters.

Somewhat heterogeneous populations may be subdivided into strata in order to increase the efficiency of sampling. For example a forest manager may wish to know the standing volume in a block of 20-year-old *Cupressus lusitanica,* part of which has been damaged by windblow. Providing that the manager can estimate the area of the two parts of the forest separately and he does not wish to set limits of precision to the volume of standing timber in the two parts separately, then each part may be designated a *stratum.* Then the selection of sampling units in each stratum is done independently of the other. However, if the object of the sampling is to provide separate estimates of the volume of timber standing in the two parts, each with a given degree of precision, then each must be considered a population and sampled accordingly. The implication of these considerations of populations and strata is that the population for one type of inventory or objective may be a stratum in a different inventory with different objectives.

The sampling unit, that is the small part of the population that is selected to represent the whole, is both defined and designed by the organizer of the inventory on the basis either of previous experience or as a result of preliminary trials. For example, depending upon the nature of the population, sampling units may be

- single logs
- piles of logs
- lorry loads of pulpwood
- single standing trees
- small plots demarcated in the forest
- narrow transects (or lines) in the forest
- points in the forest, etc.

Once the sampling unit has been selected, then the frame of the inventory is set for that inventory as the total of all such units in the population. For example in the forest block or population of 1000 ha the sampling frame is 10,000 sampling units each of 0.1 ha but 400,000 sampling units each of 0.0025 ha.

The frame of the sampling units must equal the whole of the population without overlap or omission, if the estimate is to be unbiased. Therefore circular overlapping or non-overlapping sampling units are by definition biased, because the sum of the sampling units does not equal the whole population. When they are used then the sampler assumes that the bias is unimportant.

The most efficient sampling unit will be the smallest unit that is representative of the whole; in practice the choice of the size of the sampling unit is a compromise between the degree of variation represented by a unit and the cost. The aim of the choice in the shape of the unit is to include for a given size as representative a part as possible of the variation in the whole population. Experience has shown that the larger the pattern of variation the larger is the optimal size of the sampling unit; in very heterogeneous mixed evergreen tropical high forest or miombo with large numbers of species some of which are widely dispersed, long transects, that is sampling units in the form of long, narrow strips, are efficient. In regularly spaced plantations, care has to be taken that the sampling unit reflects the growing space of the trees sited within it; in such plantations sometimes the plot boundaries are deliberately sited midway between rows of trees (see Figure 47). Otherwise plot centres must be randomly located in relation to the pattern of the trees (see Gambil & Wiant, 1985).

The sampling fraction and the multiplier or bulking-up factor. The sampling fraction is that part of the whole population that is measured. It may be expressed in terms of either

- numbers of sampling units, i.e. $\frac{n}{N}$

or

- size, i.e. $\frac{\sum^{n} a}{A}$

Where:
n = number of sampling units measured
N = number of sampling units in the frame
a = area of a sampling unit
A = total area of the population

Sampling fractions may be either constant in all subdivisions of a population or varying from subdivision to subdivision.

The multiplier or bulking-up factor is employed to raise the sum of the values of the measured sampling units to the total value for the population. It is the reciprocal of the sampling fraction, e.g.

$$\frac{N}{n} \text{ or } \frac{A}{\sum^{n} a}$$

The sampling fraction is often an unimportant feature of a sample as it affects neither the expected value of the population mean nor the variance of the sampling units. It does, however, affect the standard error and the estimate of the precision. In finite populations, as the sampling fraction increases so the standard error decreases; when the whole population is measured then the estimated mean equals the true mean and the standard error is zero. Mostly in forestry examples the sampling fraction is very low and the estimate of the precision is unaffected by the sampling fraction (see Section 5.1.1.3).

5.1.1 Elementary Statistical Calculations

This section introduces the student to the symbols and formulae used in the remainder of the chapter. More extensive treatments are given in the standard textbooks on statistics. The books by F. Freese are highly recommended

(Freese, 1960, 1967, 1984). Details are listed in the bibliography.

Recently Goulding and Lawrence have produced *Inventory Practice for Managed Forests* which is a useful practical guide covering New Zealand field techniques.

5.1.1.1 *The Mean or Average*

The mean or average of a parameter and its use to describe a population was introduced in Chapter 1. Its definition is repeated for the reader's convenience.

$$\mu = \frac{\sum^{N} x_i}{N}$$ that is the arithmetic average of all units in the sampling frame

Where:
N = number of sampling units in the population
x_i = the value of the ith sampling unit
μ = population mean value

When μ is estimated from a sample by $\bar{x}$ and the sampling units are selected with equal probability, then

$$\bar{x} = \frac{\sum^{n} x_i}{n}$$

but if the sampling units are selected with unequal but known probability then the estimated total value X for the population is estimated by

$$X = \frac{1}{n} \sum^{n} \frac{x_i}{p_i}$$

and

$$\bar{x} = \frac{\bar{X}}{N}$$

5.1.1.2 *The Variance and Standard Deviation*

The variance, that is aptly also termed *the mean squared deviation,* is the expected value of $(x_i - \mu)^2$, i.e.

$$\sigma^2 = \frac{\sum^{N} (x_i - \mu)^2}{N}$$ (symbols as above)

Where σ^2 is estimated by s_x^2

$$s_x^2 = \frac{\sum^{n} (x_i - \bar{x})^2}{n - 1}$$

N.B. Bias is introduced if s_x^2 is estimated from small samples using n as the divisor. For speed and ease of computation the variance is usually estimated using the formula in the form of

$$s_x^2 = \frac{\sum^{n} x_i^2 - \left(\frac{\sum^{n} x_i}{n}\right)^2}{n-1}$$

This formula is mathematically identical with that above but has the advantage of restricting rounding errors to the single function

$$\frac{\left(\sum^{n} x_i\right)^2}{n}$$

The square root of the variance is called the standard deviation (σ_x) estimated by s_x. Its units are those of the parameter x; i.e. if x_i is volume in m^3 then the standard deviation is also expressed in m^3.

The coefficient of variation (C) is a useful measure of variation calculated as

$$C = \frac{s_x}{\bar{x}}$$

Since s_x and $\bar{x}$ are measured in the same units, C is dimensionless and is of use in comparing the variation of two parameters expressed in different units; e.g. if the coefficient of variation of the parameter x is 0.30 and of parameter y is 0.45, then parameter y is more variable than parameter x. C can be greater than 1.

5.1.1.3 *The Variance of a Mean*

If one set of sampling units provides an estimate $\bar{x}_1$ of a population parameter, other similar sets of sampling units will provide other and varying estimates of μ. They are unlikely to be identical owing to randomly distributed variation among the sampling units selected. These estimates of μ themselves have a variance, and the larger the number of sampling units used to estimate $\bar{x}$ the smaller is the variance. In very large populations the expected value of $(\bar{x} - \mu)^2$ can be shown to be

$$\frac{\sigma^2}{n}$$

but in finite populations it is estimated by

$$s_{\bar{x}}^2 = \left(\frac{s_x^2}{n}\right)\left(\frac{N-n}{N}\right) = \frac{s_x^2}{n}(1-f)$$

Where:
f = sampling fraction

The square root of this estimated variance is called the standard error ($s_{\bar{x}}$) or, sometimes, the standard deviation of the mean. Like the standard deviation, the standard error is expressed in the same units as the mean. The factor

$(N - n)/N$ is called the finite population correction factor; it is employed to ensure that when the whole population is included in the sample, then

$$n = N$$

$$\bar{x} = \mu$$

$$s_{\bar{x}}^2 = 0$$

The finite population correction factor is usually omitted when the sampling fraction is less than 1/20 or 5%. As explained in Section 5.1.1, the sampling fraction and the finite population correction factor can be expressed either in units of sampling numbers or in units of the size of the sample compared to the size of the whole population, e.g. in terms of area where the sampling units are expressed in terms of area.

FOR THE ADVANCED STUDENT

5.1.1.4 Variances of Functions

If

$$z = (x + y)$$

then

$$\sigma_z^2 = \sigma_x^2 + \sigma_y^2 + 2 \operatorname{cov}(xy)$$

estimated by

$$s_z^2 = s_x^2 + s_y^2 + 2 \operatorname{cov}'(xy)$$

where the estimated covariance of xy is

$$\operatorname{cov}'(xy) = \frac{\sum^{n} (x_i - \bar{x})(y_i - \bar{y})}{n - 1}$$

If the covariance of xy can be assumed to be 0, then

$$s_z^2 = s_x^2 + s_y^2$$

and if the covariance $= -1$, then $s_z^2 = 0$.

If $z = (x - y)$, then the variance of z is

$$s_z^2 = s_x^2 + s_y^2 - 2 \operatorname{cov}'(xy)$$

If $z = \bar{x}/\bar{y}$, then for large samples

$$s_z^2 = z^2 \left(\frac{s_x^2}{\bar{x}^2} + \frac{s_y^2}{\bar{y}^2} - \frac{2 \operatorname{cov} xy}{\bar{x}\bar{y}} \right)$$

If $z = x_i y_i$, and $\bar{z} \neq \bar{x}\bar{y}$ and x is linearly correlated with y then

$$s_z^2 = \bar{y}^2 s_x^2 + \bar{x}^2 s_y^2 + 2\bar{x}\bar{y} \operatorname{cov} xy$$

If $z = Nx$ where N is a constant and not a variable, then

$$s_z^2 = N^2 s_x^2$$

This is the situation when the estimates from small samples are used to estimate population totals.

5.1.1.5 *Confidence Limits* (symbols as before in Section 5.1.1)

Student's t statistic was defined as

$$t = \left| \frac{(\bar{x} - \mu)}{s_{\bar{x}}} \right|$$ where the symbol $|\ldots|$ indicates the modulus of the function, i.e. disregarding sign

t is the deviation of the estimated from the true mean of a population in units of standard errors. The value of t for a particular $\bar{x}$ from a particular sample varies with

- the size of the deviation, $(\bar{x} - \mu)$
- the estimated variance of the population, s^2
- the number of sampling units, n

t, therefore, has different distributions for different sample sizes. The functions defining these distributions (see Figure 45) have been calculated and tabulated to show the boundary or limiting values of t that define stated proportions of the total frequency. These values may be used to define confidence or fiducial limits on either side of the estimated mean within which the true population mean, μ, is expected to lie at the stated probability level.

From the definition of t

$$t s_{\bar{x}} = |\bar{x} - \mu|$$

$$\mu = \bar{x} \pm (t s_{\bar{x}})$$

or

$$\bar{x} - (t s_{\bar{x}}) < \mu < \bar{x} + (t s_{\bar{x}})$$

FIGURE 45 Distribution of Student's *t*

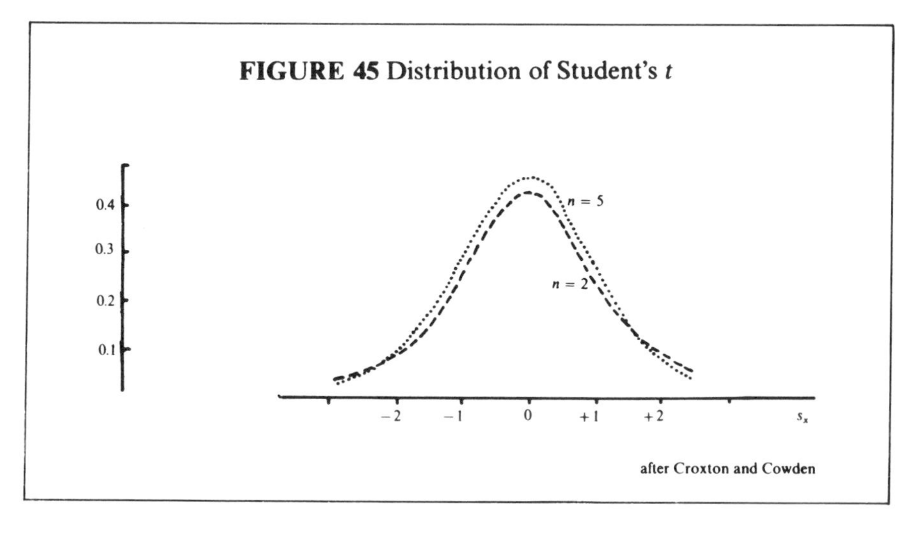

where t is the tabulated value at a given probability for the degrees of freedom of $\bar{x}$.

For example, if the tabulated value at a probability of $p = 0.05$, $t = 2.24$, then 95% of the estimated μ using the stated number of sampling units, will lie within the range of

$$\bar{x} \pm (2.24 s_{\bar{x}})$$

The value $(\bar{x} - t s_{\bar{x}})$ is called the *reliable minimum estimate* or RME.

The confidence limits of an estimated mean vary with

- the number of sampling units (or degrees of freedom of the mean)
- the estimated variance of the population
- the declared level of probability for the estimated confidence limits

EXAMPLE 35 The calculation of $\bar{x}$, s_x^2, $s_{\bar{x}}^2$, s_x, $s_{\bar{x}}$ and CL from a random sample of n units from an extensive forest population in which n/N is very small

x = tree height, m

$n = 11$

$x_i = 5,6,8,4,9,7,3,6,5,7,6$

$x_i^2 = 25,36,64,16,81,49,9,36,25,49,36$

$$\sum^n x_i = 66 \qquad \bar{x} = \frac{66}{11} = 6.0 \text{ m}$$

$$\left(\sum^n x_i\right)^2 = 4356 \qquad \frac{\left(\sum^n x_i\right)^2}{n} = 396$$

$$\sum^n x_i^2 = 426$$

$$s_x^2 = \frac{\sum^n x_i^2 - \frac{\left(\sum^n x_i\right)^2}{n}}{n-1}$$

$$= \frac{(426 - 396)}{10}$$

$$= \frac{(30)}{10}$$

$$= 3.0$$

$$s_x = \sqrt{3} = 1.7 \text{ m}$$

$$s_{\bar{x}}^2 = \frac{3}{11} = 0.3$$

$$s_{\bar{x}} = \sqrt{0.3} = 0.5 \text{ m}$$

To define CL within which μ is expected to lie in 95 such calculations out of 100, i.e. the probability of μ lying within the defined limits is 0.95 - or the probability of μ lying outside these limits is 0.05, we select the boundary value for t from the tabulation with 10 degrees of freedom = 2.2 (standard errors).

$$CL = 6.0 \pm (2.2)(0.5)\ \text{m}$$
$$= 6.0 \pm 1.1\ \text{m} \quad \text{or} \quad 4.9 < \mu < 7.1\ \text{m}$$

N.B. The t value required is the boundary or limiting value disregarding sign, so that the probability of $(\bar{x} - \mu)$ being smaller than -2.2 standard errors is 0.025 and the probability of $(\bar{x} - \mu)$ being greater than $+2.2$ standard errors is also 0.025; the value for t used is the two-tailed value at $p = 0.05$.

5.1.1.6 *Weighted Means and Their Variance* (symbols as before in Section 5.1.1)

A population may be subdivided and each part may be sampled independently. Then

$$\bar{x}_i = \frac{\sum^{n_i} x_{ij}}{n_i} \quad \text{and} \quad X_i = N_i \bar{x}_i$$

Where:

$\bar{x}_i$ = the estimated mean of the ith subdivision of the population
x_{ij} = the value of the jth sampling unit in the ith subdivision
n_i = the number of sampling units measured in the ith subdivision
X_i = estimated total value of $\sum^{N_i} x_{ij} = N_i \bar{x}_i$
N_i = number of sampling units in the sampling frame of the ith subdivision

The estimate of the total of parameter $x = X$ for the whole population of M subdivisions is

$$X = \sum^{M} X_i$$

and the estimate of the overall population mean $(\bar{x})$ is

$$\bar{x} = \frac{X}{\sum^{M} N_i} = \sum^{M} w_i \bar{x}_i$$

Where:

$$w_i = \frac{N_i}{N}$$

$$N = \sum^{M} N_i$$

and $\sum^{M} w_i = 1$

or if x refers to a sampling unit of a hectares and X_i an area of A_i hectares and X to an area of A hectares, then

$$w_i = \frac{A_i}{A}$$

Sometimes these weights are referred to as or in the form of probabilities p_i; then

$$\bar{\bar{x}} = \sum^{M} p_i \bar{x}_i$$

Pooled or weighted means derived from means of subdivisions lose a degree of freedom for each mean calculated; i.e.

$$\bar{\bar{x}} \text{ has } \left(\sum^{M} n_i - M\right) \text{ degrees of freedom}$$

The variance of the weighted mean is estimated by

$$s^2_{\bar{\bar{x}}} = \sum^{M} p_i^2 s^2_{\bar{x}_i}$$

and the standard error of the weighted mean

$$s_{\bar{\bar{x}}} = \sqrt{s^2_{\bar{\bar{x}}}}$$

If the finite population correction factor is necessary, it must be applied in the calculation of the individual $s^2_{\bar{x}}$.

The standard error of the weighted mean is needed for the calculation of its confidence limits and those of the estimate of the population total. The confidence limits of the population or weighted mean are $\bar{\bar{x}} \pm (ts_{\bar{\bar{x}}})$ where t has $\left(\sum^{M} n_i - M\right)$ degrees of freedom.

In forest inventories x_{ij} is often a volume per unit area a; then the population mean expressed in units of $m^3\ ha^{-1}$ ($\bar{\bar{X}}$) is

$$\bar{\bar{X}} = \frac{\bar{\bar{x}}}{a} \quad m^3\ ha^{-1}$$

and the estimates of the total volume in the population and its confidence limits are

$$\text{Estimated total volume} = \frac{\bar{\bar{x}}A}{a} \pm \frac{(ts_{\bar{\bar{x}}})A}{a}\ m^3$$

EXAMPLE 36 Calculation of the estimate of total volume and its confidence limits derived from a random sample in two sub-populations or strata

Sub-population 1 $A_1 = 30$ ha; $n_1 = 10$

2 $A_2 = 60$ ha; $n_2 = 15$

$a = 0.01$ ha; $x_{ij} = m^3$

$x_{1j} = 5, 6, 7, 9, 4, 7, 10, 8, 7, 7$ $\qquad \sum^{n_1} x_{1j} = 70$

$x^2_{1j} = 25, 36, 49, 81, 16, 49, 100, 64, 49, 49$ $\qquad \sum^{n_1} x^2_{1j} = 518$

$x_{2j} = 2, 3, 5, 1, 4, 6, 3, 3, 2, 1, 4, 3, 2, 5, 1$ $\qquad \sum^{n_2} x_{2j} = 45$

$x^2_{2j} = 4, 9, 25, 1, 16, 36, 9, 9, 4, 1, 16, 9, 4, 25, 1$ $\qquad \sum^{n_2} x^2_{2j} = 169$

$\bar{x}_1 = 7.0 \quad \bar{x}_2 = 3.0 \quad \bar{x} = (0.33)(7.0) + (0.67)(3.0)\ \text{m}^3$

$= 4.3\ \text{m}^3$ in 0.01 ha

Volume in subdivision 1 $= \dfrac{(7.0)(30)}{0.01} = 21{,}000\ \text{m}^3$

Volume in subdivision 2 $= \dfrac{(3.0)(60)}{0.01} = 18{,}000\ \text{m}^3$

Total volume in population $= \dfrac{(4.3)(90)}{0.01} = 38{,}700\ \text{m}^3$

N.B. The difference is a rounding error derived from expressing the weighted mean to two significant figures. Only two significant figures are justified in the total and therefore it should be rounded to 39,000 m³.

$$s_{x1}^2 = \frac{\sum^{n_1} x_{1j}^2 - \frac{\left(\sum x_{1j}\right)^2}{n_1}}{n_1 - 1} \qquad s_{x2}^2 = \frac{\left(169 - \frac{2025}{15}\right)}{14}$$

$$= \frac{\left(518 - \frac{4900}{10}\right)}{9} = 3.1 \qquad = 2.4$$

$s_{\bar{x}1}^2 = 0.31 \qquad s_{\bar{x}2}^2 = 0.16$

$s_{\bar{x}}^2 = (0.33)^2(0.31) + (0.67)^2(0.16) \qquad s_{\bar{x}} = 0.32\ \text{m}^3$

$= 0.11$

CL per plot of 0.01 ha $= 4.3 \pm (2.07)(0.32)$

$= 4.3 \pm 0.6\ \text{m}^3$ or $3.7 < \mu < 4.9\ \text{m}^3$

CL per ha $= \dfrac{4.3}{0.01} \pm \dfrac{0.6}{0.01}$

$= 430 \pm 60\ \text{m}^3\ \text{ha}^{-1}$ or $370 < \mu < 490\ \text{m}^3\ \text{ha}^{-1}$

CL of the estimate of total volume $= \dfrac{(4.3)(90)}{0.01} \pm \dfrac{(0.6)(90)}{0.01}$

$= 39{,}000 \pm 5400\ \text{m}^3$

$= 33{,}600 < \mu < 44{,}400\ \text{m}^3$

5.1.1.7 *The Number of Sampling Units Required to Achieve a Desired Precision*

(symbols as before in Section 5.1.1)

The choice of the sampling unit and sampling frame affects the variance of these units, and this is a characteristic of the population that cannot be changed by the sampler or by the choice of the sampling design. None the less the component of the variance contributing to the standard error can be altered. Also the estimate of the variance and the mean of a population or subdivision of a population is independent of the number of sampling units used and, therefore, both can be estimated from relatively few sampling units measured in a pilot or preliminary survey.

In contrast to population variance, a sampler can achieve any desired value of standard error or precision if cost is no constraint. When designing inventories, often either the standard error or the confidence limits desired are expressed as a ratio of the mean, e.g. if a ratio of standard error to mean of 0.1 is specified, then the aim is to achieve a standard error that is 10% of the mean. The number of samples (n) needed to achieve such an aim may be estimated by

$$n = \frac{s^2}{D^2\bar{x}^2} = \frac{C^2}{D^2} \quad \text{where } C = \text{coefficient of variation}$$

$$\text{and } D = s_{\bar{x}}/\bar{x}$$

(see Appendix 5).

When E = the ratio of the confidence interval ($s_{\bar{x}}t$) to the mean, i.e. $s_{\bar{x}}t/\bar{x}$, then

$$n = \frac{s^2t^2}{E^2\bar{x}^2} = \frac{C^2t^2}{E^2}$$

FOR THE ADVANCED STUDENT

Where the sampling fraction is greater than 5% and the finite population correction factor $(1 - n/N)$ is used then

$$n = \frac{Ns_x^2}{s_x^2 + ND^2\bar{x}^2}$$

where symbols are as above.

If the population falls into M subdivisions

$$n = \frac{\left(\sum^{M} p_i s_{x_i}\right)^2}{D^2\bar{\bar{x}}^2}$$

or employing the population correction factor

$$\frac{(N_i - n_i)}{N_i}$$

to each subdivision

$$n = \frac{\left(\sum^{M} N_i s_{x_i}\right)^2}{\sum^{M} N_i s_{x_i}^2 + N^2D^2\bar{\bar{x}}^2} \quad \text{or} \quad n = \frac{\left(\sum^{M} p_i s_{x_i}\right)^2}{D^2\bar{\bar{x}}^2 + \dfrac{\sum^{M} p_i s_{x_i}^2}{N}}$$

These formulae assume that the n sampling units will be allocated among the subdivisions to minimize the estimate of the population standard error. This is done by

$$n_i = \frac{n(p_i s_i)}{\sum^{M} p_i s_i}$$

If the cost of sampling varies from subdivision to subdivision, then the optimum allocation to achieve a desired standard error at minimal cost (including the finite population correction factor) is

$$n = \frac{\left(\sum^{M} p_i s_i \sqrt{c_i}\right)\left(\sum^{M} \frac{p_i s_i}{\sqrt{c_i}}\right)}{D^2 \bar{\bar{x}}^2 + \frac{\sum^{M} p_i s_{x_i}^2}{N}}$$

and the allocation to subdivision by

$$n_i = \frac{n\left(\frac{p_i s_i}{\sqrt{c_i}}\right)}{\sum^{M} \frac{p_i s_i}{\sqrt{c_i}}}$$

where c_i = variable cost per sampling unit in the ith subdivision.

If the aim is to minimize the estimated standard error for a total stated cost then

$$n = \frac{(C - c_0) \sum^{M} \frac{p_i s_i}{\sqrt{c_i}}}{\sum^{M} p_i s_i \sqrt{c_i}}$$

Where $C = c_0 + \sum^{M} c_i n_i$

c_0 = the fixed costs of the sampling

Conclusion

The following rules of thumb guide the allocation of sample numbers to subdivisions of a population: take more sampling units where

- the subdivision is larger
- the subdivision is more variable
- the cost per sampling unit is cheaper

(Cochran, 1977)

Dawkins (1985) generalized from his experience of sampling in tropical high forest and suggested that assuming a Poisson distribution so that $\sigma_x^2 = \mu$ (estimated as $s_x^2 = \bar{x}$), then

$$E\% = \frac{100 t s_{\bar{x}}}{\bar{x}} = \frac{100t}{\bar{x}}\left(\frac{\bar{x}}{n}\right)^{0.5}$$

$$= \frac{100t}{(\bar{x}n)^{0.5}}$$

or if in large samples $t = 2$, then

$$E\% = \frac{200}{(x^{0.5} n^{0.5})}$$

In practice as variance tends to increase with the extent of the forest and as trees are not randomly distributed a more general formula may be derived, as

$$E\% = K/T^b$$

Where:
$E\%$ = confidence limits expressed as a % of the mean
K = a constant lying between 200 and 250
T = $(\bar{x} \cdot n)$, i.e. the total of the sampling units
b = a constant lying between 0.4 and 0.5

Then:
$\log E = \log K - b\log \bar{x} - b\log n$
$\log E - \log K + b\log \bar{x} = - b\log n$
$\log n = (\log K - \log E - b\log \bar{x})/b$
$n = (K^{(1/b)})/(\bar{x} \cdot E^{(1/b)})$

5.1.1.8 *Relationships and Formulae in Linear Regression Analysis*

As shown in the example in Section 3.1.8.1.3, the formulae for the least squares estimates of x and y in a simple linear regression model are

$$b = \frac{\sum^{n}(x_i - \bar{x})(y_i - \bar{y})}{\sum^{n}(x_i - \bar{x})^2}$$

$$= \frac{\sum^{n} x_i y_i - \frac{\sum^{n} x_i \sum^{n} y_i}{n}}{\sum^{n} x_i^2 - \frac{\left(\sum^{n} x_i\right)^2}{n}}$$

and $a = \bar{y} - b\bar{x}$

The total variance of the parameter y can be analysed into two components: that explained by predicting y_i through the application of the regression model to the measurement of the corresponding value x_i, and the residual variance. The following tabular statement shows this:

Source	Degrees of freedom	Sum of squared deviations	Mean squared deviation	F ratio
Explained by regression	$m-1$	$\frac{\left[\sum^{n}(x_i - \bar{x})(y_i - \bar{y})\right]^2}{\sum^{n}(x_i - \bar{x})^2} = E$	$E/m = s_e^2$	$\frac{s_e^2}{s_r^2}$
Residual	$n-m-1$	$\sum^{n}\left[(y_i - (\bar{y} + b(x_i - \bar{x})))\right]^2 = U$	$U/(n-m-1) = s_r^2$	
Total	$n-1$	$\sum^{n}(y_i - \bar{y})^2 = T$	$T/(n-1) = s_y^2$	

The following mathematical relationships are derived from this table:

$$T = U + E \quad \text{or} \quad U = T - E$$

The coefficient of fit or measure of goodness of fit

$$r^2 = \frac{E}{T} = \frac{(\text{cov}_{xy})^2}{s_x^2 s_y^2}$$

The residual variance $(s_r^2) = s_y^2(1 - r^2)$ for large samples

The correlation coefficient $(r) = \dfrac{\text{cov}_{xy}}{s_x s_y}$

The standard error of the predicted value of $\bar{y} = \sqrt{\dfrac{s_r^2}{n}}$

The standard error $(s_{\bar{y}'})$ of a predicted mean value $(\bar{y}')$ corresponding to x_j

$$s_{\bar{y}'} = \sqrt{s_r^2\left(\frac{1}{n} + \frac{(x_j - \bar{x})^2}{\sum^{n}(x_i - \bar{x})^2}\right)}$$

The standard error $(s_{\bar{y}''})$ for the predicted mean value corresponding to x_j derived by applying the regression to a new sample of m sampling units from the same population

$$s_{\bar{y}''} = \sqrt{s_r^2\left(\frac{1}{m} + \frac{1}{n} + \frac{(x_j - \bar{x})^2}{\sum^{n}(x_i - \bar{x})^2}\right)}$$

Where:

x_i, y_i are paired observations on the ith sampling unit
$\bar{x}$, $\bar{y}$ are the means of the parameters x and y
E explained sum of squared deviations
U unexplained sum of squared deviations
T total sum of squared deviations
n number of paired observations of x and y
m number of sampling units in the new sample for the prediction of $\bar{y}''$

5.2 The Selection of Sampling Units

Sampling schemes may be characterized on the bases of:

- the method of selecting the sampling units, i.e. subjective or objective selection
- the degree of subdivision of the population into strata or stages
- the probability distribution employed in the selection of the sampling units
- the measurements and calculations employed to estimate the desired parameters

5.2.1 Subjectivity versus Objectivity

In subjective sampling designs the sampling units are chosen deliberately by the designer for their ability, in his opinion, to represent without distortion the whole. The designer may adopt one of several strategies. He may choose what he believes to be average units or he may deliberately include the extremes. Subjective sampling is used frequently by salesmen who may in their own interests deliberately bias the sample – for example the largest eggs are displayed for sale as if they were a representative sample. Salesmen of wood products may select samples to illustrate their wares; salesmen of tree nursery stock may select samples to illustrate the size, root development and health of their plants. Subjective sampling may be very efficient especially if the whole population can be seen at a glance and the degree of variation quickly appreciated; but it may result in a very biased representation of the whole. It is rarely used in forest inventories but is frequently used in ecological research and conservation studies. Of course no estimates of precision can be calculated from subjective selections of sampling units.

Objectivity in sampling aims to avoid bias by removing the choice of the sampling unit from the control of the observer in the field; rigid rules of selection ensure this. These rules may provide for either systematic or random selection of sampling units. In systematic sampling the sampling units are selected by a systematic routine or spatial pattern; for example a sampling unit might be every fiftieth tree in a plantation, or sampling units might be sited at the intersections of a rectangular grid of 100 m by 25 m. In random sampling the sampling unit selection is governed only by the laws of chance; for example units may be numbered and random number tables used to select the units for measurement.

5.2.2 Subdivision of a Population

An objective sampling design employing random selection of the sampling units for measurement may result in a very irregular distribution or scatter of the selected units throughout the population; parts may be intensively sampled and other parts not sampled at all. If sampling in all parts of the population is feasible, then subdivision into strata, with samples selected independently in all strata, will ensure a more even distribution. Such a design is named *stratified random sampling* in contrast to *unrestricted random sampling* where no such subdivision in the population and sample selection is applied. Alternatively, if costs or other constraints preclude sampling in every subdivision, then a preliminary selection of subdivisions in which further field sampling will be done may be made and a multi-stage design applied.

The subdivision of the population into strata aims to derive parts that each are more homogeneous than the whole; then the sampling units within a stratum are similar, i.e. the within-stratum variance is small. If the division achieves this then the precision of the overall estimate for the whole population is enhanced. For example, if aerial photographs show that a particular stand of trees has suffered windblow in the southern part beyond an extraction route that is marked on the maps, then this may form the basis of stratification for sampling to estimate the total volume remaining in the whole stand; or in montane tropical evergreen forest different vegetation associations identified on aerial photographs and delineated on maps may form different strata in order to estimate the total volume of exploitable trees.

A stratum does not have to be a single consolidated area but may consist of separate parts as long as their identity is defined without ambiguity. For example, when the basis of identifying a stratum is a particular vegetation association, then all areas with that association may be grouped as one stratum; in this case it is not necessary that all subdivisions of a stratum are sampled because the choice of sampling units within the stratum is unconstrained. Stratification is usually done before selecting the sampling units. Then the proportion or number of sampling units may be pre-determined. (N.B. Each stratum must have at least two sampling units in order to calculate the stratum mean and variance.)

Sometimes stratification is done after the selection of the sampling units, but in this case the allocation of the units to the stratum may not be optimal for the purposes of the survey. Such stratification may be very useful when employing permanent sample plots and the character of the population changes with time. For example, if a forest suffers windblow, then it may be feasible to collect all the damaged stands into one stratum and use the sample plot records in those stands to estimate the stratum mean separately from that of the undamaged stands (Whyte & Tennent, 1975).

5.2.3 The Size and Shape of Sampling Units

Sampling units generally have two characteristics

- size
- shape

The choice of sampling units concerns these characteristics and three other aspects

- the effectiveness of the unit in representing the variation in the population
- the ease of boundary definition
- the convenience and cost

Large sampling units are usually more effective in representing the variation than smaller units but are more expensive to identify and measure. In contrast, for a given sampling fraction a larger number of smaller units will provide a more precise estimate than fewer larger units.

Very many empirical studies have examined the advantages and disadvantages of sampling units of different sizes and shapes. However, these encounter the difficulty that variation due to size and shape is both confounded with sampling errors and unique to the population being studied (Yates, 1954). Modern simulation techniques have permitted more intensive studies but still generalizations are difficult. In plantation crops circular sampling units within the range of 0.01–0.10 ha are common; in tropical evergreen or semi-deciduous high forest larger sampling units – either transects or clusters – are more common. In miombo woodland, Temu (1979) found that sample plots and point samples similar to those used in mature plantations were suitable.

Such studies also have shown that the estimate of variance for a given population is a function of plot size that can conveniently be modelled by a negative exponential relationship (see Section 3.3.1.1). This is illustrated in Figure 46. However, in pattern analyses on complex vegetation types, aberrant variances associated with particular plot sizes have been taken as evidence of non-randomness of species distributions. Consequently the simple picture illustrated in Figure 46 may not be applicable in all cases.

If the relationships of cost and variance with plot size are known then it is possible to choose the plot size and numbers that will minimize the cost of achieving a desired precision. Generally small plots are cost effective but may be liable to bias arising from

- the demarcated sampling unit not representing the growing space of the trees within it
- inaccurate survey and demarcation

FIGURE 46 The relationship of estimated variance on sample plot size

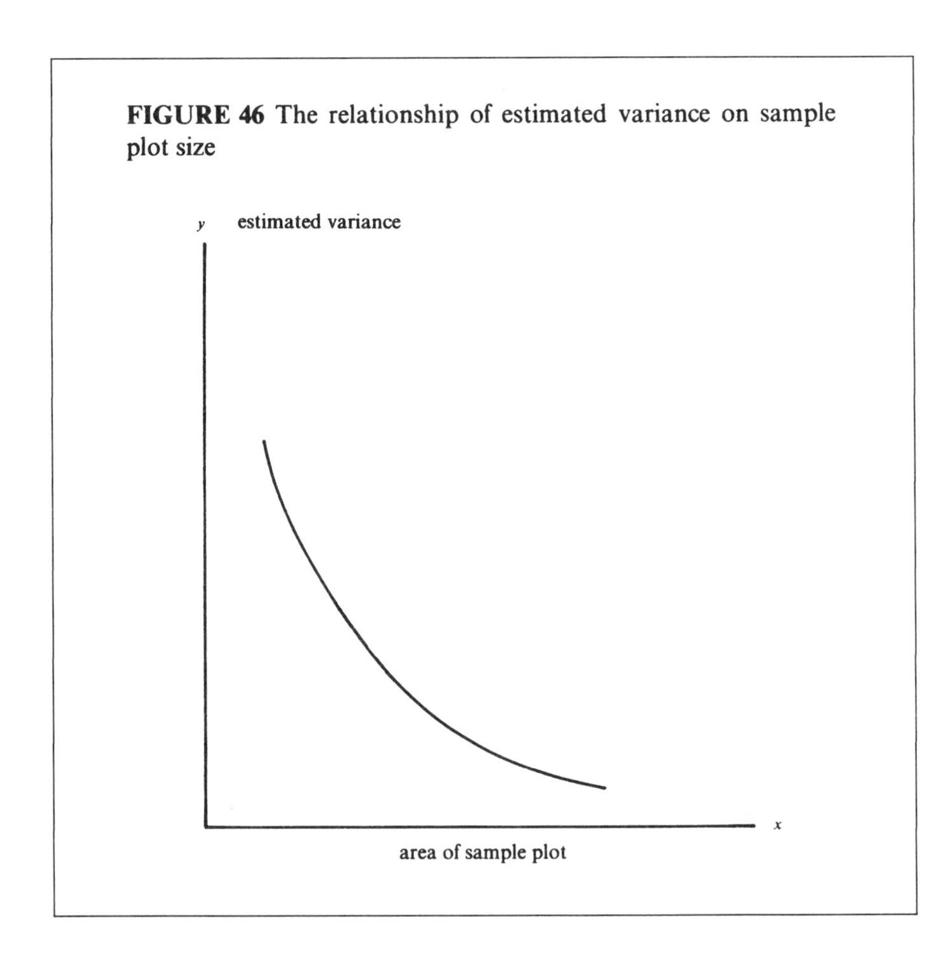

In regularly spaced plantations the danger of bias arising from the former of these two sources is great. Figure 47 illustrates three different plots of the same size site in a regularly spaced plantation but containing four, six or nine trees depending upon the position of the boundaries relative to the planting lines. Similar variation can be found using circular sample plots. The smaller the plot the greater is the risk of bias.

Therefore in regularly spaced plantations, plot centres should be sited randomly and care taken to avoid bias resulting from plot dimensions being related to tree spacing. Sometimes it is preferable to locate boundaries subjectively along the centre lines between rows – as shown in Figure 47. In this illustration the corners are located deliberately at the intersection of the two diagonals between four planting spots; then the plot boundaries joining these corners run between the rows and there are no borderline trees, so the plot area corresponds closely to the growing space of the trees within. The disadvantage is that the sampling units vary in shape and area and their survey to determine area is complex and necessitates measuring the four sides and at least one diagonal, or the four sides and included angles.

FIGURE 47 Illustration of bias in sample plots in plantations

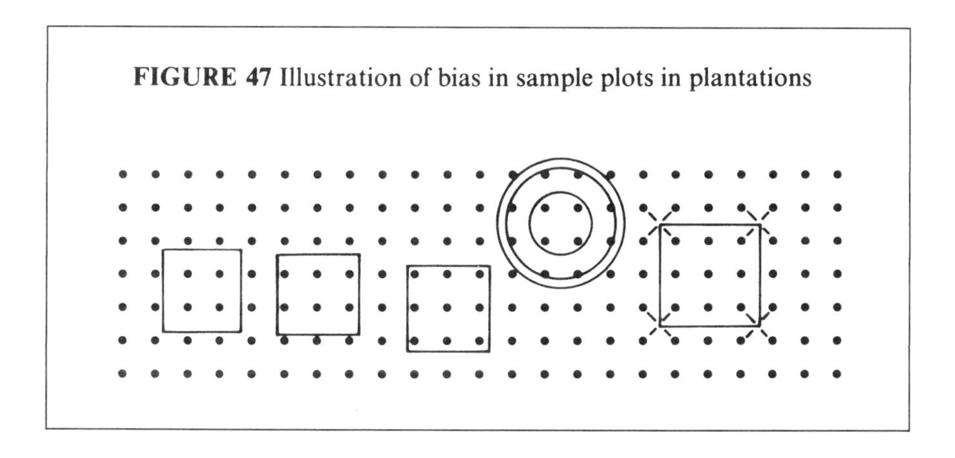

Goulding & Lawrence (1992) recommend siting the diagonal of square plots along a row of trees, but the centre of the diagonal must be random with respect to tree spacing.

In populations with linear trends such as pure, even-aged plantations growing on a hill side where the trees on the lower slopes are taller than those nearer the crest, long narrow transects located so that their long axis is parallel to the trend may be efficient. Unfortunately such transects have a very low area to perimeter ratio and are liable to serious bias from errors in demarcation and the treatment of borderline trees.

Since Bitterlich's development of sampling using a relascope, plotless, point or line, sampling units have become common in forest inventories. (Some authors now refer to these techniques as polyareal sampling units.) These have the advantage of introducing more flexible probabilities of selection by having no pre-set land boundaries; hence finite population correction factors may be omitted from calculations of variance. Nevertheless, in practice, the size of the reference angle of the gauge employed governs the probability of tree selection – for a given tree size, the larger the angle the smaller the probability of selection.

Summary on plot size and shape

- small sampling units (and larger reference angles in point and line sampling) are liable to bias unless care is taken in their siting and demarcation
- large units are effective in representing population variance but may be expensive
- in some populations long narrow transects aligned parallel to trends may be effective units, but having a low area to perimeter ratio are liable to bias unless great care is taken in their demarcation and survey
- small units are usually easier and cheaper to assess than larger units. In practice very large units may have to be subdivided in order to control and check their assessment and may be costly
- circular plots are easy to demarcate in open stands of trees with little ground cover or shrub layer
- multi-storeyed crops in which visibility is poor may be unsuited for line or point sampling

5.2.4 Probability in Selecting Sampling Units

The rules applied in the selection of sampling units for detailed measurement in random sampling designs may apply equal or unequal weights to affect the probability of the selection of a particular unit. Consequently sampling unit selection may be:

- with equal probability
- with probability proportional to size – PPS
- with probability proportional to prediction – 3P

N.B. Systematic sampling always employs equal probabilities to the selection of the first unit.

Sampling with equal probability may be achieved in many ways. In forestry the two commonest are:

- each unit in the sampling frame is given a serial number; samples are selected using random number tables. This is called list sampling
- the units in the sampling frame are drawn to scale on a map or diagram and enclosed within a rectangular grid. Each intersection in the grid is identified by a pair of rectangular coordinates. Sampling units are chosen by selecting pairs of random numbers identifying a particular intersection where a sampling unit is sited

Examples of random sampling

1. *List sampling with equal probabilities*
In a particular research plot each tree is numbered from 1 to 123 and the following ten random numbers from 1 to 123 are selected. Once selected each number is not replaced and is excluded from later selections. The trees identified by the ten numbers 91, 23, 20, 21, 55, 61, 111, 92, 121, 47 form the sample,

2. *Selecting sampling units by coordinates*
A map of the population to be sampled on a scale of 1:10,000 is overlain by a rectangular millimetre square grid. A rectangle of 56 by 72 encompasses the whole population. The grid intersections then represent the NW corner of sampling units 10 m apart and 0.01 ha in area. Sampling units are selected by pairs of random numbers within the ranges of 1–56 and 1–72 respectively; selected coordinates lying outside the population are discarded and replaced by a fresh pair of random numbers.

S No.	Random pairs North	East
1	52	18
2	7	11
3	43	10
4	6	20
5	38	10
6	10	29
7	41	55
8	30	53
9	14	40
10	36	7

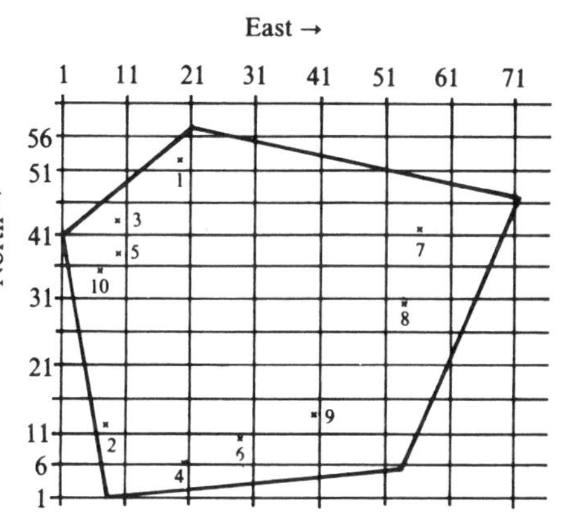

The dimensions of the rectangular grid used for selecting samples by pairs of random coordinates may impose a constraint on the number of units in the sampling frame through the distance between adjacent intersections. If the sampling unit area is not equal to the area delineated by a single grid unit then, in fact, one is sub-sampling each sampling unit in the frame. Providing that the selection of sampling units is not unduly restricted, i.e. the sampling fraction defined by the paired coordinates is not greater than, say, 2%, this is unlikely to introduce bias; nevertheless, conceptually the sampling frame must be biased as there are parts of the population outside the sampling frame and the sum of the sampling units does not equal the total population.

FOR THE ADVANCED STUDENT

Sampling with probability proportional to size is often employed when sampling units vary in size. It may be achieved by a form of list sampling in which serial numbers are allocated to subdivisions of a population in proportion to their size – see Example 37. From this example it can be seen that a small tree has a small range of numbers associated with its selection, whilst a larger tree has a larger range. For example tree no. 3, 15 cm dbh, can only be selected if a random number between 103 and 120 turns up – that is with a probability of 0.0072 – whereas tree no. 4, dbh 45 cm, is associated with the numbers 121 to 279 or a probability of 0.0639, i.e. the probability is proportional to the tree's basal area. Exactly the same results can be obtained using d^2 as the basis of the probabilities; e.g. tree no. 3 $d^2 = 225$, $p = 225/31{,}125 = 18/2490$, or $p = 0.0072$.

EXAMPLE 37 List sampling with probability proportional to size

Object: to select a sample of ten trees from a population of 50, each with probability proportional to the tree's basal area.

The dbh of all 50 trees in a population have been measured and their basal areas calculated and summed. A list is prepared showing:

- tree serial number
- dbh
- basal area
- cumulative basal area
- probability of selection: $p_i = \dfrac{g_i}{\sum^{50} g_i}$
- range of random numbers for selection of each tree as a sample

Serial no.	dbh	g_i	$\sum g$ cum.	p_i	Range of random numbers for selection
1	20	0.031	0.031	0.031/2.490 = 0.0124	1-31
2	30	0.071	0.102	0.071/2.490 = 0.0285	32-102
3	15	0.018	0.120	0.018/2.490 = 0.0072	103-120
4	45	0.159	0.279	0.159/2.490 = 0.0639	121-279
5	17	0.023	0.302	0.023/2.490 = 0.0092	280-302
6	48	0.181	0.483	0.181/2.490 = 0.0727	303-483
7	30	0.071	0.554	0.071/2.490 = 0.0285	484-554
:	:	:	:	:	:
:	:	:	:	:	:
49	13	0.013	2.433	0.013/2.490 = 0.0052	2421-2433
50	27	0.057	2.490	0.057/2.490 = 0.0279	2434-2490
				$\sum = 1.000$	

Random numbers between 1 and 2490 are selected to identify ten sample trees. The numbers were:

466, 1538, 22, 1129, 1134, 2456, 521, 1651, 1778, 1605

Rearrangement in ascending order and comparison with the ranges of random numbers results in the following selection and measurement:

Random no.	Tree no.	p_i	Volume ob m^3	v_i/p_i m^3
22	1	0.0124	0.346	27.90
466	6	0.0727	1.625	22.35
521	7	0.0285	0.751	26.35
1129	10	0.0169	0.437	25.86
1134	11	0.0820	0.310	25.83
1538	27	0.0216	0.516	23.89
1605	29	0.0305	0.729	23.90
1651	31	0.0080	0.214	26.75
1778	36	0.0149	0.362	24.30
2456	50	0.0279	0.602	21.58
				248.71

$$\text{Total volume} = \frac{1}{n}\sum^{n}\left(\frac{x_i}{p_i}\right) = 24.87\ \text{m}^3$$

Sampling with probability proportional to prediction is a relatively new technique in forestry. Instead of allocating the probabilities objectively as in the previous example, the sampler applies a probability for selection by estimation rather than measurement. To apply this technique successfully considerable experience of this type of sampling and knowledge of the population being sampled are needed.

5.2.4.1 *The Theory of 3P Sampling* (Grosenbaugh, 1952)

The probability of selecting a particular sampling unit (e.g. in Example 38, a log) is a function of the size of the total population, the magnitude of the estimated parameter on the sampling unit (volume), and the range of cards from which a random variable is selected.

The probability of selecting the ith tree with equal probability is multiplied by the probability that the estimated parameter is equal to or greater than the random variable generated by drawing a card. Therefore the probability of measuring the ith tree as a sample is

$$p_i = \left(\frac{1}{N}\right)\left(\frac{E_i}{z}\right) = \frac{E_i}{Nz}$$

Where:

E_i = estimated value on the ith sampling unit

N = total number of sampling units in the frame

z = number of cards from which the random variable is chosen so that $z \geqslant$ the maximum value of E

N.B. z and E must be in the same units.

If by mischance z is chosen as less than the maximum value of E actually encountered, then sampling units with a larger value than E are 'sure to be measured'; then their value must be kept separate from the calculations of the estimated population total and added to that estimate in the final presentation (Hetherington, 1982).

EXAMPLE 38 3P sampling

An observer wishes to estimate the total volume of 500 logs (N) lying in a forest and reckons that this is around 800 m^3. He expects the largest log to be just under 3 m^3. Past experience of the variation in such collections of logs indicates that a sample of 20 logs will be needed to estimate the total volume with a sampling error of $\pm 5\%$ ($p = 0.05$). The sampler examines each log and:

- estimates its volume to 0.1 m^3
- keeps a record of each estimate
- compares his estimate with a value on a card drawn at random from a pack. In the pack each card has a different value in the series 0.1-3.0 (or 1-30) with an additional ten cards having the symbol X. In this case there are 40 cards in all. If the sampler's estimated volume is equal to or greater than the value on the drawn card then that particular log is measured for volume as a sample. If the card drawn is an X or has a value larger than that of the estimated volume, then the log is not in the sample and is not measured

The measured logs form the sample by means of which the total volume of all 500 logs is estimated, as shown below

Log no.	Ocular estimate of volume, m^3	Random card	Decision
1	1.7	1.9	reject
2	1.9	2.7	reject
3	0.8	1.2	reject
4	2.1	2.5	reject
5	0.9	1.3	reject
6	1.4	2.0	reject
7	2.3	2.4	reject
8	0.4	0.3	select
:	:	:	:
:	:	:	:
435	2.2	1.7	select
:	:	:	:
:	:	:	:
500	1.8	1.9	reject
	892.0		

The list of those selected and measured is:

Log no.	Ocular estimate of volume, E_v, m^3	Measured volume M_v, m^3	Ratio M_v/E_v
8	0.4	0.414	1.035
15	0.6	0.527	0.878
91	2.9	3.161	1.090
94	1.7	1.543	0.908
116	0.9	1.017	1.130
175	2.4	2.049	0.854
183	1.8	2.011	1.112
204	1.6	1.564	0.978
227	2.3	2.266	0.986
241	2.0	2.053	1.026
256	1.9	3.102	1.106
270	0.8	0.641	0.801
298	1.1	0.983	0.894
361	2.8	2.954	1.055
389	2.3	2.517	1.094
392	1.7	1.820	1.071
404	2.5	2.391	0.956
435	2.2	2.148	0.976
			17.950

$$\text{Total volume} = \frac{\sum^{N} E_v}{n} \sum^{n} \left(\frac{M_v}{E_v}\right) = \frac{(892.0)(17.950)}{18}$$

$$= 889.5 \text{ m}^3$$

$N = 500$
$n = 18$

Then the estimated total population value derived from a sampling unit i is:

$$TV = \frac{M_i}{P_i} = \frac{M_i}{E_i/Nz} = \frac{Nz \cdot M_i}{E_i}$$

Where M_i = the measurement of the value to be estimated in the ith sampling unit.

Averaged over all the frame

$$TV = \frac{1}{N} \sum^{N} Nz \frac{M_i}{E_i} = z \sum^{N} \left(\frac{M_i}{E_i}\right)$$

But for all trees not measured $M_i/E_i = 0$

Therefore TV may be estimated by

$$TV = z \sum^{n} \left(\frac{M_i}{E_i}\right)$$

Where:
n = number of trees measured

In this formula z controls the sampling intensity – as z increases the sampling fraction decreases; for example, if in the calculation above z is increased to 80 by increasing the number of X, or automatic reject cards, then the expected number of samples (n_e) falls to 10 as

$$n_e = \frac{\sum^{N} E_i}{z} \quad \text{or} \quad z = \frac{\sum^{N} E_i}{n_e}$$

Similarly if any tree or part of the population is estimated to be larger than the maximum value of E, then they are sure to be measured. Their measured value must be tallied separately and added to the estimated value from trees equal to or smaller than the maximum value of E as represented by the random cards.

One of the disadvantages of the system is that n cannot be set exactly, although the expected value (n_e) can be controlled through z, provided that a reasonable guess of the total being estimated is available (Grosenbaugh, 1967, 1976).

More precise results are obtained using an adjusted formula for the estimation of total volume (TV_a). In this adjusted formula the term z is replaced by

$$\frac{\sum^{N} E_i}{n}$$

which is calculated after the sampling has been completed and both terms are known. Then

$$TV_a = \frac{\sum^{N} E_i}{n} \sum^{n} \left(\frac{M_i}{E_i}\right)$$

This formula has the disadvantage that it is biased, but in practice this bias has been found negligible. Also there is no function for the variance of this estimate, though it may be approximated by

$$s^2_{\overline{TV_a}} = \frac{\sum^{n} \left[\left(\frac{M_i}{E_i} \sum^{N} E_i\right) - TV_a\right]^2}{n(n-1)}$$

The principle of 3P sampling assumes that the ratio M_i/E_i is constant over the range of trees measured. If this is unlikely then a regression estimate is to be preferred. (H.T. Schreuder has many articles comparing 3P and regression or model sampling, e.g. Biggs *et al.*, 1985.) Often in practice a tendency to overestimate small trees and underestimate large trees has been observed. This leads to bias with the large trees under-represented in the sample.

5.2.4.2 An Application of 3P Sampling with a Spiegel Relaskop

Point sampling may be combined with 3P sampling and measuring the volume of sample trees with a Spiegel relaskop. The probability of a tree being measured for volume as a sample is

$$P_i = \frac{g_i E_i}{Fz}$$

Hence the estimate of total volume per hectare is

$$TV_a = \frac{FzM_i}{g_i E_i} \quad \text{m}^3\ \text{ha}^{-1}$$

Hence the estimate of volume obtained at one point at which the tallied trees (N) are sub-sampled with 3P and measured for volume is

$$TV_a = F \frac{\sum^{N} E_i}{n} \sum^{n} \left(\frac{M_i}{g_i E_i} \right)$$

and as an average for r points

$$TV_a = \frac{F}{r} \sum^{r} \left[\frac{\sum^{N_i} E_{ij}}{n_i} \sum^{n_i} \left(\frac{M_{ij}}{g_{ij} E_{ij}} \right) \right] \quad \text{m}^3\ \text{ha}^{-1}$$

N.B. This uses a separate ratio and sampling fraction for each sampling point. These are more usually pooled to give

$$TV_a = \frac{F}{r} \frac{\sum^{r}\sum^{N_i} E_{ij}}{\sum^{r} n_i} \sum^{r}\sum^{n_i} \left(\frac{M_{ij}}{g_{ij} E_{ij}} \right) \quad \text{m}^3\ \text{ha}^{-1}$$

The estimated or concomitant parameter E need not be a direct estimate of the measured parameter, but must be correlated with it. Hence when estimating total volume, then $(\text{dbh})^2$ is a useful estimated parameter and, in fact, may be measured. Height has been used in some studies (Rennie, 1976) and also volumes from a one parameter volume table based on a measured diameter (Bell, 1973). However, a direct ocular estimate of volume is not difficult and is usually satisfactory. In contrast, the measurements of the parameter to be estimated, M_i, must be without error.

Further reading

Useful practical manuals for 3P sampling are:

Hetherington, J.C. (1982) 3P sampling – the devil you don't know. *Scottish Forestry Journal* 36(1), 25–35.

Space, J.C. (1974) *3P Forest Inventories. Design, Procedures, Data Processing.* Metric edition. USA Forest Service, State and private forestry – SE area.

Wiant, H.V. (1976) *Elementary 3P Sampling.* Bulletin 65 OT. Agricultural and Forestry Experimental Stn, West Virginia University.

Attention is also drawn to the articles by H.T. Schreuder, e.g. Schreuder &

Anderson (1984); Schreuder *et al.* (1984); Schreuder & Wood (1986); also to Banyard (1987); Iles & Fall (1988); Iles & Wilson (1988).

5.2.4.3 Other Applications of Sampling with Unequal Probability

5.2.4.3.1 Line sampling with an angle gauge

Vertical point sampling and critical height sampling have been described in Sections 3.1.4.3 and 3.1.8.2. Both of these techniques may be used in line sampling where a line of a constant length replaces a point. The sighting of the gauge is always made at right angles to the line so that the plot size for a given diameter is a rectangle whose breadth depends upon the angle of the gauge and the tree size and the length is that of the line. Care must be taken in viewing borderline trees to ensure that the line of sight is at right angles to the transect line. The gauge may be used either on one side of the line only, or on both sides. The appropriate formulae derived from the respective ratios of tree volume to plot volume are:

For horizontal line sampling – one sided:

$$G = \frac{10^4\theta}{L}\sum^{n}\left(\frac{g_i}{d_i}\right)$$

$$V = \frac{10^4\theta}{L}\sum^{n}\left(\frac{v_i}{d_i}\right)$$

For vertical line sampling – one sided:

$$G = \frac{10^4 q}{L}\sum^{n}\left(\frac{g_i}{h_i}\right)$$

$$V = \frac{10^4 q}{L}\sum^{n}\left(\frac{v_i}{h_i}\right)$$

Where:
G = basal area per ha, m^2
L = length of line, m
q = as defined in Section 3.1.4.3, i.e. $(\tan\alpha + \tan\beta)$
g_i = basal area of subject tree i selected by the angle gauge
v_i = volume of tree i
h_i = height of tree i
θ = relascope angle
n = number of trees selected by the angle gauge

The need to measure these tree parameters is the main disadvantage of line sampling with an angle gauge. (See also McTague & Bailey, 1985; Deusen & Meerschaerrt, 1986; Lynch, 1990.)

5.2.4.3.2 Intersect sampling

In many parts of the world logging control necessitates the estimation of wood residues remaining after logging. Often these are estimated using intersect sampling. Lines of a set length are located in felled-over areas and the length and diameter at the point of intersection of each piece of wood which intersects with the sample lines are measured. Then for each line

$$V = \frac{\pi^2}{8L} \sum d_i^2 \quad \text{m}^3\ \text{ha}^{-1}$$

The derivation of this formula is based on the assumptions that:

- the logs are cylindrical – but taper does not introduce an error
- all logs are lying in a horizontal plane – sloping terrain introduces small errors
- logs are randomly oriented

The variance between sampling units:

- is inversely proportional to piece density
- increases with varying log diameter and length

and given a fixed total of sampling line to be employed

- decreases with the length of each sampling unit

Precision is a function of line length, number of lines and the orientation of both the lines and the logs. Bailey (1970), Van Wagner & Wilson (1976), Pickford & Hazard (1978) and Van Wagner (1968) provide insight into the theory and practice of line intersect sampling. Other applications include sampling wooded strips (Hansen, 1985) and assessing fire hazard.

5.2.4.3.3 Sequential sampling

This technique, often combined with 2-stage sampling, limits the sampling to a minimum in order to take a decision, e.g. that an area is or is not stocked with a minimum defined stocking. It is sometimes used in surveys of regeneration or insect attack, etc. (Fairweather, 1985). Randomly selected units are examined in sequence until the accumulated results of sampling fall into one of two or more defined classes. As a result little work is required where the result is clear cut and extensive sampling restricted to borderline situations. Its common use is in quality control in manufacturing.

Prior to sampling, three pieces of information are necessary:

- the distribution of the population being investigated; commonly these are modelled as Normal, Poisson or binomial
- the threshold levels at which a decision will be made
- the risk probabilities α and β where:
  - α = the probability of wrongly rejecting the null hypothesis (in quality control – the producer's risk)
  - β = the probability of wrongly accepting the null hypothesis (the consumer's risk)

The sampler then constructs a sampling chart depicting the zones of acceptance, uncertainty or rejection of the null hypothesis. Samples are taken one after the other until the accumulated results of sampling, plotted continuously on the control chart, fall unambiguously into a zone of certain acceptance or rejection. The structure of the chart varies with the assumption of the population distribution concerned. Details of the theory of sequential sampling are given in Wald (1947) and Wetherill (1975).

Forestry applications are not numerous – largely because of the cost of taking random samples sequentially. The design is only suitable where the cost of accessing each independent sampling unit is small compared to the cost of measuring or classifying it. Examples and comments are given by Waters (1955), Dick (1963), Loetsch & Haller (1964), Loetsch *et al.* (1973) and Roeder (1979).

5.2.4.3.4 Importance sampling

This is a further application of 3P sampling in which the measurements are collected with probability proportional to their importance in the total estimation. It has been suggested as a suitable technique for the collection of samples from a tree for the determination of volume or biomass (Valentine *et al.*, 1984; Furnival *et al.*, 1986; Gregoire *et al.*, 1986; Wiant *et al.*, 1989). The importance sampling for volume may be a second stage in a forest inventory.

In these applications, a prediction or model of the tree is postulated – either a simple taper function or an approximate weighting of the different parts reflecting their volume or weight, etc. A random number from a uniform distribution (0,1) is used to determine – in the former case – a point at which a single diameter along the bole is measured or – in the latter case – the part of the bole or branch which will represent that part of the tree under consideration. The total volume is then calculated by integration of the model from the diameter, or other sample measurements selected with a known probability. The closer the model used is to reality the more precise the estimate.

5.3 Measurements and Calculations in Inventory

The parameter of interest X (where X might be, for example, the mean volume per ha of trees in the forest or population)

- may be measured directly on the samples
- may be derived as a ratio from two measurements such as volume and area of sampling unit
- may be estimated indirectly, for example by means of a regression equation. Often in regression estimates the field measurements concentrate on a parameter x_2 which is relatively cheap to measure. Then x_2 is used to predict the parameter of interest x_1 which is more expensive to measure, using a relationship estimated from the data of a sub-sample in which both x_1 and x_2 have been measured.

Examples 39, 40 and 41 illustrate these methods. Further details are given in Section 5.4.2.1.1.

EXAMPLE 39 The calculation of a simple estimate, sampling with equal probability for all sampling units

$$\bar{x} = \sum^{n} x_i / n$$

$$X = N\bar{x} \quad \text{or} \quad X = A\bar{x}/a \quad \text{or} \quad X = \sum^{n} x_i / f$$

Where:

x_i = value of the *i*th sampling unit, $i = 1 \ldots n$
$\bar{x}$ = average or mean of the n sampling units
X = total to be estimated for the whole population
N = number of sampling units in the frame
n = number of units sampled
A = total area of population
a = area of each sampling unit
f = sampling fraction, expressed as n/N in terms of numbers of sampling units or $(na)/A$ in terms of area

1. If X is the total volume of 400 logs, and

x_i = 1.5, 0.6, 1.2, 0.9, 1.0, 0.8, 0.9, 1.1, 0.8, 0.7 m^3 per log

$$n = 10, \quad \sum^{10} x = 9.5, \quad \bar{x} = 0.95$$

$$X = \frac{(400)(9.5)}{10}$$

$$= 380 \text{ m}^3$$

2. The total volume (TV) in a forest of 120 ha is estimated using 12 sampling units, each a plot of 0.01 ha. The data collected were

v_i, m^3 per 0.01 ha

2.10 1.95 4.36 2.64 3.17 4.04
2.89 3.16 2.91 2.39 2.55 2.46

$n = 12$,

$$\sum^{12} v_i = 34.62 \text{ m}^3, \quad \bar{v} = 2.88$$

$$TV = \frac{\sum^{n} v_i A}{na} = \frac{(34.62)(120)}{(12)(0.01)} \text{ m}^3 \text{ in the forest}$$

$$= 34{,}620 \text{ m}^3 \text{ in 120 ha}$$

EXAMPLE 40 The calculation of a ratio estimate

Ratio estimates are frequently used in economic surveys where the sampling units vary in size. If one is estimating the average cost of planting in different forest projects, then the sampling units may be a compartment, stand or annual planting coupe.

N = total number of sites planted in the year (known)
n = number of sites at which planting was specially costed (selected at random after planting had been completed)
c_i = total cost of planting at site i; $i = 1 \ldots n$
a_i = area planted at site i
C = total cost of planting at all N sites (to be estimated)
A = total area planted at all N sites (known)

Example 40 continued

$$C = \left(\frac{\sum^{n} c_i}{\sum^{n} a_i}\right) A$$ which is a more efficient estimator than

$$C = \sum^{n} \left(\frac{c_i}{a_i}\right)\left(\frac{N}{n}\right)$$

c_i	10,200	18,750	17,100	30,600	80,000	23,500	5,000
a_i	17	25	19	102	200	47	5

$n = 7$; $\sum^{7} c_i = 185{,}150$; $\sum^{7} a_i = 415$; $A = 80{,}640$

$$C = \frac{(185{,}150)(80{,}640)}{415}$$

$$= 35{,}977{,}000 \text{ shillings}$$

$$\bar{c} = \frac{185{,}150}{415}$$

$$= 446 \text{ shillings per ha}$$

EXAMPLE 41 Calculation of a regression estimate

Regression estimates are frequently used to estimate standing volume using measurements made on aerial photographs combined with a sub-sample in which both the aerial photogrammetric measurements and ground measurements of volume have been taken.

In an area of 5000 ha, field trials showed that standing volume was a linear function of canopy closure as measured on aerial photographs. 500 measurements of canopy closure were made on the photographs including a subsample of 50 in which ground measurements of standing volume were also taken.

From these 50 samples a linear regression of

$$v = c + bx$$ was estimated,

Where:

v = volume in plot as measured on the ground
x = canopy closure as measured on an aerial photograph
c,b = constants estimated from the paired observations of v and x

The total standing volume in the area was estimated from

$$TV = (\bar{v} + b(\bar{x}_{ap} - \bar{x}_r))\frac{A}{a}$$

Where:

$\bar{v}$ = mean volume measured on the ground plots
$\bar{x}_{ap}$ = mean of the 500 canopy closure measurements taken on the aerial photographs
$\bar{x}_r$ = mean of the 50 canopy closure measurements taken on the plots for which volume was also measured
A = total area
a = area of a plot in which volume was measured

This formula follows from the regression as

$$c = (\bar{v} - b\bar{x}_r)$$

then the volume of a plot v_i for which no ground measurement was made is estimated by

$$v_i = (\bar{v} - b\bar{x}_r + bx_i)$$
$$\bar{v}_{ap} = (\bar{v} - b\bar{x}_r + b\bar{x}_{ap})$$
$$\text{so } TV = (\bar{v} + b(\bar{x}_{ap} - \bar{x}_r))\,\frac{A}{a}$$

$\bar{x}_{ap} = 67\ m^3$
$\bar{x}_r = 73\ m^3$
$\bar{v} = 79\ m^3$
$b = 0.47$
$A = 5000$ ha
$a = 0.25$ ha

$$TV = \left(79 + 0.47(67 - 73)\right)\left(\frac{5000}{0.25}\right)$$
$$= 1{,}523{,}600\ m^3$$

5.4 Sampling Designs

One classification of sampling designs is:

- Subjective sampling
- Objective sampling
  - — Systematic sampling[1]
  - — Random sampling
    - One stage sampling
      - Unrestricted random sampling
      - Stratified random sampling
    - Multi-stage sampling
      - Two-stage sampling
      - More than two stages in the sample section

N.B. With random sampling unit selection may be

- with equal probabilities for all sampling units
- with probability proportional to size
- with probability proportional to a predicted attribute of the sampling unit

In all designs the estimation of the parameter of interest may be direct, by proportion or ratio, or through a more complex relationship expressed in a mathematical model – often a simple linear or multiple linear regression, these latter being termed 'multi-phased sampling'. The final sampling unit may, in fact, be a cluster of sub-units in some designs (see Section 5.4.2.1.3).

Subjective sampling has been described in Section 5.2.1 and will not be dealt with further.

Whatever the sampling scheme employed, bias should be avoided. For example, when estimating standing volume of a forest crop, bias may arise from malpractice such as

[1] Systematic designs may be stratified also.

FIGURE 48 A classification of sampling designs

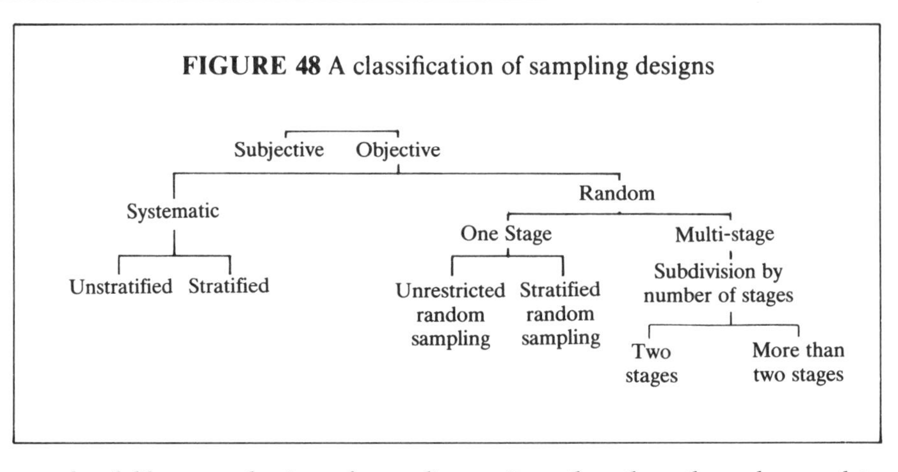

- the deliberate selection of sampling units rather than through complete objectivity; for example, moving sampling units that fall where there are many nettles may give rise to bias as such ground is then unsampled and may be more fertile and carry a higher standing volume than average
- unconscious bias arising, for example, by siting plot centres midway between trees in a regularly spaced plantation (see Section 5.2.3)
- the rejection of unstocked sampling units or sampling units difficult of access

Bias can be avoided using unrestricted random selection, or random selection subject to known restrictions; however, the instructions on plot location must be adhered to strictly. N.B. *Randomization* does not mean haphazard selection. It is quite wrong to select sampling units in a forest situation by drawing a random bearing and distance from a starting point or previous sampling unit location; it is also wrong to stick pins in or throw darts at a map. Random selection means that each sampling unit has a known probability of selection. Selection must be rigorous and no element of choice left to the field staff.

The relative efficiency of sampling designs

Sampling designs may be compared on the basis of the product of the cost per sampling unit and the estimated variance. The design with the smallest product is the most efficient by this criterion. None the less there are, of course, other factors to be considered when making the final choice.

To achieve equal precision with two designs, then

$$\frac{t_a S_a^2}{n_a} \text{ must equal } \frac{t_b S_b^2}{n_b}$$

but as t_a is approximately equal to t_b then the relative efficiency of the two designs may be compared by their costs.

$$\frac{\text{Cost of design } a}{\text{Cost of design } b} = \frac{c_a n_a}{c_b n_b}$$

but

$$n_a = \frac{t^2 S_a^2}{E^2} \text{ and } n_b = \frac{t^2 S_b^2}{E^2}$$

Therefore

$$\frac{\text{Cost of design } a}{\text{Cost of design } b} = \frac{c_a S_a^2}{c_b S_b^2}$$

Where:

t = Student's t

S_a^2, S_b^2 = estimated variances derived from design a and b

n_a, n_b = number of sampling units for design a and b to provide estimates of equal precision E

c_a, c_b = cost per sampling unit of designs a and b

EXAMPLE 42 Comparison of efficiency of two sampling designs

Design A uses small circular plots costing 3 shillings per sampling unit. The estimate of variance is 100.

Design B uses larger rectangular plots costing 5 shillings per sampling unit. The estimate of variance is 50.

Relative efficiency of design A to design B (RE)

$$\text{RE} = \frac{(3)(100)}{(5)(50)} = \frac{1.2}{1}$$

indicating that design B is the more cost effective.

5.4.1 Systematic Sampling

Systematic sampling is objective. The selected sampling units are arranged in a rigid sub-frame so that each has a common relationship with its neighbours, and the selection of the first or any one unit automatically selects the remainder.

Examples of a systematic layout are:

- when the selected sampling units lie at the intersections of a rectangular grid
- when every 10th tree in every 6th row is a sample
- when every 50th load of timber delivered each day by lorry to a pulp mill is a sample load

Even where the choice of the first sampling unit is random, if this determines the remaining units then the sample is a systematic one.

The inherent danger of bias in a systematic sample arises from the possibility of some periodic pattern of variance in the population matching or partially matching the pattern of sampling. Figure 49 illustrates and exaggerates how a systematic sample aligned across ridges and valleys might sample the forest on the ridges and fail to sample that in the valleys. Even if the coincidence of the pattern of variation of sampling units is only partial, considerable bias may still result.

Systematic samples selected from lists prepared for a separate purpose may be random provided that there is no correlation between the parameter

S_a^2, S_b^2 = estimated variances derived from design a and b

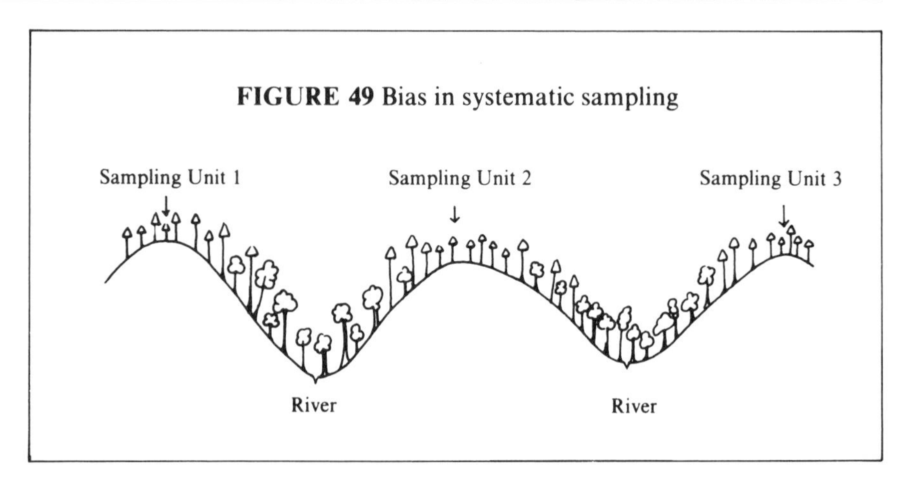

FIGURE 49 Bias in systematic sampling

used for the ordering in the list and the parameter to be estimated by sampling. For example, a systematic sample of forest reserves selected from an alphabetical list of names is unlikely to be biased and should represent the population of all reserves even although certain names may be more common in some parts of the country than in others. In contrast, a systematic sample of journeys made by a Land Rover extracted from the vehicle log book may be very biased if the vehicle has a routine that is partially correlated with the periodicity of the sampling, i.e. the systematic sample may result in selecting Saturdays when the vehicle goes on a long journey into the nearest town.

Another source of error arises in systematic samples – especially in small populations – from over- or under-representation at the margins. For example, if there are 95 rows and a systematic sample of one row in ten selects rows 9, 19, 29, 39, 49, 59, 69, 79, and 89, then the sampling fraction is not 10% but $9/95$ (assuming all rows are of equal length) or 9.5%. Also the edge rows may be under- or over-represented in the sample. The nine sampling units truly represent the population only if there is no linear trend, from row 1 to row 95 in the example above. If there is such a trend, then the imperfect representation of the rows towards the edge of the plantation will introduce bias. For example:

Given a population with a frame of sampling units having the values 1, 2, 3, 4, 5, 6, 7, 8, 9, 10

$$\text{total} = 55, \text{ mean value } \mu = 5.5$$

A systematic sample of every third unit starting at unit no. 3 gives a biased estimate

$$\text{sample } 3, 6, 9 \quad \text{total} = 18, \text{ estimated mean} = 6.0$$

A systematic sample of every fourth unit starting with the first unit is also biased

$$\text{sample } 1, 5, 9 \quad \text{total} = 15, \text{ estimated mean} = 5.0$$

EXAMPLE 43 A systematic layout and calculation of population estimate

Aim To estimate the total volume of trees in a population

Method Fell and measure a systematic sample = 2% approx.

Frame There are approximately 4000 trees arranged in approximately 80 rows with 50 trees per row. The rows are distinct but do not contain the same number of trees.

2% of 4000 trees = 80 trees. Consequently options open include:
1 tree from each of 80 rows
2 trees from every 2nd row
3 trees from every 3rd row, etc.
40 trees from every 40th row
Every 50th tree regardless of rows

This last option entails walking past every tree although only every 50th is used. Hence it is very time consuming. The penultimate option reduces the walking but at the expense of concentrating the sampling units into only two rows. A suitable compromise giving a fairly uniform spread of sampling units but reducing the time spent is to measure every 6th tree in every 10th row

Design Fell every 6th tree in every 10th row and measure its volume; start with the 4th row, this being a random selection in the range of 1-10 and the 3rd tree (random selection in the range 1-6). Count trees continuously from row to row. Count the actual number of rows beyond the last one sampled and the actual number of trees beyond the last sample tree in the last row sampled.

Results

Total number of rows sampled	8
Total number of rows	76
Total number of trees sampled	63
Number of trees beyond the last sample in the last row sampled	5
Number of trees in rows sampled = (63)(6) + (5 − 3)*	380
Fraction of rows sampled	8/76
Estimated total number of trees = (380)(76)/8	3610
Estimate of sampling fraction = 63/3610	0.01745
Total volume of the 63 samples	35.52 m^3
Estimate of total volume in population = (35.52)(3610)/63	2035 m^3

*N.B. Three is subtracted from the remainder of trees at the end of the final row sampled as the final sampling unit represents a group of six, three of which succeed the sampling unit chosen. This leaves a balance of two outside this sampling frame.

Advantages of systematic sampling

- the distribution of the selected sampling units is controlled by the designer and is regular. In forest inventories all parts of the population are both seen and represented in the sample and planning the layout of the sampling units is simple
- field teams understand the layout and can usually identify and locate the units easily

- qualitative and additional information over the whole population can be collected at regular intervals during the survey; for example, soil, ground vegetation and topographical information may be added to maps of the forest; the regularity in the distribution of the sampling units facilitates adding such information on maps
- as the distribution of the sampling units is regular and all parts of the population are sampled, the precision of the parameters estimated is usually high

Disadvantages of systematic sampling

- no valid estimates of the precision are feasible unless the pattern of variation is known. Many approximate estimates of accuracy may be calculated but, as the laws of chance have not operated on the choice of sampling unit, then an occasional, unforeseen and possibly large bias may exist
- there is no rational method of choosing the sampling intensity and layout of the sample to achieve a desired precision at minimum cost. Consequently the actual costs incurred may be higher than necessary to achieve the required precision depending on the layout and intensity chosen

FOR THE ADVANCED STUDENT

Many formulae have been suggested to estimate approximately the variance of a mean from a systematic sample. If the designer is convinced that the systematic selection has resulted in an essentially random order of sample selection, then the formula for a random sample may be applied.

Unfortunately this assumption is unrealistic in many naturally occurring populations. More suitable formulae are based on differences between successive sub-sets of the whole data.

For example: the sampling units may be considered in c sub-groups each of i sampling units. A new parameter (d_j) is calculated for each sub-group. Then, assuming independence of the sub-groups, an estimate of the variance of the mean of the original sampling units can be calculated as follows.

When

$$i = 9$$

$$d_j = (\tfrac{1}{2}x_1 - x_2 + x_3 - x_4 + x_5 - x_6 + x_7 - x_8 + \tfrac{1}{2}x_9)$$

$$c = n/i$$

$$\bar{x} = \frac{\sum^{n} x}{n}$$

$$s_{\bar{x}}^2 = \frac{\left(\sum^{c} d_j^2\right)}{\left(\sum^{i} k_h^2\right)(c)(n)}$$

where k_h = the coefficients of the h terms in d_j; so in this example where $i = 9$

$$\sum^{i} k_h^2 = (\tfrac{1}{2}^2 + 1^2 + 1^2 + 1^2 + 1^2 + 1^2 + 1^2 + 1^2 + \tfrac{1}{2}^2) = 7.5$$

Generally, but not always, this and the similar formulae overestimate the true variance (Yates, 1954; Rebner & Ek, 1983).

Often in forest inventory systematic sampling designs involve some form of two-way grid. In such cases another formula to provide an approximate estimate of the variance takes sampling units in groups of four as shown in Figure 50. In a design with 24 sampling units arranged in four rows of six units per row, an estimate of the variance is given by

$$s_{\bar{x}}^2 = \frac{\sum_i^c \left(\sum_j^4 x_{ij}^2 - \left(\sum_j^4 \frac{x_{ij}}{4} \right)^2 \right)}{4c^2(3)}$$

N.B. Three degrees of freedom are contributed by each of the c sets of four plots. This is the formula for a stratified random sample with c strata of equal area.

FIGURE 50 A systematic layout, with variance calculated from six sets of four deviations

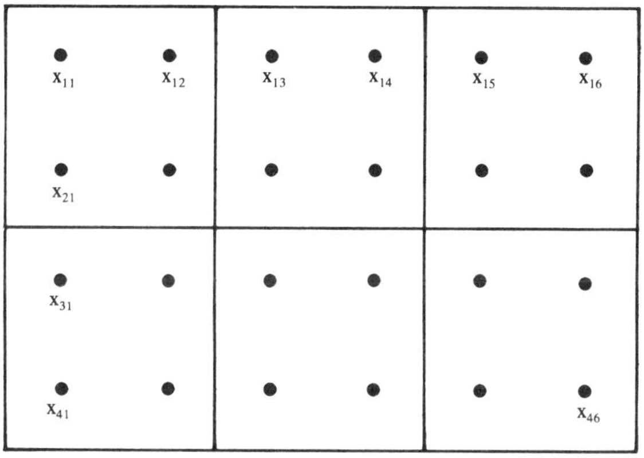

In a further modification, the marginal cells around a set of four sampling units may be used to contribute deviations from the cell mean, as illustrated in Figure 51.

The estimate of the variance is derived from the sum of 16 weighted squared deviations in each cell, but the degrees of freedom are $c - 1$.

$$s_{\bar{x}}^2 = \frac{\left(\sum^c d_j^2 - \frac{\left(\sum^c d_j \right)^2}{c} \right)}{\left(\sum^i k_h^2 \right)(4c)(c-1)}$$

FIGURE 51 Modification for the calculation of the variance for a systematic design

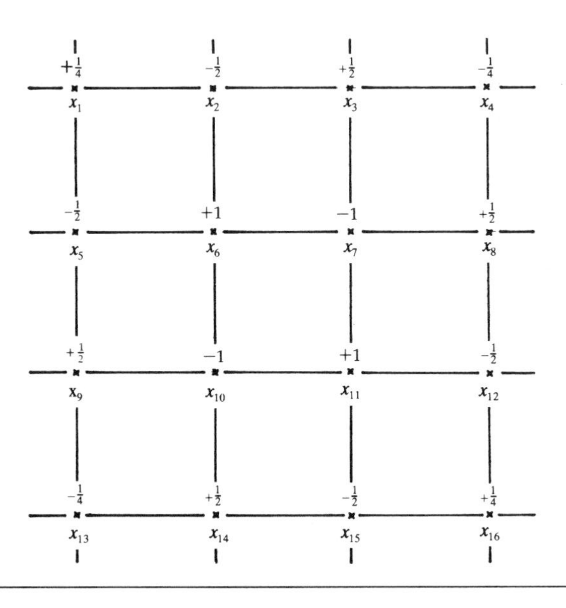

Where:

c = number of cells

$$d_j = \tfrac{1}{4}x_1 - \tfrac{1}{2}x_2 + \tfrac{1}{2}x_3 - \tfrac{1}{4}x_4$$
$$- \tfrac{1}{2}x_5 + x_6 - x_7 + \tfrac{1}{2}x_8$$
$$+ \tfrac{1}{2}x_9 - x_{10} + x_{11} - \tfrac{1}{2}x_{12}$$
$$- \tfrac{1}{4}x_{13} + \tfrac{1}{2}x_{14} - \tfrac{1}{2}x_{15} + \tfrac{1}{4}x_{16}$$

k_h = the coefficients or weights in d_j

$\sum^{i} k_h^2 = 6.25$

$j = 1 \ldots c;\ h = 1 \ldots i$

When a cell is incomplete, i.e. at the edges of the frame, it should be completed using 0 for the missing values.

5.4.2 Random Sampling

In random sampling the choice of each sampling unit for measurement is made independently of that of any other – that is the selection of any one unit gives no indication of the identity of any other selected unit. Selection must embody the laws of chance so that each unit in the sampling frame has a known probability of selection.

An improvement in the efficiency of sampling may be achieved by subdividing the population. For example, in two-stage sampling, firstly the population is divided into a sampling frame of relatively large primary sampling units from which some but not all is selected to represent the whole. Then, in the second stage, each of the selected primary sampling units is subdivided into a secondary sampling frame and some of the secondary sampling units selected to represent the whole primary unit. In multi-stage sampling the process of successive definition of sampling frames and selection of sampling units is extended – to three or more stages.

5.4.2.1 One-stage Sampling

One-stage random sampling may be either unrestricted or restricted; in unrestricted random sampling each sampling unit is available for selection whenever a sampling unit is to be chosen. This is the case where an individual sampling unit may be selected more than once; however, for practical reasons, sampling is usually done without replacement – meaning that the same sampling unit may only be selected once. In restricted random sampling the population is subdivided into strata and the sampling units in each stratum are usually selected separately using separate sub-sampling frames; consequently when the sampling units in one stratum are being selected those in other strata are excluded. Every stratum must have a minimum of two sampling units in order to have a meaningful mean and variance which are components of the population mean and variance.

In forest inventory strata are frequently defined by:

- geographical location – separate forest blocks or compartments or stands form the individual strata
- vegetation type – separate forest crop types, vegetation or site types as mapped or identified on aerial photographs or other imagery, form the individual strata
- age classes – especially in plantations

Sometimes the information on which stratification is based is collected during or after the survey. In such cases the allocation of sampling units to strata has not been done using separate frames, and the survey designer has no control over the allocation among strata. Consequently optimal allocation cannot be employed (see Section 5.1.1.7).

The term 'one stage' implies that there is only one stage in the random selection of sampling units in contrast to the successive hierarchical selections of multi-stage sampling. Stratification is a restriction on the selection but not part of a succession of stages in selection, each dependent on the previous stage.

5.4.2.1.1 Unrestricted random sampling (see Figure 52a, Example 45 and Section 5.1.1)

The direct estimate and its confidence limits

Assume that data are collected from n random sampling units, each being a circular plot of area a. The data are standing volumes, m^3. Then

$$\bar{x} = \frac{\sum^{n} x_i}{n} \quad m^3 \text{ per sampling unit, assuming selection with equal probability}$$

$$s_x^2 = \frac{\sum^{n} (x_i - \bar{x})^2}{n - 1}$$

$$\bar{X} = \frac{\bar{x}}{a} \quad m^3\ ha^{-1}$$

$$TV = \left(\frac{\bar{x}}{a}\right) A \quad m^3 \text{ in the forest}$$

Where:
$\bar{x}$ = estimated average volume per sampling
s_x^2 = estimated variance
$\bar{X}$ = estimated average volume per hectare

and the confidence limits at a stated probability level p

$$\text{for } \bar{x} \text{ are } \pm t\sqrt{\left(\frac{s_x^2}{n}\right)} \ m^3 \text{ per plot; or}$$

$$\text{for } \bar{X} \text{ are } \pm \frac{t\sqrt{\left(\frac{s_x^2}{n}\right)}}{a} \ m^3\ ha^{-1}\text{; or}$$

$$\text{for } TV \text{ are } \pm \frac{t\sqrt{\left(\frac{s_x^2}{n}\right)}A}{a} \ m^3 \text{ in the forest}$$

Where:
t = value for Student's t at a probability of p with $(n - 1)$ degrees of freedom

Unrestricted random sampling is very efficient in relatively homogeneous populations. Any desired precision in the result may be obtained by adjusting n, the number of sampling units selected; the greater the number of sampling units the greater is the precision. The number n necessary to achieve a desired precision may be estimated if the standard error of the estimate ($s_{\bar{x}}$) or the confidence limits are stated as a proportion of the mean (see Section 5.1.1.7).

EXAMPLE 44 Calculation of the number of sampling units needed to achieve a given precision

Problem 1	To estimate the number of sampling units needed to achieve a desired standard error of 5% of the mean ($D = 0.05$) using unrestricted random sampling
Formula	$n = \frac{s_x^2}{D^2 \bar{x}^2}$
Data	From a pilot sample the mean of the population ($\bar{x}$) was estimated at 5 m per sampling unit of 0.02 ha, with a variance (s_x^2) of 1
Calculation	$n = \frac{1}{(0.05)^2(5)^2} = 16$ sampling units
Problem 2	To estimate the number of sampling units needed to achieve confidence limits of 10% ($E = 0.1$) of the mean, using unrestricted random sampling
Formula	$n = \frac{t^2 s^2}{E^2 \bar{x}^2}$
Data	As in Problem 1

As the value of t depends upon the degrees of freedom in the mean, which itself depends upon n, this estimate of n has to be obtained by iteration.

A first estimate of t^2 is taken as 5; then

$$n = \frac{(5)(1)}{(0.1)^2(5)^2} = 20$$

The calculation is then repeated using a t value appropriate for 19 degrees of freedom = 2.09

1st iteration:

$$n = \frac{(2.09)^2(1)}{(0.1)^2(5)^2} = 17.47 \text{ or } 18$$

2nd iteration using t for 17 df = 2.11

$$n = \frac{(2.11)^2(1)}{(0.1)^2(5)^2} = 17.81$$

Therefore the decision to use 18 plots is accepted. N.B. The validity of this estimate depends upon the accuracy of the estimates of s^2 and $\bar{x}$; there is no guarantee that 18 plots will achieve the desired precision.

A convenient method of selecting sampling units from a population consisting of several separate blocks of forest is to trace their outline so that they lie as close together as possible (see Figure 52b). A rectangular grid is then laid over the tracing so that the intersections represent the sample frame, each representing a sampling unit. The sample may be selected using pairs of random coordinates. Any sampling unit falling outside a forest block is rejected. The chosen sampling units are then traced back to the original forest map and located from convenient points on the map by, for example, compass bearings and distances. Using this design some forest blocks may contain no samples whereas, by chance, other blocks may have more than the number expected from their proportion of the total area; the distribution is a matter of chance although the expected distribution would be in proportion to block areas. (See Section 5.2.4 for comments on bias in sampling with rectangular coordinates.)

FOR THE ADVANCED STUDENT

The ratio estimate and its confidence limits

Ratio estimates may be used when a population parameter X is known or can be estimated precisely and is proportional to another population parameter Y; we wish to estimate Y which is unknown and difficult or costly to estimate directly. Both the parameters x_i and y_i are measured in the selected sampling units and the ratio r of y to x is estimated from these measurements.

EXAMPLE 45 Unrestricted random sampling

$A = 37.0$ ha total area of population
$a = 0.02$ ha area of each sampling unit
$n = 12$ number of sampling units
v_i = volume, m³, per plot

Data:

Plot	1	2	3	4	5	6	7	8	9	10	11	12
v_i	4.7	4.4	3.8	5.1	4.0	4.6	4.0	4.6	4.8	6.1	5.6	4.3 m³

$$\sum v_i = 56.0 \qquad \bar{v} = \frac{56}{12} = 4.67$$

$$\sum v_i^2 = 266.32 \qquad s_v^2 = \frac{266.32 - \left(\frac{56.0^2}{12}\right)}{11} = 0.45 \qquad s_{\bar{v}} = \sqrt{\frac{0.45}{12}} = 0.19 \text{ m}^3$$

t at 0.05 with 11 df = 2.20

Confidence limits $= \bar{v} \pm ts_{\bar{v}} = 4.67 \pm (2.20)(0.19)$ m³

Confidence limits for estimate of volume per plot

$= 4.67 \pm 0.42$ m³ per 0.02 ha

Confidence limits per ha

$$= \frac{4.67}{0.02} \pm \frac{0.42}{0.02} = 233 \pm 21 \text{ m}^3 \text{ ha}^{-1}$$

Confidence limits for total volume

$$= \frac{(4.67)(37)}{0.02} \pm \frac{(0.42)(37)}{0.02} = 8640 \pm 780 \text{ m}^3$$

Unfortunately the ratio estimate is biased. Cochran (1977) states:

> There is no difficulty if the sample is large enough so that the ratio is nearly normally distributed and the large-sample formula for variance is valid. As a working rule, the large sample results may be used if the sample size exceeds 30 sampling units and is also large enough so that the co-efficients of variation of $\bar{x}$ and $\bar{y}$ are both less than 10%.

$$\text{that is } \frac{s_{\bar{x}}}{\bar{x}} < 0.1 \text{ and } \frac{s_{\bar{y}}}{\bar{y}} < 0.1$$

Ratio estimates are common in forestry when the sampling units have different areas and the total area of the population is known. Then

$$Y = X\left(\frac{\sum^{N} y_i}{\sum^{N} x_i}\right) \quad \text{or if} \quad R = \left(\frac{\sum^{N} y_i}{\sum^{N} x_i}\right)$$

then $\quad Y = RX$

which is estimated by

$$Y' = rX$$

Where:

$$r = \left(\frac{\sum^{n} y_i}{\sum^{n} x_i}\right)$$

or $$\bar{y}_r = r\bar{x}$$

and an alternative estimate of Y' is

$$Y' = N\bar{x}r$$

Hence

$$\sigma_y^2 = \sum^{N} (y_i - Rx_i)^2/N$$

and the population variance is estimated by

$$s_Y^2 = \frac{N^2(1-f)}{n(n-1)}\left(\sum^{n} (y_i - rx_i)^2\right)$$

$$= \left(\frac{N^2(1-f)}{n}\right)(s_y^2 + r^2 s_x^2 - 2rs_{yx}) \qquad \text{(see Appendix 6)}$$

OR as Y' may be estimated from rX and N may be estimated from $\frac{X}{\bar{x}}$

$$s_Y^2 = \left(\frac{X^2(1-f)}{\bar{x}^2 n}\right)(s_y^2 + r^2 s_x^2 - 2rs_{yx})$$

Where:

Y = population parameter to be estimated

Y' = estimate of Y

X = population parameter known

y_i, x_i = values of x and y measured in the ith sampling unit

N = total number of sampling units in the population

n = number of samples

f = sampling fraction or $\frac{n}{N}$

FIGURE 52 Examples of one-stage sampling designs

(a) Unrestricted random sampling

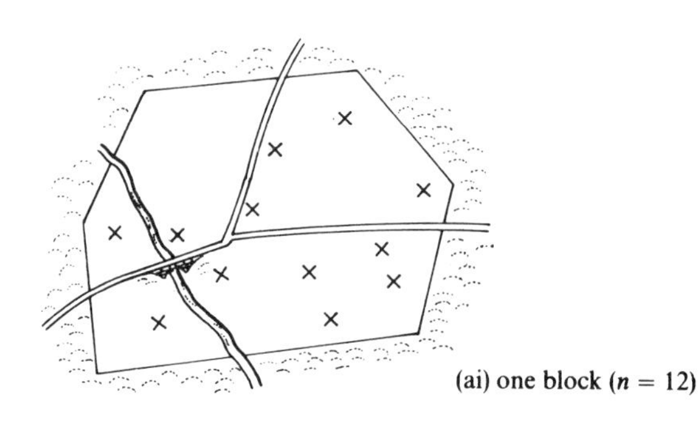

(ai) one block (n = 12)

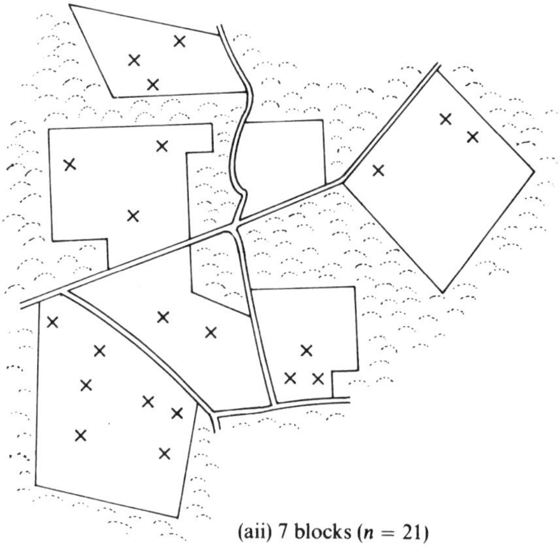

(aii) 7 blocks (n = 21)

N.B. By chance one block has no sampling units sited within it.

$\bar{x}$ = average of sample values of x_i
$\bar{y}_r$ = ratio estimate of mean value of y per sampling unit
σ^2 = population variance of sampling units
s_x^2, s_y^2 = estimate of variance of sample values of x and y
s_Y^2 = estimate of population variance – ratio
s_{yx} = estimate of covariance of y and x

An example of a calculation was shown in Section 5.3.2. Hansen (1985) gives an example of ratio sampling wooded strips.

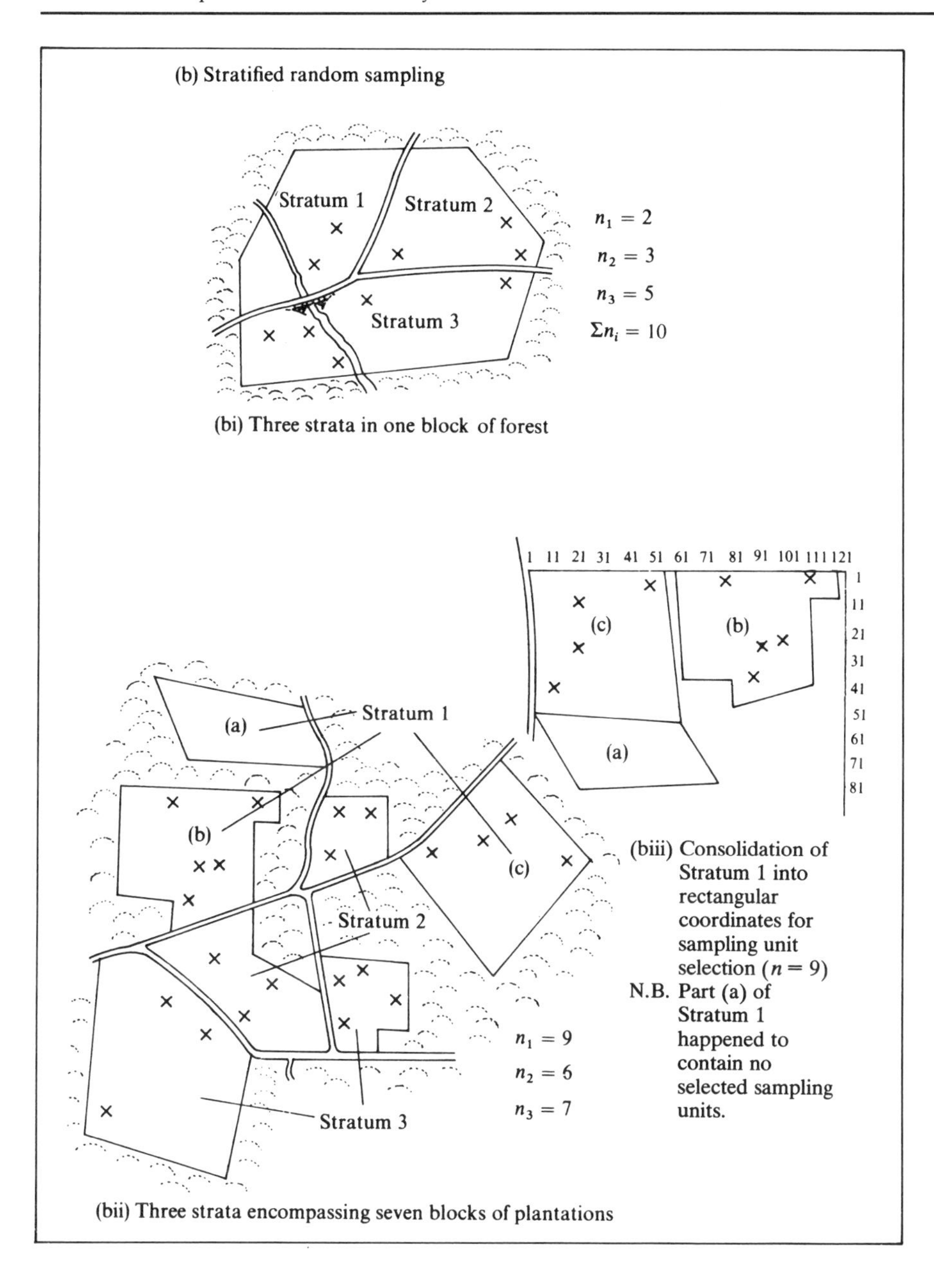

(bi) Three strata in one block of forest

(bii) Three strata encompassing seven blocks of plantations

(biii) Consolidation of Stratum 1 into rectangular coordinates for sampling unit selection ($n = 9$)
N.B. Part (a) of Stratum 1 happened to contain no selected sampling units.

The regression estimate and its confidence limits

Regression estimates are useful when a population parameter X is known or can be estimated easily and precisely and is highly correlated with, but not directly proportional to, the unknown parameter Y; usually Y is more difficult and costly to measure directly than is X. Example 41 given in Section 5.3 illustrates the scenario in which total standing volume is regressed on canopy closure measured on aerial photographs.

$$y_i = \mu + \beta(x_i - \mu_x) + e_i \quad \text{but } \mu \text{ and } \beta \text{ are estimated}$$

$$y'_i = c + bx_i$$

$$\bar{y}_{lr} = c + b\bar{x}$$

$$Y'_{lr} = N\bar{y}_{lr}$$

or
$$Y'_{lr} = \bar{y}_{lr}\frac{X}{\bar{x}}$$

or
$$Y'_{lr} = \bar{y}_{lr}\frac{A}{a}$$

where the estimated value of $\bar{y}_{lr}$ is per unit a of area.

Similarly the variance of the estimates of $\bar{y}_{lr}$ and of Y'_{lr} may be derived from:

$$s^2_{\bar{y}_{lr}} = \left(\frac{1-f}{n}\right)(s^2_y(1 - r^2))$$

where $s^2_y(1 - r^2)$ is the residual variance of the regression

$$s^2_{Y'_{lr}} = \left(\frac{N^2(1-f)}{n}\right)(s^2_y(1 - r^2)) \quad \text{if } N \text{ is known}$$

or

$$s^2_{Y'_{lr}} = \left(\frac{X^2(1-f)}{\bar{x}^2 n}\right)(s^2_y(1 - r^2)) \quad \text{if } X \text{ is known}$$

or

$$s^2_{Y'_{lr}} = \left(\frac{A^2(1-f)}{a^2 n}\right)(s^2_y(1 - r^2)) \quad \text{if } \bar{y}_{lr} \text{ is per unit } a \text{ of area and } A \text{ is known}$$

Where:

$\bar{y}_{lr}$ = linear regression estimate of population mean
$\bar{y}$ = mean of the n values of y_i measured on the sample
$\bar{x}$ = mean of measured x_i values estimating the population mean μ_x
c, b = estimated regression constant and coefficient
N = number of sampling units in the population
n = number of samples
A = total area of population
a = area of a sampling unit
Y = population value
Y'_{lr} = linear regression estimate of Y
f = sampling fraction
s^2_y = estimated variance from the n values of y_i
$s^2_{\bar{y}_{lr}}$ = estimate of variance of $\bar{y}_{lr}$
$s^2_{Y'_{lr}}$ = estimate of the variance of the regression estimate of Y
r = estimate of the correlation co-efficient of x and y

N.B. In simple regression sampling x_i and y_i are measured in all samples, whereas in double or two-phase sampling the y_i values are measured on a sub-sample only. Recent references to regression sampling in forestry can be found in Marshall & Demaerschalk (1986); Schrender & Wood (1986).

5.4.2.1.2 Stratified random sampling

The bases of subdivision of a population were described in Section 5.2.2. Figure 53 illustrates two applications of the techniques described and the example in Section 5.1.1.6 details the calculations.

Assuming that m strata have been defined and the total area (A) and the areas of each stratum (A_i) are known; then if n_i sampling units are measured in a stratum for the parameter x, so that x_{ij} is the measurement of x in the jth sampling unit of the ith stratum; then

$$\bar{x}_i = \frac{\sum^{n_i} x_{ij}}{n_i} \quad \text{that is the mean value per sampling unit in stratum } i$$

$$\bar{\bar{x}} = \sum^{m} p_i \bar{x}_i \quad \text{that is the overall mean}$$

Then

$$X = N\bar{\bar{x}} \quad \text{that is the estimate of the population total}$$

$$= \frac{\bar{\bar{x}}A}{a} \quad \text{where } \bar{\bar{x}} \text{ is a measure per unit area } a \text{ in the population}$$

Usually in forest inventory p_i – the proportion of the stratum area to the total area, i.e. $[(A_i)/(A)]$ – will be known.

The estimation of the standard error ($s_{\bar{\bar{x}}}$) of the overall mean and of the estimate for the population total (X) follows from these relationships;

$$s^2_{\bar{x}_i} = \frac{\sum^{n_i} (x_{ij} - \bar{x}_i)^2}{n_i(n_i - 1)} \quad \text{or variance of } \bar{x} \text{ in stratum } i$$

$$s_{\bar{\bar{x}}} = \sqrt{\sum^{m} p_i^2 \frac{s^2_{\bar{x}_i}}{n_i}}, \quad \text{or} \quad s_{\bar{\bar{x}}} = \sqrt{\frac{\sum^{m} A_i^2 s^2_{\bar{x}_i}}{A^2}}$$

where p_i can be expressed in terms of area.

s_X, the standard error of the estimate population total X, is calculated from

$$s_X = s_{\bar{\bar{x}}} N \quad \text{or} \quad s_X = s_{\bar{\bar{x}}} \frac{A}{a}$$

where x_{ij} is a measure per unit area a in the population. If necessary, the finite population correction factor is included in the calculation of the variance of $\bar{x}_i$.

FIGURE 53 Examples of stratification

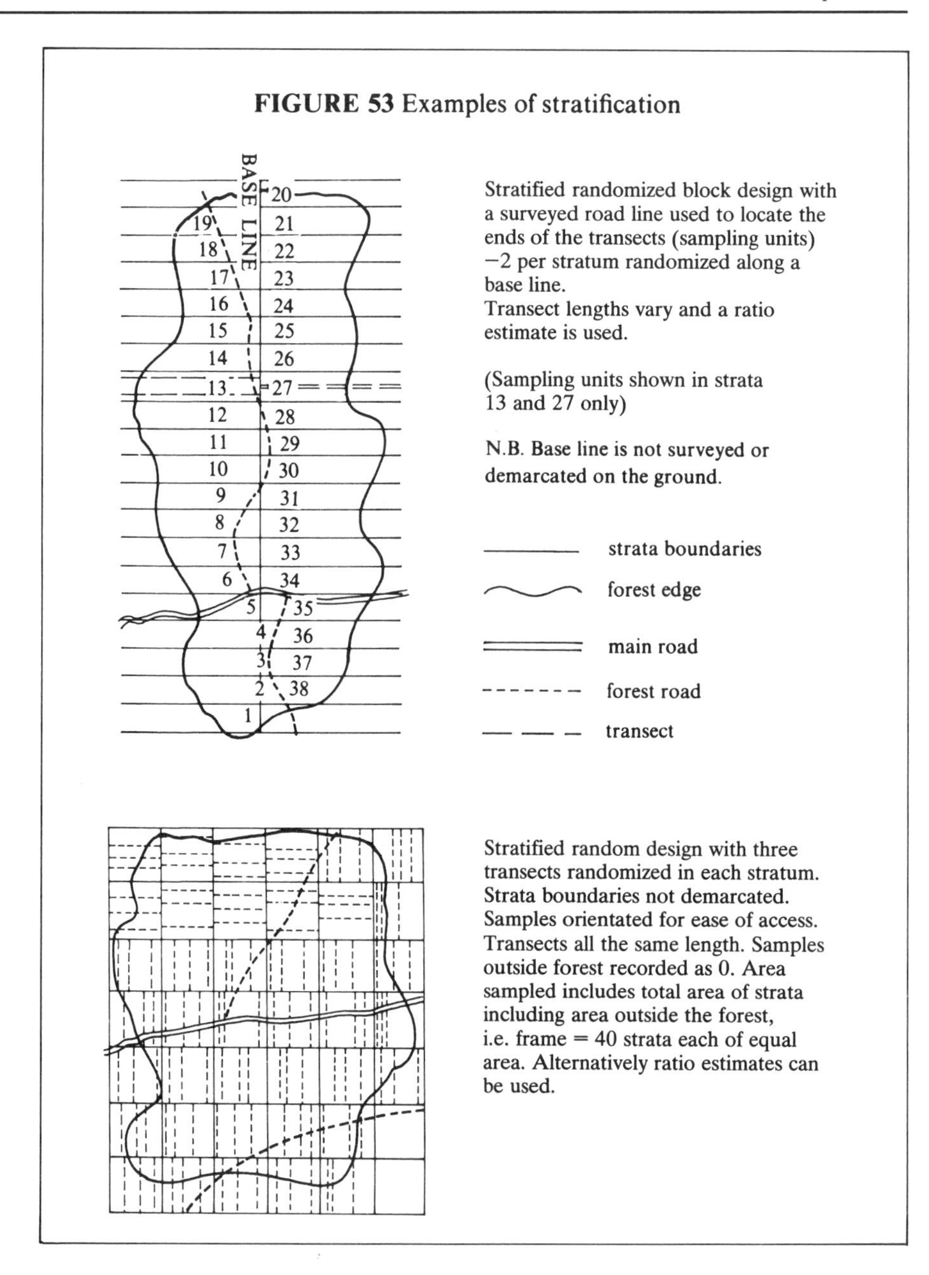

FOR THE ADVANCED STUDENT

The efficiency of stratified random sampling arises from the use of the strata means in the derivation of the population mean. Consequently only the variance within each stratum contributes to the standard errors. The differences, or deviations, between strata means are ignored because of the use of the separate means. However, this implies some loss in the total number of degrees of freedom of the t statistic used to estimate the confidence limits of the population mean; that is the maximum number of degrees of freedom is $\sum n_i - m$, where there is a total of $\sum n_i$ sampling units in m strata. Table 2 shows an analysis of the total variance into the two components

- that arising from the deviations between the means of the strata; and
- that arising from the deviations of the sampling units within a stratum from the stratum mean

Where:

$$cf = \frac{\left(\sum^{m}\sum^{n_i} x_{ij}\right)^2}{\sum^{m} n_i}$$

cf = correction factor (see Section 5.1.1.2)
SS_B = sum of squared deviations between means
SS_W = sum of squared deviations within means
SS_T = total sum of squared deviations
m = number of strata
n_i = number of sampling units in stratum i
x_{ij} = value of the jth sampling unit in the ith stratum, $j = 1 \ldots n$, $i = 1 \ldots m$

As the sums of squared deviations between the means of strata plus the

Table 2 Analysis of variance in a stratified random sampling design, assuming a common within stratum variance

Source of variation	Degrees of freedom	Sums of squared deviations	Variance	*F* ratio
Between means of strata	$m - 1$	$\sum^{m}(n_i(\bar{x}_i - \bar{x})^2)$ $= \sum^{m}(1/n_i)\left(\left(\sum^{n_i} x_{ij}\right)^2 - cf\right)$ $= SS_B$	$\frac{SS_B}{m-1} = s_B^2$	$\frac{s_B^2}{s_W^2}$
Within strata	$\sum^{m}(n_i - 1)$	$\sum^{m}\sum^{n_i}(x_{ij} - \bar{x}_i)^2$ $= \sum^{m}\left(\sum^{n_i} x_{ij}^2 - \frac{\left(\sum^{n_i} x_{ij}\right)^2}{n_i}\right)$ $= SS_W$	$\frac{SS_W}{\sum^{m}(n_i - 1)} = s_W^2$	
Total	$\sum^{m} n_i - 1$	$\sum^{m}\sum^{n_i}(x_{ij} - \bar{x})^2$ $= \sum^{m}\sum^{n_i} x_{ij}^2 - cf$ $= SS_T$	$\frac{SS_T}{\sum^{m} n_i - 1} = s_{total}^2$	

sums of squared deviations within strata must equal the total sum of squared deviations, then as s_B^2 increases s_W^2 decreases and the F ratio increases. Hence the aim of stratification is to make the division of the population so that the strata, and the strata means, are very different from each other, while the strata themselves are very homogeneous.

The expected value of s_W^2, or the within strata variance, is σ_W^2, whereas the expected value of s_B^2, the between strata variance, is $m\sigma_B^2 + \sigma_W^2$. Hence the F ratio $= \dfrac{m\sigma_B^2 + \sigma_W^2}{\sigma_W^2}$ and is expected to equal 1 if σ_B^2 equals 0, in which case the stratification has been ineffectual. (N.B. σ_B^2 is the stratum component of the variance.)

Sampling units should be designed to represent the residual variance within the strata and minimize the variation between these units. Stratified random sampling is most efficient when the source of variation is known and the stratification can be designed accordingly; for example, where a known and mapped part of a stand has been thinned and the remainder is unthinned, or where on aerial photographs differences in stand stocking or vegetation associations can be identified and mapped.

Even where such detailed information is lacking, extensive forest populations may be stratified into regularly shaped subdivisions in the belief that the variation of adjacent parts of the forest will be smaller than that between more distant parts. Two such designs are illustrated in Figure 53.

5.4.2.1.3 Cluster sampling

The term 'cluster sampling' has been used in different ways by different authors. Many authors include multi-stage designs, but this misleads as cluster sampling concerns the arrangement of units in a sampling stage.

In simple cluster sampling the sampling unit is selected at random in an unstratified or stratified population; the unit is then subdivided into smaller units comprising the cluster. These smaller units may all be measured – but this is rare; more usually only a sub-sample within the cluster is measured. The reason for sub-sampling is to economize by applying the assumption that a systematic sub-sample will sample efficiently and fully represent the variance within the sampling unit; then the estimated value of the sampling unit derived from the sub-sample will be so close to the true value for that sampling unit that any difference may be ignored. If this is incorrect then the estimation of the confidence limits will be optimistic.

Common designs for the sub-units in a cluster are illustrated in Figure 54:

- relascope sweeps at intervals around a traverse such as a square
- line plots at intervals along transects
- transects around the perimeter of the sampling unit

Transects arranged as radii around a camp site or groups of plots have also been used. Cluster sampling implies the acceptance of the hypothesis that the pattern of variance is such that a very large sampling unit, i.e. the cluster, is necessary to represent the population variance, and that a large sampling fraction at the cluster level will reduce the sampling error within the cluster to a negligible amount. Hence the value for the sampling unit is derived from the sub-sample and treated as if it had been measured rather than estimated. This implies that the total value for each cluster is calculated and used as the sampling unit datum from which the sample mean, its standard error and population estimates are calculated. In forestry, especially in forest inventory, the sampling unit values are often expressed per hectare and population values derived from the total area of the forest.

FIGURE 54 Common designs of clusters

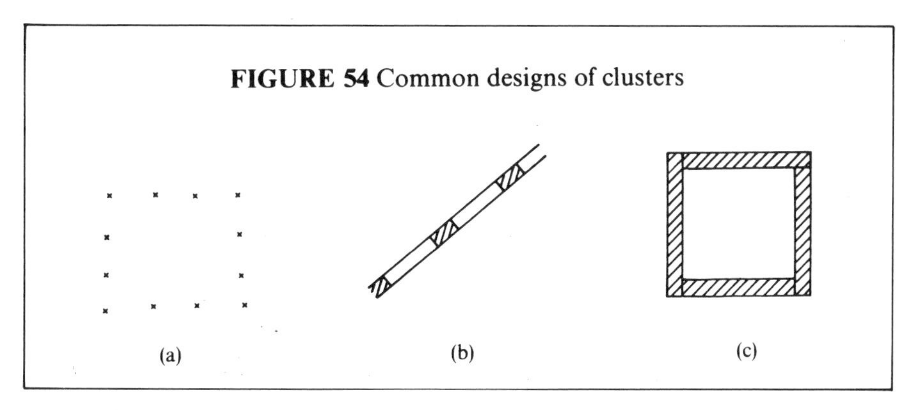

The calculations of sample and population characteristics are shown below.

$$x_i = \frac{\sum^{n'} x_{ij}}{f'} \text{ estimate of value for cluster } i$$

$$\bar{x} = \frac{\sum^{n} x_i}{n}$$

$$\bar{X} = \frac{\bar{x}}{a} \text{ mean value per hectare}$$

$$\text{Total } X = \bar{X}A$$

$$s_{\bar{x}} = \sqrt{\frac{\sum^{n} (x_i - \bar{x})^2(1-f)}{n(n-1)}}$$

Where:

x_{ij} = value of the jth sub-sampling unit in the ith sampling unit or cluster, $j = 1 \ldots n'$, $i = 1 \ldots n$

n' = number of sub-sampling units in a cluster

f' = sampling fraction within cluster

n = number of sampling units or clusters

f = sampling fraction of clusters

x_i = estimate of value for the ith sampling unit derived from the sub-sample

$\bar{x}$ = mean of the sampling units

a = total area comprising a cluster
A = total area of population
$\bar{X}$ = mean value per hectare
Total X = estimate of total in population
$s_{\bar{x}}$ = estimate of standard error of mean of sampling units (N.B. Ignoring error due to sub-sampling)

The within cluster variance has been omitted from the calculation of the standard error, assuming that f' approaches 1 and the within cluster variance approaches 0 so that, therefore, the component is small in comparison to the between cluster variance.

5.4.2.1.4 Double or two-phase sampling

A two-phase regression estimate is frequently used in forest inventory. An example was given in Section 5.3. It is suited to the situation where the object of the inventory – parameter y – is difficult, expensive, or destructive to measure, but an associated parameter x is both highly correlated with y and easier to measure. Unlike single-phase regression estimation, no prior knowledge of the population value (X) is needed. The values of x are measured on each sampling unit and the regression parameters for predicting y from x are estimated from measurements of x and y taken on a sub-sample. This enables the surveyor to predict Y as a function of the estimated X from the main sample. The application of this technique implies that the residual variance of the regression is very much smaller than the variance of the x values in the population. There are, therefore, two sources of error contributing to the error of the estimate:

- the error in the estimate of the regression parameters
- the error in the estimate of X

The calculations of sample and population characteristics are shown below and a worked example follows.

A linear regression in the form

$$y = c + bx$$
$$c = \bar{y}_r - b\bar{x}_r$$

is estimated from a sub-sample of n_1 sampling units in which both x_i and y_i are measured, where

$\bar{x}_r$ = mean of the x_i measurements made on the sub-sample
$\bar{y}_r$ = mean of the y_i measurements made on the sub-sample
s_y^2 = variance of y estimated from the n_1 measurements
$\bar{x}_2$ = mean of the x measurements made on the full sample of n_2 sampling units
$\bar{y}_{lr}$ = estimated mean of y for the population, using the linear regression
Y = estimate of population value for the parameter y

Then

$$\bar{y}_{lr} = c + b\bar{x}_2$$
$$= \bar{y}_r + b(\bar{x}_2 - \bar{x}_r)$$

$Y = N\bar{y}_{lr}$ or if y is a value per unit area a in a population of total area A

$Y = \frac{Ay_{lr}}{a}$ where Y is the population value of the parameter y

The variance of this linear regression estimate from two-phase sampling is derived from the sum of the variances of the two phases. The variance of the n_1 measured values of y_i can be used to estimate the residual variance of the regression and the variance of the y predicted from the n_2 sampling units, provided that the sub-sample was selected at random. Hence

$$s^2_{\bar{y}_{lr}} = s^2_{\bar{y}_2} + \frac{\sum^{n_1} (y_i - (\bar{y}_r + b(x_i - \bar{x}_r)))^2}{n_1(n_1 - 2)} = \frac{s^2_y}{n_2} + \frac{s^2_y(1 - r^2)}{n_1}$$

or with the finite population correction factor

$$\begin{aligned} s^2_{\bar{y}_{lr}} &= \left(\frac{N - n_2}{N}\right)\frac{s^2_y}{n_2} + \left(\frac{n_2 - n_1}{n_2}\right)\frac{s^2_y(1 - r^2)}{n_1} \\ &= \left(\frac{1}{n_2} - \frac{1}{N}\right)s^2_y + \left(\frac{1}{n_1} - \frac{1}{n_2}\right)s^2_y(1 - r^2) \\ &= \frac{s^2_y}{n_1}(1 - r^2) + \frac{s^2_y r^2}{n_2} - \frac{s^2_y}{N} \quad \text{(often the final term is omitted)} \end{aligned}$$

Also

$$\begin{aligned} s^2_Y &= N^2 s^2_{\bar{y}_{lr}} \\ &= \frac{A^2}{a^2} s^2_{\bar{y}_{lr}} \quad \text{where } y \text{ is a value per unit area } a \end{aligned}$$

Cunia & Michelakackis (1983) refer to the error calculation in tree biomass table construction using two-phase sampling.

EXAMPLE 46 Calculation of inventory results using a two-phase design

An inventory of a large area of miombo woodland was carried out to estimate the total standing volume of firewood. The design was simple random sampling with a two-phase regression estimate. An initial investigation showed that the model

$$y = c + b(cc)$$

was a useful predictor equation, where y = volume of firewood, $m^3\ ha^{-1}$; cc = canopy cover as measured on aerial photographs, using a scale of 1–10 to quantify the degree of canopy closure.

The inventory yielded the following data:

Area (A) 24,700 ha
n_1 50 – number of plots in which y, volume per ha, was measured
n_2 400 – number of sampling units in which cc was measured
$\bar{x}_r$ 6.4 – average canopy cover score in the sub-sample of n plots
$\bar{y}_r$ 184.3 m^3 – average volume per ha in the n_1 sample plots

Example 46 continued

c 115 - regression constant, estimated
b 10.2 - regression coefficient, estimated
r^2 0.5184 - square of the correlation coefficient, estimated
s_y^2 3674 - estimated variance of y
$\bar{x}_2$ 5.7 - mean of n_2 sampling units

Then

$$\bar{y}_{lr} = \bar{y}_r + b(\bar{x}_2 - \bar{x}_r)$$

$$= 184.3 + (10.2)(5.7 - 6.4), \text{ m}^3 \text{ ha}^{-1}$$

$$= 184.3 + (10.2)(-0.7), \text{ m}^3 \text{ ha}^{-1}$$

$$= 177.2 \text{ m}^3 \text{ ha}^{-1}$$

$$Y = \bar{y}_{lr}\frac{A}{a} = (177.2)\frac{24,700}{1}$$

$$= 4,380,000 \text{ m}^3$$

$$s_{\bar{y}_{lr}}^2 = \frac{3674}{50}(1 - 0.5184) + (3674)\frac{0.5184}{400} - \frac{3674}{24,700}$$

$$= 35.39 + 4.76 - 0.15 = 40.00$$

$$s_{\bar{y}_{lr}} = 6.32 \text{ m}^3 \text{ ha}^{-1}$$

Confidence limits of $Y = 4,380,000 \pm (2.0)(6.32)(24,700)$
$(t)\ (s_{\bar{y}_{lr}})\ (A)$
$= 4,380,000 \pm 310,000 \text{ m}^3$

5.4.2.2 *Multi-stage Sampling*

Multi-stage sampling is characterized by the division of the population into a frame of primary sampling units, only some of which are sampled. The primary units are selected and each is subdivided into a frame of secondary sampling units. Either final sampling units are selected or units are selected for further subdivision into tertiary stage sampling frames. This hierarchical successive subdivision of the sampling units may proceed further, although frequently only two stages are employed.

Multi-stage sampling differs from stratified random sampling in that in the former only some of the subdivisions are sampled whereas in stratified random sampling all the strata must be sampled with at least two sampling units in each. The distinction between two-phase and two-stage sampling is in the nature of the work done; in two-stage sampling the primary units are selected although no measurements are taken using these as units – they form the frame from which the second stage units are selected. Measurements are only taken on these second or final stage sampling units. In two-phase sampling, however, one complete set of measurements is made on the main sample and an additional supplementary measurement is made on the sub-sample in order to estimate the regression parameters.

The calculations depend upon

- the relative size of the primary units

- the method of selection
- the probability of selection
- the sampling fractions

In the simplest situation with

- equal sized primary units
- selected either with or without replacement
- equal probability of selection
- equal sampling fractions within the primary units

then

$$X_i = \frac{M_i}{m_i}\sum^{m_i} x_{ij} = \sum^{m_i}\frac{x_{ij}}{f_2} \qquad \bar{x}_i = \frac{\sum^{m_i} x_{ij}}{m_i}$$

the estimate of the population total X is

$$X = \frac{N}{n}\left(\sum^{n} X_i\right) = \sum^{n}\left(\frac{X_i}{f_1}\right) \quad \text{and} \quad \bar{\bar{x}} = \frac{\sum^{n} M_i\bar{x}_i}{\sum^{n} M_i}$$

$$= \frac{N}{n}\left(\sum^{n} M_i\bar{x}_i\right)$$

$$= \frac{N}{n}\left(\sum^{n} M_i\left(\frac{\sum^{m_i} x_{ij}}{m_i}\right)\right)$$

but in this design M_i and m_i are constant for all i; and

$$\frac{1}{f_2} = \frac{M}{m} \qquad \bar{\bar{x}} = \frac{\sum^{n}\bar{x}_i}{n}$$

Then

$$X = \left(\frac{N}{nf_2}\right)\left(\sum^{n}\sum^{m} x_{ij}\right) = \frac{1}{f_1 f_2}\left(\sum^{n}\sum^{m} x_{ij}\right)$$

The variance of X is given by

$$s_X^2 = \frac{N^2(1-f_1)}{n}\frac{\sum^{n}(X_i-\bar{X})^2}{(n-1)} + \frac{N}{n}\sum^{n} M^2(1-f_2)\frac{\sum^{m}(x_{ij}-\bar{x}_i)^2}{m(m-1)}$$

or in terms of the mean values of the secondary sampling units, the formula above can be rearranged to give the following:

When $m_i = m$, the variance of $\bar{\bar{x}}$ is given by

$$s_{\bar{\bar{x}}}^2 = \frac{(1-f_1)}{m^2 n}\frac{\sum^{n}(X_i-\bar{X})^2}{n-1} + \frac{1}{Nn}(1-f_2)\frac{\sum^{n}\sum^{m}(x_{ij}-\bar{x}_i)^2}{m(m-1)}$$

or

$$s^2_{\bar{\bar{x}}} = \frac{(1-f_1)}{n}\frac{\sum^{n}(\bar{x}_i - \bar{\bar{x}})^2}{n-1} + \frac{n}{N}\frac{(1-f_2)}{nm}\frac{\sum^{n}\sum^{m}(x_{ij}-\bar{x}_i)^2}{n(m-1)}$$

Where:

X = estimate of total population value of parameter x
X_i = estimate of total in primary unit i
N = number of primary units
n = number of primary units sampled
$\bar{\bar{x}}$ = estimate of population mean value per secondary sampling unit
x_{ij} = value of x in the jth secondary unit of the ith primary unit
m_i = number of secondary units selected in the ith primary unit (= m if all m_i are equal)
$\bar{x}_i$ = mean of the j secondary units in the ith primary unit
$f_1 = \frac{n}{N}$ $f_2 = \frac{m}{M}$
X_i = estimate of total value of x in the ith primary unit
$\bar{X}$ = average of X_i
M_i = number of secondary sampling units in the ith primary unit (= M if all M_i are equal).

This last formula indicated that when $\frac{n}{N}$ is small, i.e. $f_1 \rightarrow 0$, then the second component may be omitted. In which case,

$$\text{variance of } \bar{\bar{x}} = \frac{\sum^{n}(\bar{x}_i - \bar{\bar{x}})^2}{n(n-1)}$$

This also follows from the fact that the formula for the variance of the mean including the second component may be re-written, omitting the finite population correction factors, as

$$\text{variance of } \bar{\bar{x}} = \frac{s_b^2}{n} + \frac{s_w^2}{nm}$$

Where:

s_b^2 = the block component of variance, and
s_w^2 = the within block component of variance

Therefore

$$\text{variance of } \bar{\bar{x}} = \frac{ms_b^2 + s_w^2}{nm}$$

but this is the variance between the primary sampling units (see Section 5.4.2.1.2).

$$= \frac{\sum^{n} m(\bar{x}_i - \bar{\bar{x}})^2}{nm(n-1)}$$

$$= \frac{\sum^{n}(\bar{x}_i - \bar{\bar{x}})^2}{n(n-1)}$$

If the primary units are unequal in size and they are chosen with probability proportional to their size, then usually sampling is with replacement. There are several methods of dealing with the situation when a primary unit is selected more than once; in this case the selection of secondary units within the primary units is repeated independently on each occasion the primary unit is selected. Then using the symbols as above:

$$\bar{x}_i = \frac{\sum^{m_i} x_{ij}}{m_i}$$

$$X_i = M_i \bar{x}_i$$

$$X = \frac{1}{n}\sum^{n}\left(\frac{X_i}{p_i}\right) \qquad \bar{x} = \frac{X}{\sum^{N} M_i}$$

$$s_x^2 = \frac{\sum^{n}\left(\frac{X_i}{p_i} - X\right)^2}{n(n-1)}$$

if $p_i = \frac{M_i}{\sum^{N} M_i}$, and $s_x^2 = \left[\frac{\left(\sum^{N} M_i\right)^2 \sum^{n}}{n(n-1)}\right]\left[\bar{x}_i - \left(\frac{X}{\sum^{N} M_i}\right)\right]^2$

The finite population correction factor is unnecessary as the sampling is with replacement. The relative simplicity of this formula encourages the use of this design provided that it does not result in too great a loss of precision compared to sampling without replacement.

Multi-stage sampling is particularly useful when simple random or stratified random sampling would be too costly because of the costs of access to the sampling units, and where the between primary unit variation is small compared to that within primary units.

Further reading

The recent literature on multi-stage sampling is voluminous. An introduction into some of the topics may be found in Jeyaratnam *et al.* (1984), Furnival *et al.* (1986) and Murchison & Ek (1989).

5.5 Recurrent Forest Inventory

Synott (1979, 1980), Adlard (1990) and Alder & Synott (1992) have provided detailed comments and recommendations concerning the techniques for recurrent forest inventory in the tropics – both in plantations and in mixed forests. Also the literature on the calculation and analysis of data from recurrent forest inventory is voluminous. Ware & Cunia (1962), Cunia (1965), Sullivan & Reynolds (1976) and Cochran (1977) provide early references; Loetsch *et al.* (1973) provide a very useful general account of recurrent forest inventory principles. Most recent literature has been reviewed in Malla *et al.* (1984), and especially that dealing with points rather than plot sampling units is to be found in Flewelling (1981), Iles & Beers (1983) Martin (1983), Scott (1984) and Yang & Chao (1987). The proceedings of the joint meeting of IUFRO in April 1989 (Adlard & Rondeux, 1989) are also relevant.

In this introduction, only three cases are considered:

- using only permanent sample plots
- using only temporary sample plots
- using both temporary and permanent sample plots

The sampler is interested in obtaining the best estimate of the current growing stock and the change since the last inventory. For the sake of the simplicity only measurement on two successive occasions is considered.

5.5.1 Recurrent Forest Inventory Using Only Permanent Sample Plots

When:
$\bar{x}_p$ = mean of all plots on the first occasion
$\bar{y}_p$ = mean of all plots on the second occasion
$\bar{z}$ = best estimate of the change between the two measurements, i.e. $\bar{z} = \bar{y}_p - \bar{x}_p$
r = estimated correlation coefficient of x_p and y_p
n_p = number of permanent sample plots remeasured

then (see Section 5.1.1.4)

$$s_{\bar{z}}^2 = \frac{(s_{y_p}^2 + s_{x_p}^2 - 2rs_{x_p}s_{y_p})}{n_p}$$

Consequently the closer that r approaches the value of +1 the smaller is the standard error of $\bar{z}$. Normally the finite population correction factors may be ignored as the sampling fractions are very low.

EXAMPLE 47 Calculations in a two-stage sampling design using primary units selected with probability proportional to area, with replacement

A forest was divided into 30 primary units of varying size and totalling 24,000 ha. Six primary units were selected, with replacement with probability proportional to their area. In each primary unit the volume per ha was measured in three secondary sampling units. The calculation of the estimate of the total standing volume and its confidence limits ($p = 0.05$) was as follows:

Data

Id. no. of primary unit selected	Area	p_i	x_{ij} m³ ha⁻¹	$\bar{x}_i$	X_i m³	X_i/p_i = Total X, m³
6	1000	10/240	20,30,50	33.3	33,300	800,000
9	900	9/240	10,30,40	26.7	24,000	640,000
13	600	6/240	15,25,60	33.3	20,000	800,000
17	700	7/240	20,15,75	36.7	25,700	880,000
26	800	8/240	20,40,60	40.0	32,000	960,000
28	1400	14/240	10,70,10	30.0	42,000	720,000
						4,800,000

$$X = \frac{4{,}800{,}000}{6}$$

$$= 800{,}000 \text{ m}^3$$

$$\bar{x} = \frac{800{,}000}{24{,}000} = 33.3 \text{ m}^3 \text{ ha}^{-1}$$

$$s_X^2 = \sum^{n}\left(\frac{X_i}{p_i} - X\right)^2 \Big/ (n(n-1))$$

$$= \frac{(640.00)(10^8)}{(6)(5)} = (21.33)(10^8)$$

$$s_X = \sqrt{s_X^2} = 46{,}000 \text{ m}^3$$

Confidence limits $= 800{,}000 \pm t(46{,}000)$ m³
$= 800{,}000 \pm 120{,}000$ m³

N.B. t has $n-1$ df $= 2.57$ at $p = 0.05$

If the same data had been collected in primary units of equal size, i.e. 800 ha each, selecting without replacement and with equal probability, then the calculations would be:

Id. no. of primary unit selected	$X_i = M_i\bar{x}_i$	$\sum_j^{m_i}(x_{ij} - \bar{x}_i)^2$	$(X_i - \bar{X})^2$
6	26,700	467	0
9	21,300	467	$(29.2) \cdot 10^6$
13	26,700	1117	0
17	29,300	2217	$(6.8) \cdot 10^6$
26	32,000	800	$(28.1) \cdot 10^6$
28	24,000	2400	$(7.3) \cdot 10^6$
	160,000	7468	$(71.4) \cdot 10^6$

Example 47 continued

$\bar{X}_i = 26{,}700\ m^3$

$X = (30/6) \cdot 160{,}000 = 800{,}000\ m^3$

$$s_X^2 = N^2(1-f_1)\frac{\sum^n (X_i - \bar{X})^2}{n(n-1)} + \frac{1}{f_1} M_i^2 (1-f_2)\frac{\sum^n\sum^m (x_{ij} - \bar{x}_i)^2}{m(m-1)}$$

$$= 30^2\left(1-\frac{6}{30}\right)\frac{(71.4)(10^6)}{(6)(5)} + \frac{30}{6}800^2\left(1-\frac{3}{800}\right)\frac{(7468)}{(3)(2)}$$

$$= (17.1)(10^8) + (39.7)(10^8)$$

$$= (56.8)(10^8)$$

$s_X = ((56.8)(10^8))^{0.5} = 75{,}000\ m^3$

Confidence limits $= 800{,}000 \pm t(75{,}000)\ m^3$

$= 800{,}000 \pm 190{,}000\ m^3$

5.5.2 Recurrent Forest Inventory Using Only Temporary Sample Plots

Using the symbols

$\bar{x}_t$ = mean of the temporary sample plots on the first measurement
$\bar{y}_t$ = mean of the temporary sample plots on the second measurement
$\bar{z}$ = best estimate of the change between the two measurements

i.e.

$\bar{z} = \bar{y}_t - \bar{x}_t$

then

$s_{\bar{z}}^2 = s_{\bar{y}}^2 + s_{\bar{x}}^2$, ignoring the finite population correction factor

As the two sets of sampling units are independent, the expected value of the correlation coefficient is 0 and so the third term appearing in the corresponding equation above in Section 5.5.1 can be omitted. In the normal situation and the limits on cost, these variances are large in relation to $\bar{z}$ and hence the design is little used, but has been included here to emphasize the importance of the correlation between measurements in permanent sample plots in reducing the standard error of the change in volume.

5.5.3 Recurrent Forest Inventory Using Both Permanent and Temporary Sample Plots

Very frequently the cost of using all permanent sample plots is prohibitive; also at times the structure of the forest varies through felling, regeneration and accidents so that a very small sample of permanent plots may be inadequate to represent the growing stock even though it is adequate to represent the growth relationships.

Consequently a combination of the two types of sampling unit has advantages. The procedure is to carry out an inventory using temporary sample plots in a normal manner, except that random sub-samples are designated for re-measurement at a later date and are, therefore, carefully demarcated and their location accurately surveyed so that they may be relocated. The extra costs of demarcation, survey and maintenance are the major contribution to the expense of permanent sample plots. Also, if the objects of the inventory include the estimation of removals, mortality and recruitment, then the individual trees within the permanent sample plots must be identified so that their fate can be followed from measurement time to re-measurement time. This further increases the costs.

The change in the population over the time interval between successive re-measurements makes the estimation of the optimal combination of numbers of permanent and numbers of temporary sample plots inaccurate. None the less a guide can be given.

This optimal combination depends upon the objects set, e.g.

- to select the combination of temporary and permanent plots that will minimize the standard error of the change $\bar{z}$ for a given cost
- to select the combination of temporary and permanent sample plots that will minimize the variance of both $\bar{z}$ and the estimate of the current growing stock on the occasion of the second assessment, etc.

In this second case, and assuming the cost function in the form of

$$C = c_t n_2 + c_p n_p$$

Where:

C = total variable cost of permanent and temporary plots
c_t = variable cost per temporary plot at the time of re-measurement
c_p = variable cost of a permanent sample plot
r = correlation coefficient of the measurements in the permanent sample plots at the two occasions
n_2 = number of temporary sample plots at the time of re-measurement
n_p = number of permanent sample plots
N_1 = total permanent and temporary sample plots used on the first measurement

Then

$$n_p = \frac{N_1\sqrt{(1-r^2)}}{r^2}\left[\sqrt{\frac{c_t}{c_p}} - \sqrt{(1-r^2)}\right]$$

$$n_{t_2} = \frac{s^2_{\bar{v}2}}{s^2_{\bar{v}2}} - \frac{N_1}{r^2}\left[1 - \sqrt{\frac{c_t(1-r^2)}{c_p}}\right]$$

Loetsch *et al.* (1973)

As in sampling with only permanent sample plots, the closer that r approaches to 1 the more precise are the estimates. Ware & Cunia (1962) suggest that permanent sample plots are only worth while if their cost is less than twice that of temporary plots and the period of re-measurement is short so that the value of r is near 1. However, they were not speaking of tropical forest conditions where permanent sample plots are the only feasible source of data on growth, recruitment and mortality.

The complexity of the formulae disguises the simplicity of the principle behind the technique. In essence the formula for $\bar{z}$ re-estimates both $\bar{y}$ and $\bar{x}$ using all the information supplied by both inventories. Through the regression of x on y estimated from the permanent sample plots, the temporary sample plots of the second inventory are provided with predicted values for their situation had they been measured at the time of the first inventory. $\bar{x}_b$ is then calculated from all $n_p + n' + n''$ plots. Similarly, using the regression of y on x the n' temporary plots of the first inventory are provided with predicted values for their situation at the time of the second inventory; $\bar{y}_b$ is again calculated from data for all temporary and permanent sample plots. Therefore the permanent sample plots represent the rate of change, whereas the sum of both the permanent and temporary plots at both occasions represents the growing stock.

When

$\bar{x}_p$ = mean of permanent plots on the first measurement
$\bar{y}_p$ = mean of permanent plots on the second measurement
$\bar{x}_t$ = mean of temporary plots on the first measurement
$\bar{y}_t$ = mean of temporary plots on the second measurement
n_p = number of permanent sample plots
n' = number of temporary sample plots at the first measurement
n'' = number of temporary sample plots at the second measurement
n_1 = $n_p + n'$
n_2 = $n_p + n''$
$\bar{x}_b$, $\bar{y}_b$ and $\bar{z}_b$ = the best estimates of μ_x, μ_y and μ_z
r = correlation coefficient of x_i, y_i measured in the permanent sample plots
b_{yx} = regression coefficient of y and x estimated from the paired x_i and y_i measured in the permanent sample plots
b_{xy} = the corresponding regression coefficient of x on y

then

$$\bar{z}_b = \bar{y}_t(1 - A) - \bar{x}_t(1 - B) + \bar{y}_p A - \bar{x}_p B$$

where

$$A = \frac{n_p(b_{xy}n'' + n_1)}{(n_1 n_2 - n'n''r^2)}$$

$$B = \frac{n_p(b_{yx}n' + n_2)}{(n_1 n_2 - n'n''r^2)}$$

therefore

$$s^2_{\bar{z}_b} = \frac{(A^2 s_y^2 + B^2 s_x^2 - 2ABrs_x s_y)}{n_p} + \frac{(1 - A)^2 s_y^2}{n''} + \frac{(1 - B)^2 s_x^2}{n'}$$

$$\bar{y}_b = a_y(\bar{x}_t - \bar{x}_p) + c_y\bar{y}_p + (1 - c_y)\bar{y}_t$$

and

$$s_{\bar{y}_b}^2 = a_y^2 s_x^2\left(\frac{1}{n'} - \frac{1}{n_p}\right) + \frac{c_y^2 s_y^2}{n_p} + \frac{(1 - c)^2 s_y^2}{n''} - \frac{2a_y c_y r s_x s_y}{n_p}$$

when

$$a_y = \frac{\left(\dfrac{rs_y}{s_x}\right)(n_p n')}{(n_1 n_2 - n'n''r^2)}$$

$$a_x = \frac{\left(\dfrac{rs_x}{s_y}\right)(n_p n'')}{(n_1 n_2 - n'n''r^2)}$$

$$c_y = \frac{(n_p n_1)}{(n_1 n_2 - n'n''r^2)}$$

$$c_x = \frac{(n_p n_2)}{(n_1 n_2 - n'n''r^2)}$$

An explanation of these formulae is given in Appendix 7.

Point sampling in recurrent forest inventory has been strongly advocated by many authors but introduces a complication in calculating increment due to the fact that 'recruits' include both trees that have regenerated in the period between inventories and trees that have grown large enough to be included in the angle gauge counts, the latter differing from the former in that only their increase in volume – not their total volume – contributes to growth. Increment is not simply

$$v_2 - v_1$$

but must be adjusted to exclude volume in v_2 that was present at the time of v_1 but not recorded. Recent references on this topic that provide an entry to the literature are Van Deusen *et al.* (1986) and Yang & Chao (1987).

6 SITE ASSESSMENT

6.1 REASONS FOR ASSESSING THE POTENTIAL PRODUCTION OF A SITE

The complex of factors such as rock, soil, climate, topography and vegetation that characterize an area of ground are referred to collectively as the *site*. Foresters usually wish to assess the potential production of a site for one of three rather distinct reasons:

- as a criterion in land use allocation and development planning
- as the basis for the choice of species for planting
- as the basis for forecasting the growth of a managed forest – especially plantations – in order to plan investment and production

6.2 A CLASSIFICATION OF THE METHODS USED IN SITE ASSESSMENT

The methods of site assessment differ with the objects of the assessment.

6.2.1 Assessment for Land Allocation and Development Planning

The most common bases for such assessments are a combination of climate, soil groups and vegetation associations which together reflect the growth potential and constraints on use of the site. There are many examples of this type of assessment; on the global and regional scale that of Holdridge *et al.* (1971) is a relatively recent example. He based subdivisions of 'life zones and associations' on the inter-play of altitude, latitude, mean annual bio-temperature[1], precipitation and the ratio of potential evapotranspiration to precipitation.

More local examples are those applied in the land classification maps of Trapnell & Griffiths (1960) in East Africa and of the Overseas Development Administration (1972) in Northern Nigeria. A great deal of the information for

[1]Bio-temperature is the average of the temperature each day of the year with the substitution of 0 for values below 0°C or above 30°C.

such land classification systems is derived from aerial photography and other remote sensing imagery, but it has to be used in conjunction with information gathered from ground surveys.

6.2.2 Assessment to Aid in the Choice of Species and Prediction of Growth

In Europe, Scandinavia, Canada and elsewhere the Patterson-Weck CVP index has been used to define homoclimes or areas of similar climate; these are useful when searching for species to introduce in order to raise the useful yield from the forest. The CVP index is defined by:

$$I = \left(\frac{T_v}{T_a}\right)(P)\left(\frac{G}{12}\right)(E)$$

Where:

I = CVP index ranging from 0 to 30,000 with forest growth possible in areas with an index greater than 25
T_v = mean monthly temperature of the hottest month, °C
T_a = difference between the mean monthly temperatures, °C, of the hottest and coldest months
P = mean annual precipitation, mm
G = length of growing season in months
E = R_p/R_s
R_p = radiation at the pole, 10^3 g cal cm^{-2} min^{-1}
R_s = radiation at the site, 10^3 g cal cm^{-2} min^{-1}

Bose (1988) refers back to his own work of the 1970s and advocates a simple thermodynamic approach to site productivity, using

$$V = f(m, T, R, r, d, s, p)$$

Where:

V = volume productivity, per hectare per year
m = water requirement of the species per unit of dry matter production
T = maximum temperature
R = mean annual rainfall
r = run-off, so $R - r$ = net soil adsorption
d = density of biomass
s = soil bulk density
p = base rock bulk density

Bose claims good correspondence between predicted and actual production for sal (*Shorea robusta*) forest in North India, for moist deciduous forest of the Eastern Ghats, and for eucalypts and explains the occurrence of hill grasslands in the Eastern Ghats of Andhra Pradesh.

In the seasonal tropics one of the critical site factors is the degree of the soil water deficit in the dry season. Walter *et al.* (1975) developed klima-diagrams to illustrate this aspect of the soil water relations. Similar water balance diagrams (Thornthwaite & Mather, 1963) show:

- monthly or periodic precipitation
- potential evapotranspiration, calculated from the Penman or other empirical formula
- actual evapotranspiration

The balance between these indicates the seasons when:

- precipitation exceeds potential evapotranspiration, and the surplus first fills the soil water storage in the profile, and then
- drains away from the site contributing to deep underground storage or stream flow
- precipitation is less than the potential evapotranspiration and the difference results in
  - first, depletion of the water available in the soil profile, and
  - then, water stress in the vegetation

Examples of such diagrams are shown in Figure 55. Comparisons are extremely useful in matching sites and assessing their suitability for different species, etc.

The oldest method of matching species to site was through indicator plants – that is, common plants which are associated with successful growth of particular species. In Europe these were members of the ground flora.

In Finland, Cajander (1926) established that the productivity of a site was closely correlated with the species association of the ground flora, and was able to predict the maximum mean annual increment of the site under the indigenous forest of Scots pine (*Pinus sylvestris*). However, this was in a country with a harsh climate and a relatively simple forest flora with few tree species. Similar methods have been much used in Europe especially in the more mountainous areas where the forest, though managed for wood production, is regenerated naturally. There the Braun-Blanquet classification of site types by vegetation associations is used to estimate the growth of the forest for the purpose of yield prediction and control. The vegetation associations used are defined by the trees, the shrub layer and the ground flora.

Very much less work on site classification has been done in the tropics – either in the indigenous forest or in plantations. In Uganda at Budongo forest, *Khaya anthotheca* regenerated successfully along with the light-demanding species *Trema guineensis*; the grass *Andropogon gayanus*, indicating montmorillonitic clays with a high pH, shows sites more suited to afforestation with *Eucalyptus camaldulensis* than with *Pinus caribaea*. Most effort has been directed at the selection of species to match the site that also minimize the need for costly weeding operations. Often the invasion of sites under agriculture by rhizomatous grasses such as *Imperata cylindricum* or *Digitaria scalarum*, or by weeds such as *Lantana camara* or *Eupatorium* affects the early growth of trees and raises the cost of plantation establishment. Their effects often mask other inherent site differences.

In West Africa considerable progress has been made in correlating the growth of teak (*Tectona grandis*) plantations with soil characters, notably the exchangeable calcium. Similarly the distribution of species in the natural forest is associated with geology and topography (Hall, 1977). Water shedding, water receiving and shedding, and water receiving sites all have markedly different soil formation processes, soil types and forest associations. For example, in

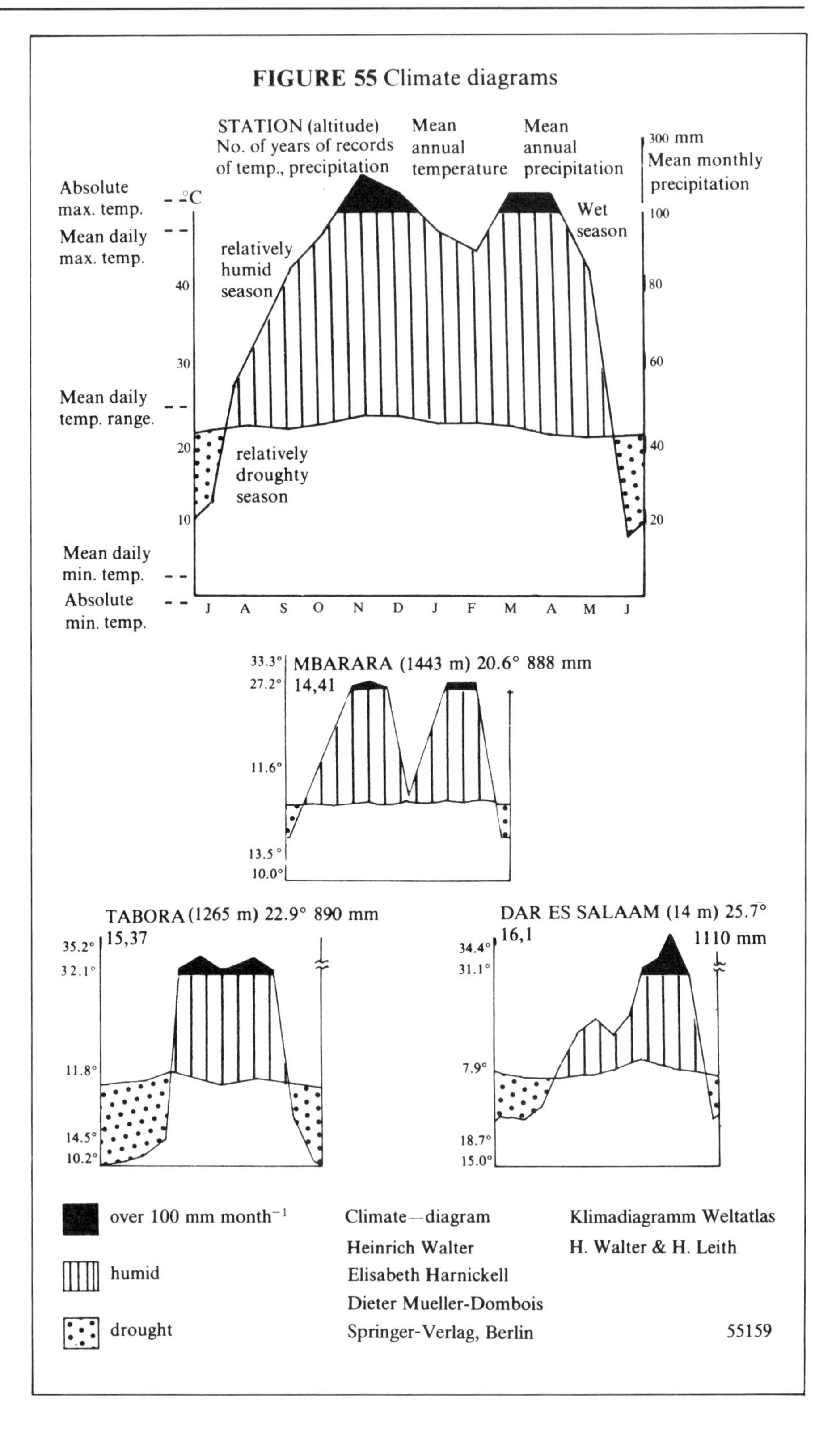

FIGURE 55 Climate diagrams

mixed semi-deciduous forest, the understorey on the ridges may wilt during the dry season when that of the lower slopes remains turgid. Often the middle slopes are the most productive, having soils of adequate depth and a favourable soil water regime at all seasons; soil on the ridges tend to be shallower and drier whereas those in the valley bottoms are less well drained and with clays of poorer physical properties for tree growth. The recognition of site differences and indicator species or associations has proceeded to only a limited extent in East Africa; in Uganda *Eucalyptus maculata* was planted in areas where the ground vegetation indicated seasonal flooding, while *E. saligna* was restricted to the better drained areas. At higher altitudes where softwoods were used, *Cupressus lusitanica* was confined to old forest sites whereas *Pinus patula* could be established easily in grasslands. *Tectona grandis* is restricted to the high base status alluvial soils with good ground water supplies on sloping ground. *Balanites aegyptiaca* grows on soils with a high pH often containing free calcium carbonate nodules. Around Lake Victoria *Loudetia kagarensis* grows on the extremely leached, white sands.

In the miombo woodlands of central Tanzania the better drained soils have associations dominated by *Brachystegia spiciformis* and *Julbernardia globulifera*, whereas the flood plains are characterized by *B. longifolia* and *Pericopsis angolensis*. Similarly in the more open grasslands of northern Uganda *Hyparrhenia filipendula, H. dissoluta, H. rufa* and *Themeda triandra* are characteristic of the better drained slopes and *Andropogon gayanus* of seasonally flooded areas.

Generally, however, in the species-abundant associations of the tropics the detailed ecological studies on which site classification could be based have yet to be done; therefore the examples quoted are obvious and matched by easily recognizable features of the sites, such as the topographical position. Little work has been done to recognize indicator species capable of differentiating sites with different growth rates of, say, *Pinus patula* within an afforestation scheme. In part this is because the plantations tend to be managed at stocking levels that preclude a well-developed ground flora, so that the species that appear tend to reflect the light conditions rather than the site quality.

In the temperate areas a great deal of work has been done to develop mathematical models predicting crop growth from parameters easily measured on sites either with or without trees. The most used predictive variables are

- altitude
- relative exposure
- temperature regime
- rainfall
- rooting depth
- stoniness
- pH and other variables associated with the base exchange system

Unfortunately most of the results have only local applications because the predictive equations are empirical and unstable, depending for their precision and accuracy on the particular set of data used and the method of constructing the equations. Extrapolation outside the range of the data is unwarranted and potentially unreliable and misleading. A recent study in Scotland is outlined in Worrell & Malcolm (1990).

EXAMPLE 48 A multivariate predictive equation for the growth rate of *Eucalyptus* species *A* in a mountainous area

$$I_A = a + b_1X_1 + b_2X_2 + b_3X_3 + b_4X_4$$

Where:

I_A = maximum mean annual volume increment of species *A*, $m^3\ ha^{-1}\ yr^{-1}$
X_1 = mean annual rainfall, mm
X_2 = number of months with less than 50 mm of rain
X_3 = relative elevation with reference to main drainage channel and watershed
X_4 = % clay in the 300-500 mm soil layer

The constants in the equation were estimated to be

$a = 2.7$
$b_1 = 0.023$
$b_2 = -1.4$
$b_3 = -0.82$
$b_4 = 0.15$

and at a particular site the variables

$X_1 = 1000$
$X_2 = 5.0$
$X_3 = 0.5$, i.e. mid-slope
$X_4 = 20$

Then

$$\begin{aligned} I &= 2.7 + 0.023(1000) - 1.4(5.0) - 0.82(0.5) + 0.15(20) \\ &= 2.7 + 23.0 - 7.00 - 0.4 + 3.0 \\ &= 21.3\ m^3\ ha^{-1}\ yr^{-1} \end{aligned}$$

This model is obviously very sensitive to rainfall and the length of the dry season (as might be expected in an area where local rain shadows provide a major source of variation in tree growth and site potential). N.B. This example is for illustration only and is not based on field data.

Perhaps the most helpful progress towards the selection of species through matching characteristics of site is the interactive computer program described by Webb *et al.* (1980) which provides details of 125 species.

In certain circumstances foliar and soil analysis can be employed to diagnose the nutrient status of a crop and to predict its response to the application of fertilizers. For success a great deal of knowledge about nutrient levels normal in the species and site must be known. Nutrient levels in plants and their response to fertilizers depend upon species, season, position of the foliage samples in the crown, and the nature of the soil. For example, pines and eucalypts growing on the same site may have vastly different concentrations of nitrogen and magnesium in their leaves. Also the response to fertilizer depends upon its availability in the soil; it may be leached away, mopped up by other vegetation or locked up in the chemical system of the soil and soil organisms.

6.2.3 Site Assessment in Plantations for Forecasting Growth and Yield

Most of the current projects on site assessment are in established plantations aiming at accurate growth prediction. Following the practice in many countries, attention has been directed towards establishing the relationship of total volume production and dominant height.

As the tallest trees are difficult to identify, dominant height has often been defined as the average height of the 100 trees per hectare with the largest diameter at breast height.

6.2.3.1 Site Classification by the Dominant Height on Age Relationship

In Europe the first attempts to establish a method of classifying plantations through their growth potential was based on subjectively defined relationships of top height on age. Data on height development were obtained from both temporary and permanent sample plots and stem analysis, most temperate species exhibiting relatively easily distinguished annual rings. Large numbers of growth patterns from dominant trees of a single species from as wide a range of sites as possible were collected and are illustrated in Figure 56. Then a suitable reference age was chosen – often 50 years – and the total range of dominant height at that age was divided into a number of classes each having an equal but limited range of dominant height – usually 2 or 3 m. These classes were termed *quality classes*. Subsequently an average top height on age relationship was illustrated for each quality class. For example, in Figure 56, plantations on quality class I sites are expected to have a top height of 30 m at age 50 whereas those on quality class IV sites will have a top height of only 15 m at the same age. Thus the quality class of a site was established for older crops and predicted for crops younger than the reference age. Obviously the site classification was for the one species only, and the productivity of a quality class I site for species *A* was not the same and had no relationship with that of a quality class I site for species *B*.

Site quality was frequently defined as an index, for example with reference to the dominant height at a particular age. Alder (1977) refers to a system of site classification of *Pinus patula* using a reference age of 20 years. Then a site on which a plantation reaches 25 m dominant height at that age is said to have a *site index* of 25. Site index is, therefore, a continuous variable that is useful in growth modelling. Most site indices refer to dominant height at a particular reference age, but this is not essential.

6.2.3.2 Methods of Defining Dominant Height on Age Relationships

Attempts have been made to reduce the subjectivity involved in the definitions of the top height on age relationship. More detailed and objective methods must consider:

- whether all sites have similar patterns of height growth

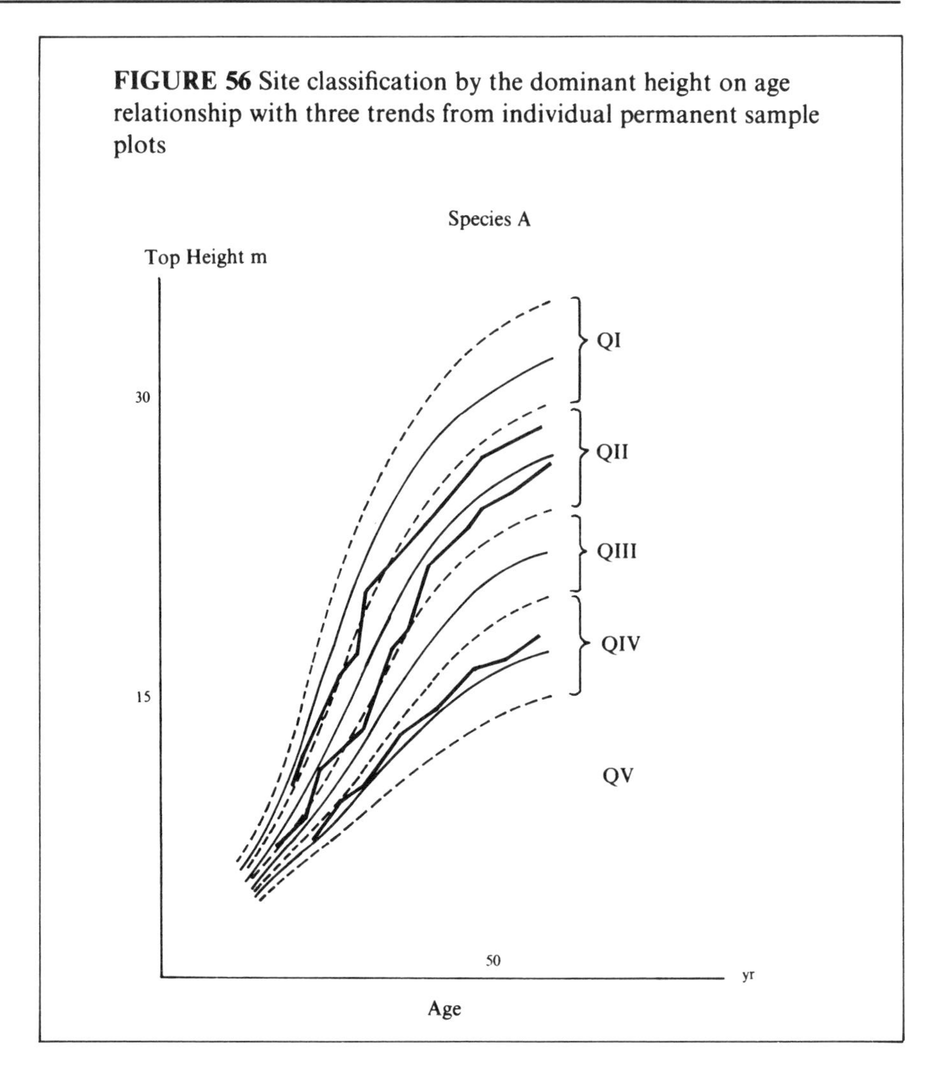

FIGURE 56 Site classification by the dominant height on age relationship with three trends from individual permanent sample plots

- whether height growth patterns have altered over time along with the practices of plantation establishment. Are the growth patterns the same now as they were 10 or 20 years ago? Can growth data from recently established plantations be combined with that from the early years of plantations established long ago?
- if it is accepted that all sites have similar patterns of height growth, then how should the sites be represented in the data so that the family of growth curves is undistorted and representative?
- how should the range of growth patterns be subdivided into classes each of which can be represented by an average relationship?
- how should the relationships be established from the data collected?

Alder (1980) gives a useful classification of methods.

6.2.3.2.1 The representation of growth patterns on different sites

The similarity of growth patterns of even-aged plantations of a single species on different sites is commonly assumed. This assumption is usually on the basis of a subjective, visual appraisal of illustrations made from data collected in permanent or temporary sample plots representing a wide range of sites. Tweite (1969, 1977) in Norway, Greaves (1978) for *Gmelina arborea* in Nigeria, Mathu & Philip (1979) for *Cupressus lusitanica* in Kenya and Alder (1977) for exotic softwoods in East Africa used this method. In contrast Waichuri (Waichuri & Wanene 1975) used different mathematical models for different sites for *Cupressus lusitanica* in Kenya, and recent Russian literature also supports this approach (see Figure 57b; also Biging, 1985; Hamlin & Leary, 1987; Newnham, 1988).

If growth results from the interaction of a large number of site factors then continuous variation about a central model would be expected. However, in certain circumstances, for example on soils with a marked limit to rooting depth, distinctive growth patterns have been established (Davidov, 1978; Griffin & Johnson, 1980).

Frequently data are collected together from records spanning a considerable period of time; consequently variation in growth at a given age includes variation due to climate, differing silvicultural regimes and techniques, and seed source as well as site. Also frequently, as plantation establishment progresses, the representation of different site types within the population changes. For example, often the best sites are planted first and afforestation spreads on to what originally were considered less suitable and even marginal sites. The early records from older stands, therefore, may refer to a smaller range of site type and have a different average growth trend compared with records obtained from recently established stands.

The aim of sampling must be to establish the central trend of the growth pattern as accurately as possible. If it is to be undistorted, the records should be symmetrically distributed around the population mean growth pattern. Often such data simply do not exist; then distortion almost inevitably results. Ideally the site of the average and extreme rates of growth should be identified by a set of extrinsic site characters and not by crop growth itself – but again this is often impracticable. Nevertheless great care must be taken to select objectively the sites from which the data for growth analysis are derived, with the aim of faithfully representing the full range of growth patterns and rates. This representation should be maintained at all ages. If tending techniques of other practices have been introduced and resulted in different patterns of growth, then separate relationships should be derived for crops established at different times. There is little point in representing sites in proportion to their area; in fact the extremes in terms of site quality are likely to be sampled far more intensively than the average site.

If, as frequently happens, the range of sites on which a species is established is extended either by the introduction of new provenances or more intensive silvicultural techniques, then new permanent sample plots must be established to:

- check that such sites or plantations have similar growth patterns to those established earlier
- provide data either for the revision of the central trend or for the establishment of a separate growth model

6.2.3.2.2 Constructing dominant height on age curves

The problem is how to define the non-linear top height on age relationship – remembering that successive measurements made in the same research plot are not independent in the statistical sense. The main problem, assuming that the data from different sites and ages are poolable, is how to define the central trend, but the derivation from this of a family of curves also must be considered.

Tweite (1969) pooled data and calculated the mean height at the beginning and end of a series of equal intervals of age; he used identical plots for the two measurements for one interval, but the data for successive intervals were not from an identical set of sample plots. Generally the number of plots available decreased as the age increased. The mean trend was then defined graphically by summing and smoothing the segments. The method has the advantage of using all the data but has an element of subjectivity and does not lend itself to precise mathematical statement for use in growth models. Obviously the designer of the final graphical model has to interpolate a mean trend through a sample containing an element of random variation.

Schumacher (Schumacher, 1939; Bruce & Schumacher, 1942) long ago recommended extracting at random one pair of data (top height and age) from each research plot and using this sub-set of data to determine objectively the average trend – using a curvilinear mathematical model:

$$y_t = a\mathrm{e}^{(b/k^t)}$$

Where:

y_t = size at time t

e = base of Napierian logarithms

a, b, k = constants

The disadvantage is that only part of the data is used and that on growth rates is completely discarded. Nevertheless it does eliminate the interdependence of the data.

There are two parts to the problem of fitting mathematical models:

- the choice of the model
- the estimation of the population average relationship

There are also two approaches to the choice of the model:

- either a model may be chosen by the researcher from a range suggested by experience and preliminary analyses of the data. In this choice an objective criterion such as the minimum sum of squared deviations between the predicted and measured values should be used, i.e. an empirical approach
- or a suitable growth model may be chosen on the basis of a hypothesis concerning the growth function that is independent of the set of data (*a priori*), i.e. an analytical approach

The former is likely to provide the better fit for the particular data set, but extrapolation would be hazardous. The latter is more suited to preliminary studies with a small amount of data and where extrapolation is needed because a more intensive study based on more complete data is infeasible.

Common growth models applied to height curves are the linear, the logistic and Gompertz functions.

Linear functions, though rigid, have been used frequently and may fit a set of data with a limited range in the dependent variable; commonly used functions are:

simple linear models $Y_t = a + b(t)$

linear transformations $Y_t = at^b \rightarrow \log Y_t = (\log a) + b(\log t)$

exponential models $Y_t = ae^{bt} \rightarrow \log Y_t = (\log a) + t(b \log e)$

Where:
Y_t = prediction of dependent variable at time t, e.g. height at age 50 years
t = predictor variable of time
e = base of natural logarithms
a, b = constants to be estimated

However, growth studies in many branches of science have demonstrated that more complex functions are justified if the range of age encompasses juvenile, adolescent, mature and senescent stages. Then a function with a sigmoid form, ideally with its origin at (0, 0), a point of infection occurring early in the adolescent stage and either approaching a maximum value (asymptote), or peaking and falling in the senescent stage, is justified.

Examples of such functions are:

- Polynomial models as employed by the British Forestry Commission (Christie, 1970)

$$Y_t = a + bH + cH^2 \ldots$$

Where:
Y_t = stand parameter, e.g. total volume production at time t
H = stand dominant height at time t

- The logistic model

$$i_t = \frac{abce^{-ct}}{(1 + be^{-ct})^2}$$

or

$$Y_t = \frac{a}{(1 + be^{-ct})}$$

Where:
i_t = growth rate
e = base of Napierian logarithms
a, b, c = constants
other symbols as above

- The Chapman–Richards growth model

$$Y_t = a[1 - e^{-b(t - t_0)}]^{-(1 - c)}$$

Where:
Y_t = value of parameter Y at time t
a = asymptote
b, c = constants
t_0 = initial age such that when $t \to t_0$, $Y_t \to 0$
and $i_{Y_t} \to 0$

- The Gompertz function in the form (Nokoe, 1978)

$$Y_t = ae^{-e^{-b(t - c)}}$$

Where a, b, c = constants

Other recent references may be found in Monserud (1984), Gregoire (1987) and Newnham (1988).

Once a model has been defined then the method of defining the trends representing the subdivisions of the population, that is the quality classes, has to be chosen. Alder (1980) describes in detail the techniques of site index curve construction using both graphical and mathematical models, and illustrates them with an example using Schumacher's model.

The main methods using inter-dependent models for the different site classes are:

- simple models using equal intercepts of dominant height at the reference age
- nested regression models with a common intercept
- multiple regression models in which an *a priori* estimate of site index is an independent variable

Simple models may be based either on an average trend, or by fitting nested regression lines with a common slope to individual permanent plot data sets. When an average trend is assumed, the definition of site quality classes may use arbitrarily defined boundaries at convenient class intervals. Alternatively, if the variance of dominant height is a function of age and the deviations around the mean are normally distributed at each age, the class boundaries may be defined as:

$+2s$	to	$+0.67s$	where s = standard deviation
$+0.67s$	to	$-0.67s$	
$-0.67s$	to	$-2s$	

with undefined classes greater than $+2s$ and less than $-2s$.

Nested regression models assume either a common slope or a common intercept. Using either method a family of site quality class curves may be constructed through pre-selected values at a given reference age.

Multiple regression models are of two types – constrained and unconstrained.

Constrained models, e.g.

$$(H - S) = b_i(A - A_i) + b_2(A - A_i)^2$$

Where:
H = dominant height
S = site index
A = age corresponding to H, years
A_i = reference age for site index

In such a model dominant height is forced to equal site index S when age equals the reference age, A_i. There is no intercept.

Unconstrained models with an intercept term. The curves must be conditioned after fitting to ensure that the dominant height predicted corresponds to the site index at the reference age. An example of such a model is

$$H = b_0 + b_1A + b_2S + b_3AS + b_4A^2$$

symbols as above (Alder, 1980).

Kirkpatrick & Savill (1981) illustrate the use of the Gompertz function to define a set of height curves to fit a range of site indices defined as top height at age 30. They conditioned their equations to provide the site indices that they wished to represent. Tweite (1977) followed a somewhat similar procedure. He first generated a constrained model for a central growth trend obtained as described above. This had the form

$$H = \left[\frac{T + a}{b + c(T - a)}\right]^d$$

Where:
H = dominant height at age T
a, b, c, d = constants to be determined and constrained so that when $T = 40$, $H = 17$ m

Then the difference in dominant height ($H_{diff.}$) at a given age (T) for different site indices was derived from a polynomial function of the variable ($T - 40$), where age 40 years was the reference age for site index definition.

6.2.3.2.3 Site classification by maximum mean annual increment classes

The British Forestry Commission maintain an extensive series of permanent research plots representing the range of species and sites afforested with even-aged plantations of a single species. These plots provide data on age, dominant height, total volume production and maximum mean annual increment. The data sets derived from these plots have been used to

- define a total volume production on dominant height relationship; this has been found to have limited variation so that a single model represents all sites
- define a model to predict maximum mean annual increment as a function of age and dominant height (see Section 7.3.1.2)

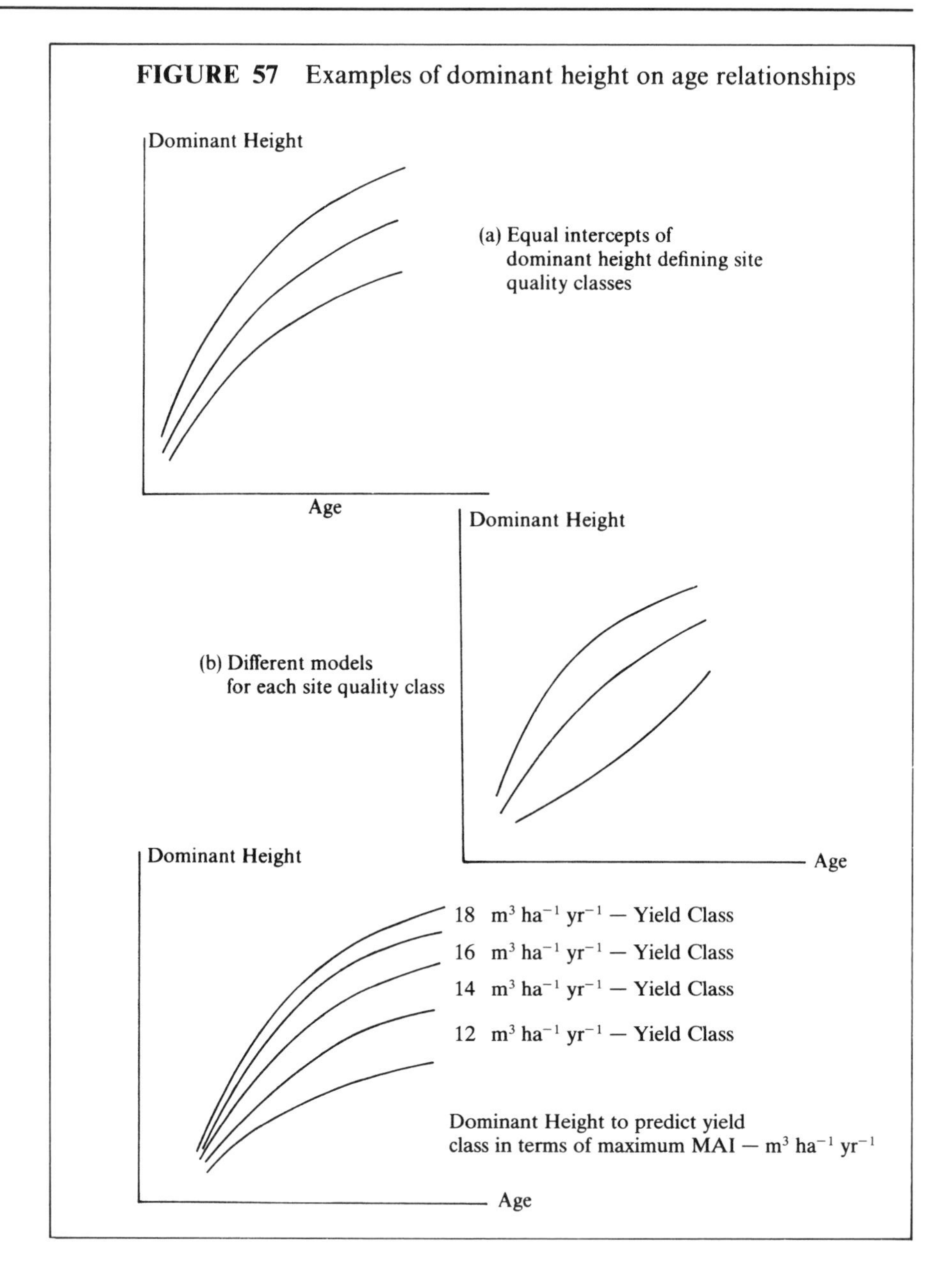

- derive a graphical model and tabulation of the development of dominant height with age for yield classes with intervals of 2 m^3 ha^{-1} yr^{-1} in maximum mean annual increment

The graphical model is used for site classification in existing plantations. The yield classes have unequal intercepts of height at a given age – the higher the yield class the narrower the intercept (see Figure 57) – that is, the more productive the site, the smaller is the increase in dominant height associated with a 2 m^3 ha^{-1} yr^{-1} increase in the maximum mean annual increment. The curves have a degree of subjectivity involved in their construction but for the common species this is believed to be small as the data set was large and the variance between plots relatively small. Later the relationships were represented by polynomials fitted to data taken off the smoothed graphical curves. Details of the Forestry Commission's methods and uses of site classification may be found in Edwards & Christie (1981), Rollison (1986), Worrell (1987) and Worrell & Malcolm (1990).

7 Forest Growth Models

7.1 Uses of Growth Models

The top height on age relationship discussed in the previous chapter is an example of a growth model; a forest growth model describes or portrays the development of tree crops as they increase in age, or as time changes. The term *forest growth model* may be misleading as it is applied here because we attempt to model the growth of the trees only and not the whole ecosystem. Hence the shorter term, growth model, will be used with the understanding that the reference is to tree crops, and growth is in terms of area – usually per hectare – even though it may be synthesized from that of individual trees.

The design of a growth model for use in a particular situation depends on the resources available, the uses to which it will be put and the structure of the tree crops – whether even- or uneven-aged – of a single species or mixed species. The four commonest uses are:

- to predict the growth of the forest so that the manager may match his harvesting and selling plans against the prediction of growth and conclude whether he is cutting more or less than, or an amount equal to, growth
- to predict the growth on a particular site to enable the land manager to make rational decisions. Often the growth model is required to provide information for conversion into economic measures to facilitate comparisons of a number of feasible investment options. Usually a single rigid crop management regime is used in these comparisons
- to predict the growth of crops under different management regimes and silvicultural practices in order to make comparisons and a choice; for example to choose the best original spacing, rotation timing and intensity of thinnings, etc.
- to predict work programmes when budgeting costs and revenues

Not all models are equally suited to these different uses. Titus & Morton (1985) reviewed models and the uses of models but only up to 1983. Avery & Burkhart (1983) is a general text with an extensive account of growth modelling for advanced students.

7.2 A Classification of Growth Models of Forest Crops

The simplest models are those for crops with the least variation in growth, that is even-aged stands of a single species; the most complex are those for very variable crops such as tropical evergreen forest, uneven-aged and with very many species. Figure 58 illustrates one form of classification of growth models based on the degree of variation in growth within the stand. See also Munro (1973).

A *stand growth model* describes the stand and predicts growth through general parameters, such as total basal area per hectare, mean values, such as mean volume per tree, and definitions of frequency distributions, such as their form and variance, etc. In contrast, a *single tree growth model* predicts the growth of individual trees and synthesizes stand growth from the sum of a representative sample of individuals.

A *distance independent model* reflects stocking and competition through average and summed terms such as numbers of stems per hectare, basal area per hectare, angle counts, etc. A *distance dependent model,* however, uses the distances from the subject tree to its competitors as one of the independent variables to predict growth.

A *static model* predicts stand or tree volume at a stated time (or age) and infers growth by subtracting the previous from the current standing volume. A

FIGURE 58 A classification of growth models

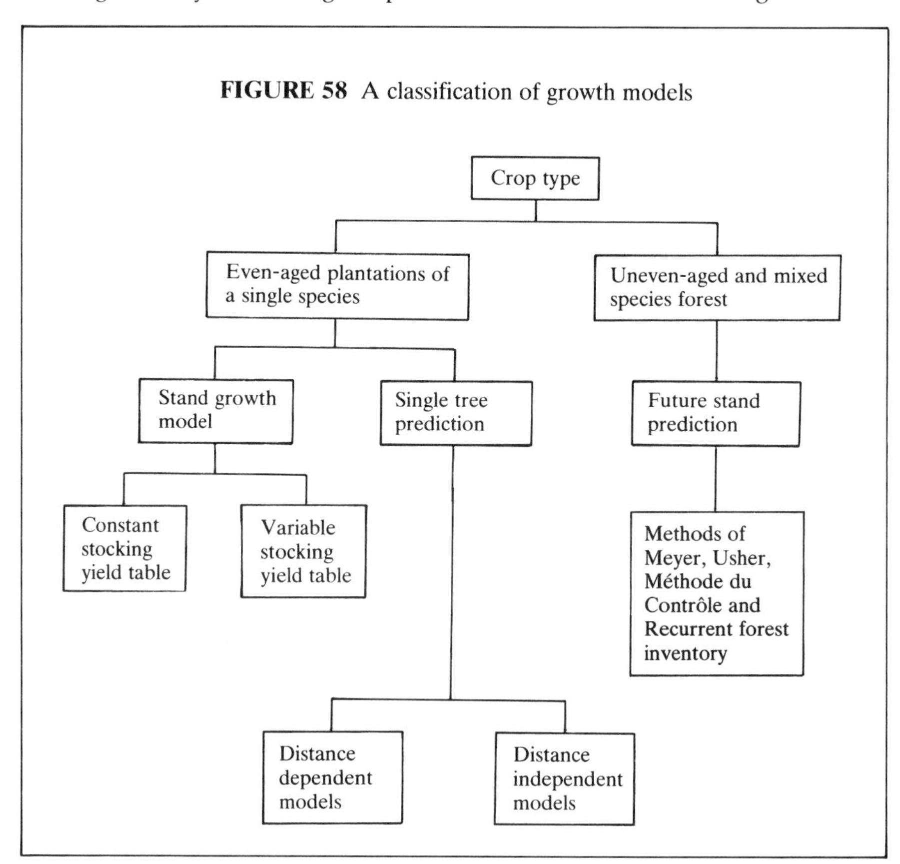

dynamic model predicts volume increment directly and deduces cumulative volumes by summing growth.

A *deterministic model* predicts the expected values under a given set of conditions, but a *stochastic model* incorporates uncertainty in the outcome by generating a random variable or variables from a prescribed probability function, and adjusts the prediction by including the effect of this stochastic element. This confers on the prediction a degree of variation to match reality. For example, a very sophisticated growth model might incorporate a variable representing the occurrence of abnormally dry periods. Then the prediction of growth and survival would be adjusted by using a value for the degree of drought in a particular period drawn from a probability distribution.

Adlard (Adlard *et al.*, 1988) has presented a wider classification including physiological and ecological approaches – see Botkin *et al.* (1972) and Brandiani *et al.* (1988).

7.3 Growth Models for Even-aged Crops of One Species Based on Average Stand Parameters

Even-aged crops of a single species have relatively little variation in growth rates and this can be further reduced by division into site, quality or yield class as described in Chapter 6. Then growth at a defined age can be described meaningfully by average stand parameters such as:

- current annual volume increment, $m^3\ ha^{-1}\ yr^{-1}$
- current annual basal area increment, $m^2\ ha^{-1}\ yr^{-1}$
- current annual increment in top height, $m\ yr^{-1}$

Often such growth models are tabulated in the form shown in Table 3. These are known as yield tables.

Many mathematical models have been used to summarize and define growth trends. They have the advantage over subjectively drawn graphs in that they can be used directly in computer programs to predict growth, whereas the handling of tabulated information derived from such graphs, though feasible, is both cumbersome and more expensive. The most frequently used models are those already described in Section 6.2.3.2.2 for representing height growth, i.e. linear transformations, polynomial, logistic, Weibull and the Gompertz function and the Chapman–Richards and Schumacher's growth models.

7.3.1 Constructing a Simple Growth Model or Yield Table

A simple growth model may be constructed from data collected in permanent research plots. Such simple models are usually independent of stocking within the range resulting from the application of a single management regime. The table may refer to a single site but, more usually, a collection of similar tables covers a range of sites on which the species is normally established.

Table 3 An example of a simple yield table for *Eucalyptus camaldulensis* Q.C.I.

	Main crop after thinning						Thinnings					Total production		Growth rates			
														CAI		MAI	
Age	N	h_d	G	$d_{\bar{g}}$	$\bar{v}$	V	N'	G'	$d'_{\bar{g}}$	v'	V'	G	V	G	V	G	V
(1)	(2)	(3)	(4)	(5)	(6)	(7)	(8)	(9)	(10)	(11)	(12)	(13)	(14)	(15)	(16)	(17)	(18)
1	1500	2.0	6.7			6						6.7	6	8.4	15	6.7	6.0
2	1500	5.0	16.7	11	0.020	30						16.7	30	10.8	35	8.4	15.0
3	1500	8.0	26.0	15	0.047	71						26.0	71	8.2	45	8.7	23.7
4	1500	10.0	33.0	17	0.079	119						33.0	119	6.6	50	8.2	29.8
5	1000	12.0	31.5	20	0.137	137	500	7.5	14	0.060	30	39.0	167	5.6	43	7.8	33.4
6	1000	13.5	36.8	22	0.176	176						44.3	206	5.0	37	7.3	34.3
7	1000	15.0	41.6	23	0.221	211						49.1	241	4.5	33	7.0	34.4
8	600	16.0	32.2	26	0.277	166	400	14.0	21	0.190	76	53.7	272	4.4	30	6.7	34.0
9	600	16.4	36.6	28	0.323	194						58.1	300	4.3	27	6.5	33.3
10	600	16.6	40.8	30	0.365	218						62.3	324	4.2	24	6.2	32.4

N.B. For example only – not based on published data.
Thinnings done at beginning of 5th and 8th years.

At each re-measurement the following set of data is collected from each research plot:

t age
N number of stems per ha
G basal area at bh, $m^2\ ha^{-1}$
V volume at stated upper diameter limits or total volume, over or under bark, $m^3\ ha^{-1}$
$\bar{h}_d$ average height of the 100 trees per hectare with greatest dbh, m
$\bar{h}_L$ Lorey's average height, m
N' numbers of stems removed in thinning, per ha
G' basal area removed in thinning, $m^2\ ha^{-1}$
V' volume removed in thinning, $m^3\ ha^{-1}$

From these sets of data *four* basic relationships are established

1. $\bar{h}_d$ on age
2. ΔG on age, or total basal area production on age
3. ΔV on age, or total volume production on age
4. N on age

Where:
ΔG = increase in total basal area production or CAI – current annual increment in basal area
ΔV = increase in total volume production or CAI in volume

These relationships express the changes in the crop and its rate of growth as it grows older. As has been discussed, they may be expressed graphically or mathematically. These and additional details for the yield table can then be calculated and entered into the table as in Table 3 in which the derivation of the columns is as below:

Parameter	Derivation	Column in Table 3
Age	pre-set	1
h_d	Relationship 1	3
CAI basal area	Relationship 2	15
CAI volume	Relationship 3	16
Total production basal area	Integration of Relationship 2	13
Total production volume	Integration of Relationship 3	14
MAI basal area	col. 13/age	17
MAI volume	col. 14/age	18

The division of the total production between main crop and thinnings has to be imposed; to a degree this is subjective. The thinning regime, defined by age and the number of stems N, volume V, and basal area G, remaining, is chosen based on the practice in the research plots. Then the numbers N', volume V', and basal area G', of the thinnings are calculated. The main crop is then augmented by the current annual increments that are assumed to be independent of the thinning regime, until the next thinning.

Parameter	Derivation	Column in Table 3
N	Defined by thinning regime	2
G	Defined by thinning regime and augmented by CAI	4
V	Defined by thinning regime and augmented by CAI	7
$\bar{v}$	$\frac{V}{N}$	6
$\bar{g}$	$\frac{G}{N}$	–
$d_{\bar{g}}$	$\sqrt{\frac{4\bar{g}}{\pi}}$	5
N'	(Line 1, col. 1) Σ (col. 8)	8
G'	Col. 13 – Σ (col. 4)	9
V'	Col. 14 – Σ (col. 7)	12
$\bar{v}'$	$\frac{V'}{N'}$	11
$\bar{g}'$	$\frac{G'}{N'}$	–
$d_{\bar{g}'}$	$\sqrt{\frac{4\bar{g}'}{\pi}}$	10

Figure 59 illustrates the relationships tabulated in Table 3.

7.3.1.1 *Uses of Dominant Height as a Predictor Variable*

There are many causes of variation in the relationships of stand parameters with age; some of the variation is due to

- site
- provenance or genotype
- early tending regimes
  — such as draining, fertilizing, etc. affecting site potential
  — such as weeding, spacing etc. affecting tree growth

In Britain dominant height (defined as the average height of the 100 trees per hectare with the greatest dbh) has been used as an alternative predictor variable in place of age. Crop parameters such as total volume production have very much less residual variation when expressed as a function of top height than when expressed as a function of age. Consequently top height has been used as a predictor of other parameters reflecting the development of the crop (see Figure 60). The outline of the method is to form up a master yield table for all sites as before except that column number one refers to top height, the age

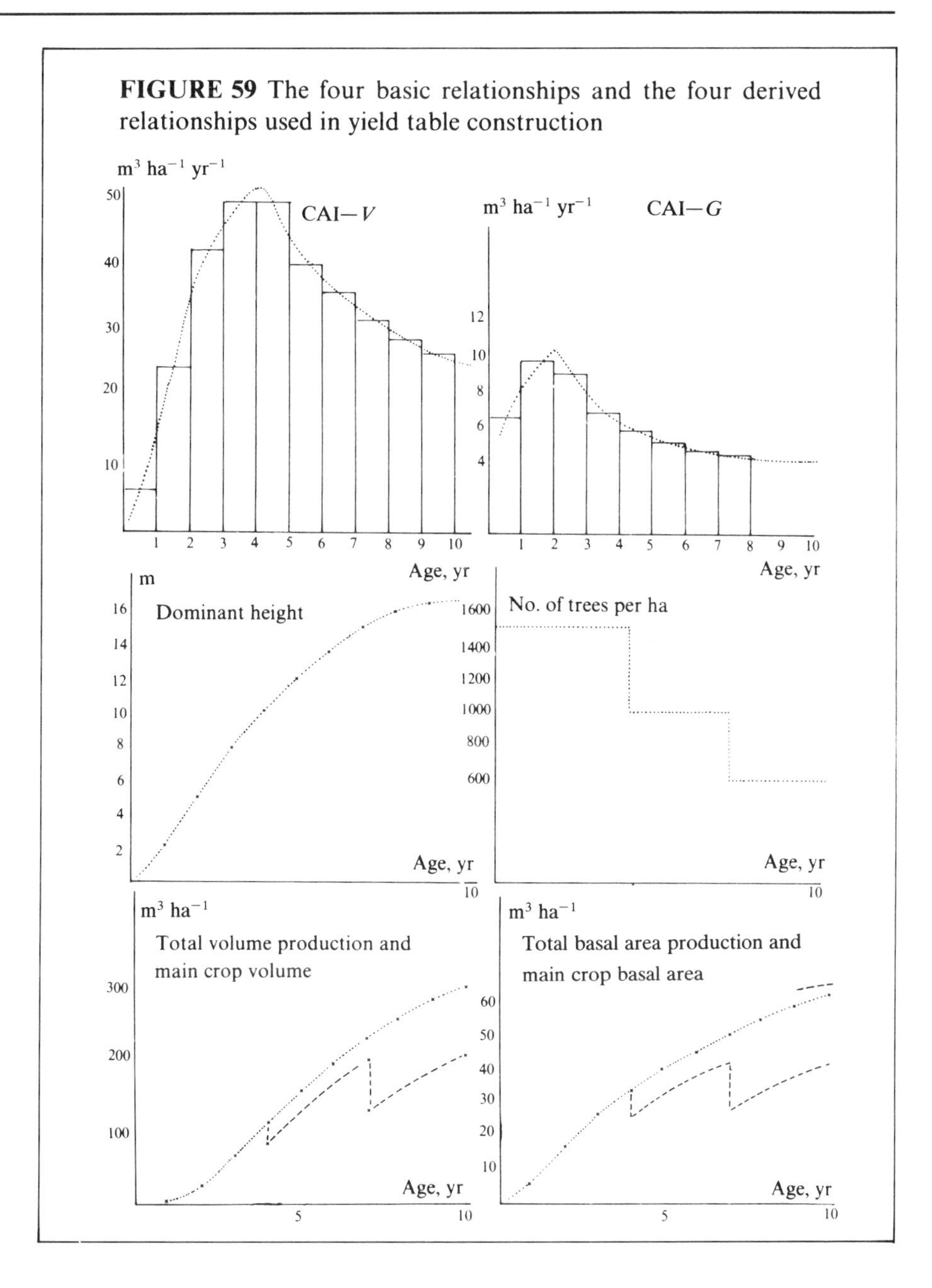

FIGURE 59 The four basic relationships and the four derived relationships used in yield table construction

column is omitted, and increments are expressed per metre of dominant height. Then, using this master table and the collection of dominant height/age curves, one for each site quality class or index, a yield table employing age as the predictor variable is constructed for each site quality class from the master table. An example showing a portion of the system involving only dominant height, total volume production and age is shown in Example 49. From such a table, the continuous relationship of total volume production on age may be derived by either graphical or mathematical procedures.

For example if

$$V = a_1 + b_1\bar{h}_d + c_1\bar{h}_d^2$$

and

$$\bar{h}_d = a_2 + b_2A + c_2\,(1/A)$$

Where a_2, b_2 and c_2 are site specific, then

$$V = a_1 + b_1\left(a_2 + b_2A + c_2\left(\frac{1}{A}\right)\right) + c_1\left(a_2 + b_2A + c_2\left(\frac{1}{A}\right)\right)^2$$

$$= a_1 + b_1a_2 + b_1b_2A + \frac{b_1c_2}{A} + c_1a_2^2 + 2c_1a_2b_2A$$

$$+ \frac{c_1c_2^2}{A^2} + c_1b_2^2A^2 + 2c_1c_2b_2 + 2\frac{c_1c_2a_2}{A}$$

$$= a_1 + b_1a_2 + c_1a_2^2 + \frac{1}{A^2}(c_1c_2^2) + \frac{1}{A}(b_1c_2 + 2c_1c_2a_2)$$

$$+ A(b_1b_2 + 2c_1a_2b_2) + A^2(c_1b_2^2)$$

$$= b_3A^{-2} + b_4A^{-1} + b_5A^0 + b_6A + b_7A^2$$

Where
$b_3 = (c_1c_2^2)$
$b_4 = (b_1c_2 + 2c_1c_2a_2)$
$b_5 = (a_1 + b_1a_2 + c_1a_2^2)$
$b_6 = (b_1b_2 + 2c_1a_2b_2)$
$b_7 = (c_1b_2^2)$
A = age all of which are site specific.

FIGURE 60 Variation in total volume production on age and on dominant height—all sites

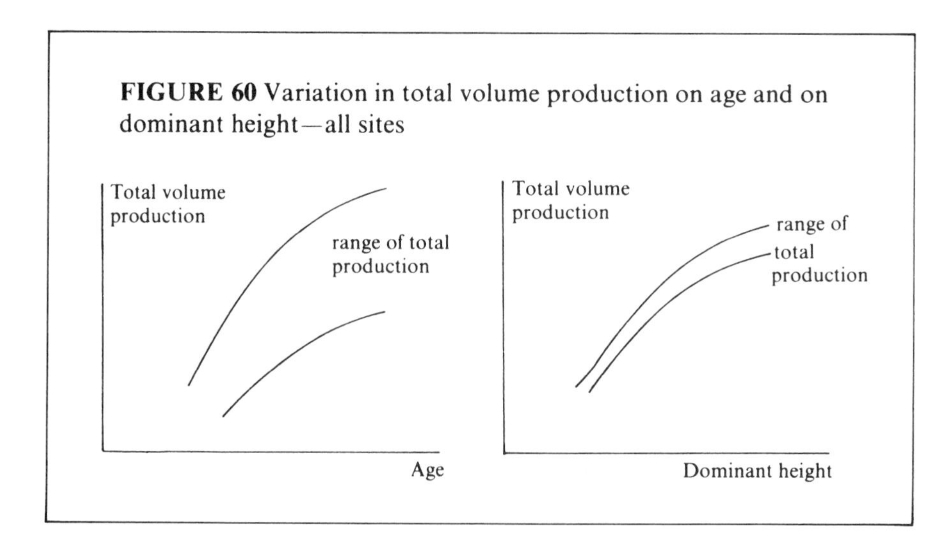

EXAMPLE 49 The use of dominant height as a predictor variable in a master yield table

Dominant height, m	Total volume production, $m^3\ ha^{-1}$	Age, years		
		Site class		
		I	II	III
4	–	1.9	2.2	2.7
6	48	2.4	3.0	3.6
8	70	3.0	3.9	4.8
10	120	4.0	5.1	6.2
12	166	5.0	6.8	8.5
14	220	6.4	8.3	–
16	275	8.0	–	–

FOR THE ADVANCED STUDENT

7.3.1.2 Use of Maximum Mean Annual Increment as an Index of Site Quality

In the previous chapter, Section 6.2.3.3, an illustration of site classification using maximum mean annual increment as a site index was given. The British Forestry Commission has used this method and constructed simple yield tables in which the different sites are defined by top height on age curves; each curve refers to a growth model differing by intervals of 2 $m^3\ ha^{-1}\ yr^{-1}$ in maximum mean annual increment (Edwards & Christie, 1981). The method of construction was analogous to that described above, though the curves in the master yield tables were defined by polynomials (Hamilton & Christie, 1973).

Kirkpatrick & Savill (1981) have outlined another procedure for constructing such models; they suggest

- fitting a Gompertz function of the form

$$\ln H_{ij} = a_i + br^{t_j}$$

Where:
$\ln H_{ij}$ = natural logarithm of top height at age t_j in the ith set (or quality class)
a_i = a constant representing the asymptote in the ith set
r, b = constants
t = age in years
i = 1 ... m sets of site quality classes
j = 1 ... n_i age in the ith quality class

- conditioning these relationships to provide a family of curves such that H_{iJ} can be used as a site index where J is the reference age. See Section 6.2.3.2.2.

Then

$$a_i = \ln S_i - br^{t_J}$$

or

$$\ln H_{ij} = \ln S_i + b(r^{tj} - r^{tJ})$$

Where:
$\ln S_i$ = natural logarithm of the site index

- if total volume production is a function of site index and age, then the corresponding total volume production and maximum mean annual increment may be calculated for each site index defined above. Alternatively the above equation may be conditioned to provide the site indices equivalent to intervals in a range of maximum mean annual increments. In turn, the top height on age relationship for each of the new site indices referring to a particular value of a maximum mean annual increment may be determined

The advantage of using site quality classes that refer to maximum mean annual increment is in the ease of comparing directly the productivity of one species or site with another, and in summarizing the growing stock in order to forecast outturn.

7.3.1.3 A Critical Assessment of Simple Yield Tables Based on Average Stand Parameters

Such simple yield tables have been a basic tool of forest plantation managers since the nineteenth century. Their limitations must be discussed in relation to their uses (see Section 7.1) the most important of which are:

- to predict growth and harvests
- to provide a conceptual model of crop development as a basis for comparative studies of crop management techniques; especially to assist the manager to decide on
  - — rotation length
  - — investment levels and timing for operations such as roading, fertilizing, pruning
  - — investment in new plantations
  - — type and intensity of thinning
  - — original spacing
  - — to provide the basis for work programming and budgeting

EXAMPLE 50 Programming and budgeting in a small *Eucalyptus* plantation, showing the application of yield table forecasts in forest management

Outline of a 5-year programme and budget

Programme

Cpt. no.	Area, ha	Age, yrs	Programme by years 1	2	3	4	5
1	10	4	–	–	–	1st thin	–
2	9	6	–	1st thin	–	–	2nd thin
3	11	7	1st thin	–	–	2nd thin	–

4	14	8	–	–	2nd thin	–	fell
5	6	10	2nd thin	–	fell	replant	tend
6	8	12	fell	replant	tend	–	–
7	21	1	tend	–	–	–	–
8	17	2	–	–	–	–	–

Forecast (from yield tables)

Year	Operation	Cpt.	Area	Cost per ha sh	Total cost sh	Volume per ha m^3	Total volume m^3	Value per m^3 sh	Total revenue sh
1	1st thin	3	11	10	110	56	616	30	18,480
	2nd thin	5	6	10	60	75	450	40	18,000
	fell	6	8	–	–	188	1504	50	75,400
	tend	7	21	200	4200	–	–	–	–
	Total yr 1				4370		2570		111,880
2	1st thin	2	9	10	90	56	504	30	15,120
	replant	6	8	800	6400		–	–	
	Total yr 2				6490		504		15,120
3	2nd thin	1	14	10	140	75	1050	40	42,000
	fell	5	6	–	–	188	1128	50	56,400
	tend	6	8	200	1600	–	–	–	–
	Total yr 3				1740		2178		98,400
4	1st thin	1	10	10	100	56	560	30	16,800
	2nd thin	3	11	10	110	75	825	40	33,000
	replant	5	6	800	4800	–	–	–	–
	Total yr 4				5010		1385		49,800
5	2nd thin	2	9	10	90	75	675	40	27,000
	fell	4	14	–	–	188	2632	50	131,600
	tend	5	6	200	1200	–	–	–	–
	Total yr 5				1290		3307		158,600

It can be seen that this programme which sticks rigidly to a thinning and felling schedule by age of plantation results in large fluctuations in work load and revenue from year to year; normally these would be unacceptable and the programme would be amended to provide a more even flow of produce.

Predictions of growth and the amounts to be harvested from existing plantations are needed in order to plan both the marketing and later processing of the wood, and the harvesting operations and transport of the wood to the processing plant (see Example 50). For these purposes only fairly crude figures are needed and simple yield tables provide adequate predictions.

The results of calculations comparing the results of different management decisions using simple yield tables usually provide adequate information to decide on:

- a suitable planning rotation
- the allocation of capital investment among comparable forest projects
- investment in new plantations

In contrast simple yield tables do not provide information that is adequate to decide upon such operations as the type and intensity of thinning or original spacing that depend upon the response of the crops to changes in the growing conditions. They are insensitive to change in stocking and provide little information about production unless accompanied by volume or diameter distributions.

Two examples from Tanzania are given in Appendix 8.

FOR THE ADVANCED STUDENT

7.3.2 Variable Stocking Growth Models

The assumptions of the simplest form of growth model are:

- variation in growth can be predicted by using different models for each site – or site quality class

and within one site quality class, then:

- within the limits of normal plantation management, crop growth (for example $m^3\ ha^{-1}\ yr^{-1}$) is relatively independent of stocking (whether measured by volume, basal area or number of stems per hectare)
- as a corollary of the above, crop growth is independent of either initial espacement or crop treatment

Obviously these assumptions are not axioms but useful simplifications to allow flexibility in the application of the model. The first assumption dates back to the work of Dr Wiedemann, a German professor, working in Scots pine (*Pinus sylvestris*) plantations managed on very long rotations. It was refuted by Professor Assmann, who presented evidence gathered from relatively old beech (*Fagus sylvaticus*) stands to support the hypothesis that there was an optimal stocking (in terms of basal area per hectare) for each site quality class that maximized basal area growth per hectare. Assmann suggested that departures from this optimal stocking resulted in marked growth reduction (10%). Since this time numerous writers have presented evidence in support of one or the other hypothesis, but much has been based on data from unreplicated research plots open to various interpretations; also often the definitions of growth and stocking, especially in the definition of the lower limit to measurements of diameter or volume, have varied. Such evidence that has been collected in a more rigorous manner has not clearly refuted either hypothesis. Consequently, for reasons of convenience and simplicity and because predictions based on such yield tables are known and accepted to be imprecise, Wiedemann's hypothesis is frequently applied. However, modern intensive methods of plantation management and recent silvicultural practices in plantation establishment, spacing and thinning result in growth responses and stockings outside the range experienced by these early workers, and their results may not be applicable in such circumstances.

Attempts have been made to introduce growth models in which stocking itself is a predictor variable. A few examples from among many that have been used are hypotheses of the form:

$$\Delta V = f(A, h_d, N, V, G, S)$$

Where:
ΔV = volume increment
f = function of
A = age
N = number of stems per ha
V = volume per ha
G = basal area per ha
S = site index

Kirkpatrick (1978) gives an example of a yield table made using such a function. His expressions were:

$$\ln V = b_0 + b_1 \ln G + b_2 S + b_3 A^{-1}$$

$$\ln I_G = c_0 + c_1 \ln G + c_2 \ln A + c_2 \ln A + c_3 \ln S + c_4 S + c_5(\ln A)(\ln S) + c_6(S)(\ln A) + c_7(S)(\ln G)$$

$$\ln N = d_0 + d_1 \ln G + d_2 S + d_3 A^{-1} + d_4(\ln G)(S) + d_5(S)(A^{-1}) + d_6(A^{-1})(\ln G) + d_7(S)(A^{-1})(\ln G)$$

Where:
ln = natural logarithm
I_G = increment in basal area per ha
A, G, V, N, S = as before

Grut (1971) used the following model for thinned stands of *Pinus radiata* in S. Africa:

$$\Delta G = f(A, h_d, N)$$

Where:
ΔG = basal area increment
other symbols as above

Vuokila (1966) advocated the use of relative growth

$$\frac{\Delta V}{V} = f(h, V, RT, t)$$

Where:
ΔV = volume increment
V = initial volume
RT = relative weight of thinning $= \frac{V'}{V}$

V' = thinned volume
t = thinning cycle
h = height

Dynamic models predicting the response of a crop to changing conditions controlled by the manager have either to use mathematical integrals of growth functions over time, or to be iterated – that is the prediction has to be repeated at intervals with the output from one prediction or iteration becoming part of the input for the following one, up to the end of the rotation. Consequently errors may accumulate unless the model is constrained, or calibrated so that it provides predictions consistent with independent data. Nevertheless, in so far as the original model is unbiased, variable stocking yield tables predict the response of crops to treatment.

However, the simple models illustrated above only predict average stand values and, therefore, translations from volume to value are crude unless volume or diameter frequency distributions are generated. Various more sophisticated models are presented by Fries (1973) in *Growth Models for Tree and Stand Simulation.*

7.4 Stand Prediction Growth Models for Even-aged Crops

Crop response to treatment has been modelled synthesizing stand growth by summing the growth of groups or individual trees. This is now a feasible tool of management and operations research because the necessary calculations can be performed quickly and cheaply on electronic computers. The method has the advantage that it generates diameter and volume distributions that provide more precise estimates of value and end use; also conceptually it should permit more sensitive analyses of treatment effects. Therefore, for example, they should predict the differential effects of early, late or no thinning on different original spacings and assist the manager to take decisions on thinning in a rational manner. The data base, nevertheless, defines the range of feasible prediction, and extrapolation beyond this range of experience may be misleading.

There are two main lines of approach:

- that which uses **stand** parameters of N, G, V and original or generated diameter or volume distributions – *distance independent models*
- that which incorporates measurements of **individual tree** basal area and/or volume and predictors of tree competition. Normally these indicators of competition attempt to integrate the sum of the competition from the neighbours into a measure such as basal area per hectare or angle summation. These are also *distance independent models.* Other models that include the distances of the subject tree to its immediate neighbours are termed *distance dependent*

7.4.1 Distance Independent Models

Stand models

Alder (1977) used a distance independent model to simulate the growth of conifers in East Africa. He generated a diameter distribution before first thinning using a Weibull function, or input an actual stand distribution. The diameter distribution was then divided into percentiles each of which was 'grown' using a function

$$\Delta d = f(RB, h_d, TD)$$

Where:
Δd = diameter increment
RB = relative basal area stocking
h_d = dominate height
TD = tree dominance

FIGURE 61 Abbreviated flow chart for Alder's model

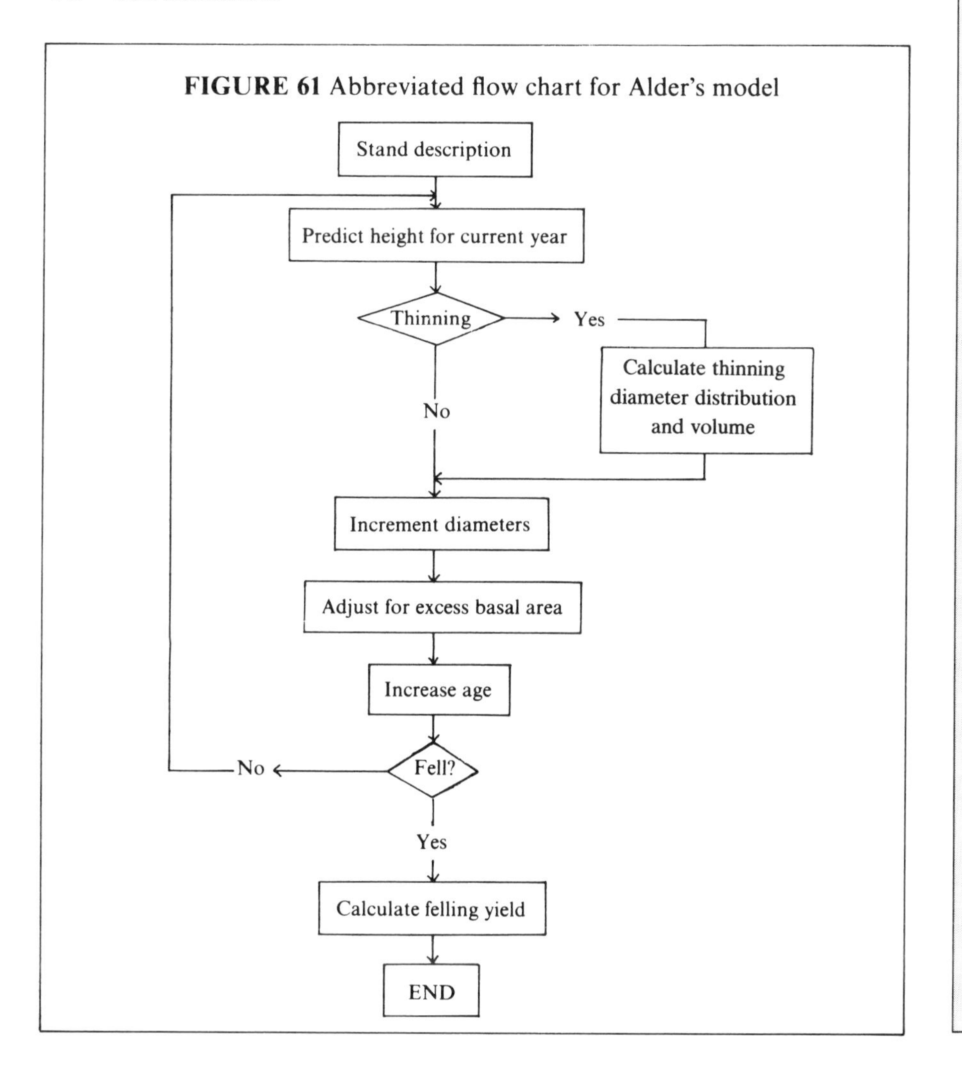

Relative basal area itself was stated in terms of a maximum basal area for the site predicted from a Chapman–Richards function of h_d. Tree dominance varied with the percentile in this distribution. Thinnings were generated by a probability function with a bias towards the smaller diameters. After thinning the diameter distribution was reassembled and the percentiles re-formed for the next iteration. Diameter increments were adjusted to prevent the accumulation of unrealistically high stockings. An abbreviated flow diagram is shown in Figure 61. Stand volumes were derived using a single tree one parameter volume table and assortments generated using a taper table. He validated the model against an independent set of data from research plots.

With Alder's model a manager can, within the limits defined by the original data, examine and predict results from different spacings and thinning regimes in terms of total volume production and volumes of trees and logs of different specifications.

This is a very considerable advance as an aid to management. Although the model performed well in the validation trials the generation of original diameter distributions for unthinned stands was a poor reflection of reality due to the very great variation in the stocking and diameter distributions occurring in the plantations. More research is needed to determine the causes of this variation. Also, and possibly as a result, the precision of the prediction of basal area increment per hectare was low.

More recently Burkhart & Knoebel (1983) have developed compatible dynamic and static growth models for *Liriodendron tulipifera*.

Individual tree models

Other models have depended upon taking representative individual trees as samples of the crop and synthesizing crop growth from that of the samples. Opie (1968) used a function of the form:

$$I_{g_i} = f(g_i, c_i)$$

Where:

I_{g_i} = basal area increment of tree i

g_i = basal area of the subject tree i

c_i = measure of competition around tree i

Opie compared the efficiency of circular plots, angle counts, and angle summation as independent measures of competition. Though improving on the precision of the estimate, like Alder's model they explained only about 65% of the total variation in basal area increment.

7.4.2 Single Tree Distance Dependent Models

Opie (1968) and Bella (1971) attempted to improve on the models described above using a distance dependent variable. This was the sum of the areas of competition overlap of the subject tree by all the immediate neighbours. The radius of the zone of influence of each tree was assumed to be circular and related to the tree's diameter at breast height. The relationship of the radius of the zone of influence and dbh was deduced from field trials. Bella refined Opie's model by weighting the overlap by functions of the relative sizes of the two trees. Although the models predicting diameter increment were highly significant, the gain in precision was small. Also the data needed for the calculation of the zone of influence are time consuming to collect and frequently unavailable from records gathered in the past. Hence the method has to have specially collected data.

Chamshama[1] supported Alder's (1977) observation and displayed the large differences in the total volume production to 7 cm top diameter in the even-aged plantations of *Pinus patula* at Sao Hill in southern Tanzania. Even in unthinned plantations matched and of the same age, top height and number of stems, the diameter distributions and total volume production varied, the high volume production being associated with heavily skewed diameter distributions. One suggestion is that the source of the variation may be in the spatial distribution of the trees; in young, very regularly spaced and even crops, volume production over 7 cm top diameter appears to be very much less than in less regular crops with a proportion of relatively very large trees. The effect in young crops appears to be sustained at least up to about age 16; this may be of considerable significance in pulpwood rotations.

This example illustrates the need for more sensitive growth models to account for the variation in growth due to the spatial relationships and relative sizes of the trees in the crop. Mitchell (1975) has described an analytically based, single tree, distance dependent growth model for Douglas fir. The disadvantages are that owing to the empirical nature of the models, each has only limited application to the population used in its derivation, and so far too few models have been developed to permit generalizations and determine the relative effects and importance in crop-development of:

- genotype
- microsite variation
- random chances at the time of establishment
- inter-tree spacing effects
- inter-tree size effects

Recent literature includes Smith (1983); Smith & Bell (1983); Chang (1984); Arney (1985); Borders & Bailet (1986); McTague & Bailey (1987). Most recent models stress the need for compatibility between the estimates of growth and cumulative yield, with more and more emphasis on the prediction of diameter and volume distributions.

7.5 Stand Prediction in Crops of Mixed Species and Age

The development of stand prediction models for even-aged plantations of a single species has progressed rapidly in the last 20 years, but the underlying

[1]Chamshama, S.A.O. (1979) *Pinus patula* plantations at Sao Hill Forest Project, Southern Tanzania. MSc Thesis, University of Dar es Salaam, Division of Forestry.

knowledge of inter-specific and inter-size growth relationships is as yet totally inadequate to permit the adoption of similar techniques in uneven-aged forest of mixed species arising from natural regeneration. The methods of measuring the growth of uneven-aged crops are evolving from those developed in France and Switzerland in the last century. Then the Swiss foresters realized that the estimation of growth as the difference between the growing stock at two different dates was liable to serious error as the change was small in relation to the growing stock and the likely errors in measurement.

$$I = V_2 - V_1 + F \text{ only if } V_2, V_1 \text{ and } F \text{ are without error}$$

Where:
V_2 = volume of growing stock on second occasion
V_1 = volume of growing stock on first occasion
F = removals by felling; mortality being assumed 0
I = periodic growth

A famous Swiss forester, M. Henri Biolley, introduced his Méthode du Contrôle (control or check method) which is well documented in Knuchel's *Planning and Control in the Managed Forest* (1953). Briefly, he established the increment of the forest over successive 100% inventories by three categories – the large, the medium and the small sized trees. These data were then used in planning the fellings over the next period. Example 51 shows the method of calculation.

Notes on the calculation in the example of the Méthode du Contrôle

Column 1 the classes of diameter at breast height were grouped into three major categories of large, medium and small trees. A separate increment % was calculated for each

2 data derived from a continental-type tariff table (see Section 2.7.4.7) or a one parameter volume table

3 from an inventory of the compartment made in 1970

4 col. 2 × col. 3

5 from an inventory of the compartment made in 1975

6 col. 2 × col. 5

7 from the compartment records of outturn for the period 1970–1975

8 col. 2 × col. 7

9 only to be completed in the line of the totals for the large and medium trees = total of col. 5 + total of col. 7 – total of col. 1, i.e. $V_2 + F - V_1$ or the number of trees of medium size in 1970 recruited to the large tree category, etc.

10 col. 2 × col. 9 for totals only

11 has to be completed for the total line of the large trees first and must equal the corresponding total in col. 3. Then starting with the largest diameter class of the large trees, col. 11 = col. 5 + col. 7 until the sum of these totals equals the figure previously entered in the total line. The balance of trees in the large tree diameter classes has been recruited from the medium category and this balance is entered in the blank line at the head of the medium tree

EXAMPLE 51 The calculation of increment in the Méthode du Contrôle

		1970		1975		Felled		Recruits		Revised 1975 total		Volume inc./ha in 5 yr	Annual volume increment per ha	
Dbh (1)	Vol./tree Silve (2)	No./ha (3)	Vol./ha Silve (4)	No./ha (5)	Vol./ha Silve (6)	No./ha (7)	Vol./ha Silve (8)	No./ha (9)	Vol./ha Silve (10)	No./ha (11)	Vol./ha Silve (12)	Silve (13)	Silve (14)	% (15)
75	5.56	–	–	1	5.56	–	–			1	5.56		*Large trees*	
70	4.93	1	4.93	–	–	–	–			–				
65	4.32	–	–	–	–	–	–			–				
60	3.72	2	7.44	5	18.60	–	–			5	18.60			
55	3.20	5	16.00	6	19.20	2	6.40			2	6.40			
Total large trees		8	28.37	12	43.36	2	6.40	6	19.20	8*	30.56	2.19	0.44	1.6
									Recruits	6	19.20		*Medium trees*	
50	2.70	11	29.70	7	18.90	1	2.70			8	21.60			
45	2.22	13	28.86	17	37.74	5	11.10			22	48.84			
40	1.66	27	44.82	27	44.82	1	1.66			28	46.48			
35	1.14	32	36.48	41	46.74	4	4.56			19	21.66			
Total medium trees		83	139.86	92	148.20	11	20.02	26	29.64	83*	157.78	17.92	3.58	2.6
									Recruits	26	29.64		*Small trees*	
30	0.12	66	47.52	52	37.44	5	3.60			57	41.04			
25	0.37	69	25.53	83	30.71	13	4.81			96	35.52			
20	0.16	117	18.72	94	15.04	19	3.04			73	11.68			
Total small trees		252	91.77	229	83.19	37	11.45	40		252	117.88	26.11	5.22	5.7
Total		343	260.00	333	273.75	50	37.87			343	306.22	46.22	9.24	3.6

Excluding recruits

*This total is made the same as that in the earlier inventory. If no fellings had taken place there would have been 12 + 2 = 14 trees in the large-size class in 1975. Therefore 14 – 8 = 6 trees had been recruited from the medium-size class. Similarly, if there had been no fellings or recruits there would have been 92 + 11 + 6 = 109. Medium-sized trees: 1975. Therefore 109 – 83 = 26 trees had been recruited from the small-size class.

category in col. 11 and labelled 'recruits' in the adjoining space in col. 10
12 col. 2 × col. 11
13 completed for the total line of the large, medium and small tree category only = col. 12 — col. 4
14 col. 13 divided by the period of years between the inventories
15 column 14 expressed as a % of col. 4

The important features of the system were:

- the elimination of sampling errors by 100% inventories
- the minimization of measuring errors by using a marked point of measurement on each tree
- the calculation of standing volume using a simple one dimension tariff table
- the calculation of removals using the same tariff
- the calculation of growth in the units of the tariff, even though sales were done on felled measure
- the control of removals by the tariff volume
- the calculation of growth separately by size classes

However the system was only feasible in small intensively managed forest.

Meyer (1953) reviewed the stand prediction method and its assumptions. Essentially the system was similar to the Méthode du Contrôle but predicted the future structure of a stand either from increment measured from successive inventories, or in sample plots or by increment borings. Normally no account of mortality and recruitment was taken. The example below illustrates the type of calculation. N.B. It assumes a rectangular distribution of the trees within each class. A worked example is shown in Example 52. (Information on growth measurement in mixed uneven-aged crops is given under recurrent forest inventory in Chapter 5.)

Notes on the calculation in the example of the stand prediction method

Column 1 classes of diameter at breast height
2 from a one parameter volume table
3 from an inventory
4 from increment borings or repeated measurements on sample trees
5 col. 4 divided by the diameter class interval
6, 7, 8 if the increment is 1/2 of the class interval, then on average 1/2 of the trees in the class will move out of the class into the next larger diameter class; similarly if the increment is only 1/10 of the class interval only 1/10 of the trees will grow up a diameter class. However, if the increment is more than the class interval, then all the trees will move up. If the increment is $1\frac{1}{4}$ times the class interval, all the trees will move up and 1/4 will move beyond into the next but one diameter class. Thus:

EXAMPLE 52 The calculation of increment using the stand prediction method

Dbh cm (1)	Vol. per tree m^3 (2)	Inventory 1 No. of stems ha^{-1} (3)	Diameter increment in 5 years cm (4)	Ratio i/c* (5)	Number of stems moving: Stationary (6)	Number of stems moving: 1 diam. class (7)	Number of stems moving: 2 diam. class (8)	Future stand No. of stems ha^{-1} (9)	$\bar{V}_1$ $m^3\ ha^{-1}$ (10)	Volume prediction $m^3\ ha^{-1}$ (11)	Species e.g. mangrove (12)
42	1.80	–	–	–			–	1.10	–	1.98	
40	1.58	2	1.1	0.55	0.90	1.10	–	2.70	3.16	4.27	
38	1.38	3	1.2	0.60	1.20	1.80	–	5.45	4.14	7.52	
36	1.19	5	1.7	0.85	0.75	4.25		10.55	5.95	12.55	
34	1.02	8	2.0	1.00	–	8.00		8.50	8.16	8.67	
32	0.88	9	2.4	1.20	–	7.20	1.80	11.70	7.92	10.30	
30	0.76	13	2.2	1.10	–	11.70	1.30	17.10	9.88	13.00	
28	0.65	18	1.9	0.95	0.90	17.10	–	13.70	11.70	8.90	
26	0.55	16	1.6	0.80	3.20	12.80	–	22.80	8.80	12.54	
24	0.46	28	1.4	0.70	8.40	19.60	–	28.90	12.88	13.29	
22	0.38	41	1.0	0.50	20.50	20.50		58.20	15.58	22.12	
20	0.31	58	1.3	0.65	20.30	37.70		70.90	17.98	21.98	
18	0.25	92	1.1	0.55	41.40	50.60		99.45	23.00	24.86	
16	0.18	129	0.9	0.45	70.95	58.05		70.95	23.22	12.77	
	Total	422						422.00	152.37	174.75	

*i = diameter increment; c = class interval = 2 cm
e.g. when i = 1.1, then i/c = 1.1/2.0 = 0.55
5 years' total increment 22.38; Current annual increment 4.48 per yr or 2.9%

	col. 6 = 0, if col. 5 is greater than 1, or else = (1 − col. 5)(col. 3)
	col. 7 = col. 3 − (col. 6 + col. 8)
	col. 8 = 0, if col. 5 is less than 1, or else = (col. 5 − 1)(col. 3)
Column 9	col. 6 + (the entry in col. 8 of 2 diameter classes lower) + (the entry in col. 7 of 1 diameter class lower), e.g.
	dbh 42 col. 9 = 0 + 1.10 = 1.10
	dbh 40 col. 9 = 0.90 + 1.80 = 2.70
10	col. 2 × col. 3
11	col. 2 × col. 9

N.B. the total of column 3 must equal that of column 8.

Even today in these forests with complex species and age structures, increment is being estimated using simple forms of stand prediction little changed from that of the classical Swiss and French methods developed in the last century and described above. However, their implementation is much simplified by the use of simple computer programs and spreadsheets (Kofod, 1982; Korsgaard, 1988; Alder, 1992). The data have to be derived from recurrent forest inventory with permanent sample plots. The current practices and state of the art were reviewed at the IUFRO meeting of Sections 4.01/4.02/1.07 held in Malaysia, June 20–24, 1988 – *Growth and Yield in Tropical Mixed/Moist Forests.* At this meeting P.G. Adlard (Adlard *et al.*, 1988) presented a paper summarizing the current thinking on modelling the tropical high forest. This stressed the need for empirical models using existing data for the planning and control of the forest; it also recognized the need for a better understanding of the growth processes of tropical high forest and a cadre of competent modellers to apply and develop the most recent techniques in modelling.

One feasible method available is a modification of Meyer's graphical method proposed by Usher (1966) and used by Bruner & Moser (1973). The following data are needed and can be obtained from recurrent measurement of a representative permanent sample – usually individually identified trees in permanently demarcated sample plots (see Alder & Synott's *Permanent Sample Plot Techniques for Mixed Tropical Forest,* 1992).

For each species:

- diameter and diameter increment for all size classes as well as mortality, and recruitment
- a one parameter volume table

From these data the probability of a tree of a given species and diameter, dying, or growing less than an amount needed to change diameter class, or sufficient to change 1 cm, 2 cm, 3 cm diameter classes within a 5-year period is calculated. Perhaps 100 permanently measured individually identified trees of each species or species group will be needed.

In addition an inventory to provide information on the current diameter frequency or stand structure of each species is needed. Then using matrix algebra and handling the mass of individual tree data on electronic computers, it is feasible to predict future stand structures and calculate the increment. A partial illustration of the calculations, much simplified, for one species using a limited range of diameter classes and a discontinuous probability matrix is given in Example 53. No account of recruitment is included. The system involves somewhat heroic assumptions – not without precedent in growth modelling – that past growing conditions will be repeated in the future and that the data on probabilities of growth or mortality, especially in the classes with low frequencies, are representative. Trials using data from permanent plots re-measured over long periods have proven somewhat disappointing over more than one or two growth periods. Also growth rates, and especially mortality rates, are liable to depend upon the frequency of gaps in the canopy that affect the crown position of neighbouring trees and the ground level conditions for regeneration and recruitment (Brandiani *et al.*, 1988; Wadsworth *et al.* 1988).

EXAMPLE 53 Calculation of increment for a mixed species uneven-aged tropical semi-evergreen forest

Species: *Maesopsis eminii*

(a) *Calculation of probabilities of changes in diameter class for the 26 cm class (25.5–26.5 cm)*

Data from permanent sample plots

dbh cm (1)	dbh after 5 yrs cm (2)	New diameter class (3)	Change in stand structure, no. of trees: Death (4)	No. change (5)	Diameter classes: +1 cm (6)	+2 cm (7)	+3 cm (8)
25.5	26.4	26		1			
25.6	26.6	27			1		
25.7	26.7	27			1		
25.7	26.6	27			1		
25.8	27.2	27			1		
25.9	dead	–	1				
26.0	26.4	26		1			
26.2	27.0	27			1		
26.2	27.8	28				1	
26.3	28.1	28				1	
26.3	28.8	29					1
26.4	28.3	28				1	
26.4	28.4	28				1	
		Total	1	2	5	4	1
$n = 13$							
		Probabilities	0.077	0.154	0.385	0.308	0.077

Notes on the calculation:

Cols 1 and 2, from individual tree measurements in permanent sample plots
Col. 3 from col. 2 in accordance with the class boundaries
Cols 4, 5, 6, 7 and 8, one entry per line according to the revised diameter class

$$\text{Probability of death} = \frac{\text{number dead}}{n} = \frac{1}{13} = 0.077, \text{etc.}$$

Example 53 continued

(b) *Calculation of increment 35–44 cm dbh class Species A*

			Probability					Number of stems/10 ha							
				Diameter class					Diameter class						
dbh cm (1)	No. of stems in 10 ha (2)	Vol/ stem m³ (3)	Death (4)	0 (5)	+ 1 cm (6)	+ 2 cm (7)	+ 3 cm (8)	Dead (9)	unchanged (10)	+ 1 cm (11)	+ 2 cm (12)	+ 3 cm (13)	Predicted no. of stems in 10 ha (14)	V_2 in 10 ha m³ (15)	V_1 in 10 ha m³ (16)
44	–	2.35		–	–	–	–	–	–				0.60	1.41	–
43	–	2.15		–	–	–	–	–	–				2.00	4.30	–
42	4	1.96	0.06	0.29	0.50	0.15	–	0.24	1.16	2.00	0.60	–	6.25	12.25	7.84
41	7	1.78	0.05	0.30	0.65	–	–	0.35	2.10	4.55	–	–	8.89	15.82	12.46
40	9	1.64	0.07	0.32	0.55	0.06	–	0.63	2.88	4.95	0.54	–	15.83	25.96	14.76
39	13	1.53	0.07	0.28	0.55	0.10	–	0.91	3.64	7.15	1.30	–	19.20	29.38	19.89
38	18	1.42	0.06	0.20	0.46	0.25	0.03	1.08	3.60	8.28	4.50	0.54	13.22	18.77	25.56
37	26	1.30	0.05	0.25	0.37	0.28	0.05	1.30	6.50	9.62	7.28	1.30	6.50	8.45	33.80
Total	77							4.51	19.88	36.55	14.22	1.84	72.49	116.34	114.31

Increment on 10 ha for 5 yrs = $V_2 - V_1$ = 2.03 m³

Current annual increment for species A 37-44 cm dbh = $\frac{V_2 - V_1}{n}$ = 0.41 m³ in 10 ha = 0.4%

Notes on the example of calculation of the increment for a mixed species, uneven-aged tropical semi-evergreen forest:

Col. 1, breast height diameter classes
Col. 2, from the current inventory
Col. 3, from a single parameter volume table
Cols 4 to 8, as in the 1st table in the example
Col. 9, col. 2 × col. 4
Col. 10, col. 2 × col. 5
Col. 11, col. 2 × col. 6
Col. 12, col. 2 × col. 7
Col. 13, col. 2 × col. 8
Col. 14, col. 10 + entry in col. 11 one line below + entry in col. 12 two lines below + entry in col. 13 three lines below, as indicated in the table for the 41 cm diameter class
Col. 15, col. 3 × col. 14
Col. 16, col 3 × col. 2

Ek & Monserud (1973) described a simulation model – FOREST – for mixed species, even- or uneven-aged stands. The inputs were a set of tree coordinates describing the spatial pattern of a stand and associated tree characteristics describing the individual trees. The trees are then grown for a number of projection periods using growth functions including individual tree competition indices. Mortality is modelled using a threshold value of growth below which a tree dies. Optional routines to model recruitment and different silvicultural treatments affecting the stand development may be included. Increment prediction in the complex botanical associations of mixed tropical evergreen forest may follow this development. Wan Razali (1988) has modelled mortality in the forests of peninsular Malaysia.

Synott (1979) has described two programs for constructing stand tables and calculating increment. The earlier model, called GROWTH, used simple regressions of increment on diameter and incorporated a random element to ensure that the predicted increments show a variability comparable with that of the observed values.

The second model, called GROPE, was far more sophisticated. GROPE uses a generalized multivariate model which projects rates of change of the commonly measured tree or crop variables. These variables include mortality, recruitment, increment, height, crown position, etc. This, also, is a stochastic model incorporating random elements.

Lieberman & Lieberman (1985), and Lieberman *et al.* (1985) derived a simple simulation technique to predict the growth of individual trees from data, derived from recurrent inventory, giving distributions of diameter increment for different diameter classes of each species. The simulation covered the period from the age of the smallest diameter class to that of the largest. Trees were grown using the increment for successive size classes drawn at random from the distributions provided. The output was a family or envelope of growth patterns from the most optimistic to the most pessimistic; from this envelope an expected average trend could be identified.

In the 1980s, Kofod and Korsgaard working in Sarawak also programmed and implemented interactive stand prediction (simulation) techniques that employ:

- the initial stand in terms of species, species groups, canopy levels, diameter classes and diminution quotients
- growth rates by species or species group, crown illumination categories and size class
- mortality rates
- stand improvement rates that provide enhanced growth as a result of factors – such as silvicultural treatments that improve crown illumination or position
- ingrowth into the first size class, i.e. established regeneration

Prediction is done by time periods – e.g. 5 years – with harvesting and silvicultural treatments as options between periods, these reflecting the management system and objectives. The input of variables controlling crop development allows great flexibility and versatility but at a cost of empiricism. Adlard (Adlard *et al.* 1988), however, also sees the need for process models, especially those that model the ecosystem such as the JABOWA model (Botkin *et al.*, 1972) and those that take special account of the mosaic structure and species richness of the tropical mixed forests (Brandiani *et al.*, 1988).

Further reading

ON STAND FORECASTING

Fries, J. (ed.) (1974) *Growth Models for Tree and Stand Simulation.* IUFRO, Stockholm.

Proceedings of the IUFRO meeting of Sections 4.01/4.02 & 1.07, June 20–24, 1988. Forest Research Institute of Malaysia, Kuala Lumpur, Malaysia.

Synott, T.J. (1979) *Tropical Rainforest Silviculture.* CFI Occasional Paper No. 10, Oxford.

APPENDICES

APPENDIX 1: The Graduation of a Biltmore Stick

ΔABC $\equiv$ ΔADE because all angles are the same

$$\therefore \frac{BC}{DE} = \frac{AB}{AD} \quad \text{and} \quad \frac{s/2}{d/2} = \frac{L}{x}$$

i.e.

$$\frac{\frac{s}{2}}{\frac{d}{2}} = \frac{s}{d} = \frac{L}{\sqrt{\left(L+\left(\frac{d}{2}\right)\right)^2 - \left(\frac{d}{2}\right)^2}} = \frac{L}{\sqrt{L^2 + Ld + \left(\frac{d}{2}\right)^2 - \left(\frac{d}{2}\right)^2}} = \frac{L}{\sqrt{(L^2 + Ld)}}$$

$$\text{or} \quad s = \frac{Ld}{\sqrt{L^2 + Ld}} = \frac{d\sqrt{L}}{\sqrt{(L+d)}}$$

FIGURE 62 The graduation of the Biltmore Stick

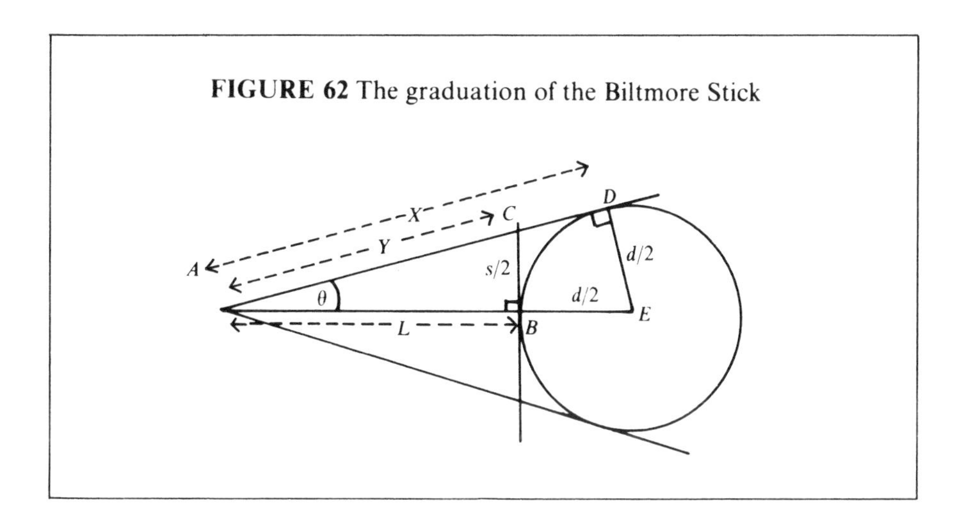

APPENDIX 2: The Systematic Errors in Applying Huber's and Smalian's Formulae to a Frustum of a Cone

N.B. $$d_m = \frac{(d_1 + d_2)}{2}; \quad d_m^2 = \frac{(d_1^2 + 2d_1d_2 + d_2^2)}{4}$$

True volume (v_t) by Newton's formula

$$v_t = \frac{L\pi(d_1^2 + 4d_m^2 + d_2^2)}{24}$$

by Huber:

Volume calculated (v_c) $$v_c = \frac{(L\pi d_m^2)}{4}$$

Error $$v_t - v_c = \frac{L\pi(d_1^2 + 4d_m^2 + d_2^2 - 6d_m^2)}{24}$$

$$= \frac{L\pi(d_1^2 - 2d_m^2 + d_2^2)}{24}$$

but

$$d_m = \frac{(d_1 + d_2)}{2} \quad \text{and} \quad d_m^2 = \frac{(d_1^2 + 2d_1d_2 + d_2^2)}{4}$$

Error $$v_t - v_c = \left[\frac{L\pi\left(d_1^2 - 2\left(\frac{d_1^2 + 2d_1d_2 + d_2^2}{4}\right) + d_2^2\right)}{24}\right]$$

$$= \frac{L\pi(d_1^2 - 2d_1d_2 + d_2^2)}{48} = \frac{L\pi(d_1 - d_2)^2}{48}$$

by Smalian:

$$v_c = \frac{L\pi(d_1^2 + d_2^2)}{8}$$

$$v_t - v_c = \left[\frac{L\pi(d_1^2 + 4d_m^2 + d_2^2 - 3d_1^2 - 3d_2^2)}{24}\right]$$

$$= \frac{-L\pi(2d_1^2 - 4d_m^2 + 2d_2^2)}{24}$$

but

$$d_m = \frac{(d_1 + d_2)}{2}$$

$$= \frac{-L\pi(2d_1^2 - d_1^2 - 2d_1d_2 - d_2^2 + 2d_2^2)}{24}$$

$$= \frac{-L\pi(d_1^2 - 2d_1d_2 + d_2^2)}{24}$$

$$= \frac{-L\pi(d_1 - d_2)^2}{24}$$

This proves that Huber's formula underestimates the volume of a frustum of a cone, whereas Smalian's formula overestimates its volume. The overestimation by Smalian's formula is twice the underestimation by Huber's. Also the errors are proportional to the length of the log and the square of the difference between the diameters of the two ends. Similar results may be proved for other solids of revolution other than the cylinder and the quadratic paraboloid.

Conclusion

Errors in the estimation of tree and log volumes are expected to be reduced by using Huber's formula and summing the volumes of sections which should be as short as is practicable.

APPENDIX 3: Volume by Height Accumulation

Consider the frustum of the cone DEFG in Figure 63(a) revolving around the X axis, i.e. the ith frustum from the apex A.

$$V = \frac{\pi}{4}\int_{x_i}^{x_{(i+1)}} Y^2\,dx$$

but within these limits

$$Y = T\left(i + \left(\frac{x - x_i}{L_i}\right)\right)$$

Let $(x - x_i) = x_z$

$$V = \frac{\pi}{4}T^2\int_{x_i}^{x_{(i+1)}} \left(i + \left(\frac{x_z}{L_i}\right)\right)^2 dx$$

$$V = \frac{\pi}{4}T^2\int_{x_i}^{x_{(i+1)}} \left(i^2 + \left(\frac{2ix_z}{L_i}\right) + \frac{x_z^2}{L_i^2}\right) dx$$

but the limits defining x_z are identical with the limits of integration.

$$V = \frac{\pi}{4}T^2\left[i^2x_z + \frac{ix_z^2}{L_i} + \frac{x_z^3}{3L_i^2}\right]_{x_i}^{x_{(i+1)}}$$

$$V = \frac{\pi}{4}T^2\left[i^2(x_{(i+1)} - x_i) + \frac{i(x_{(i+1)} - x_i)^2}{L_i} + \frac{(x_{(i+1)} - x_i)^3}{3L_i^2}\right]$$

but $\quad (x_{(i+1)} - x_i) = L_i$

$$V = \frac{\pi}{4}T^2(i^2L_i + iL_i + L_i/3)$$

Similarly the sum of frustra 1 . . . n (excluding the end cone)

$$V = \frac{\pi}{4} T^2 \sum^{n} (i^2 L_i + iL_i + L_i/3)$$

Now the expansion of $\sum^{n} (i^2 L_i + iL_i)$ can be transformed thus

$= (2L_1 + 6L_2 + 12L_3 + 20L_4 \ldots)$ a difference between coefficients increasing by + 2

$= 2(L_1 + 3L_2 + 6L_3 + 10L_4 \ldots)$

This series can conveniently be transposed from log lengths into heights from ground level giving the sum of the series because

$$\sum^{n} (i^2 L_i + iL_i) = 2 \sum^{n} a_i$$

Where:

$a_1 = L_n$
$a_2 = (a_1 + (L_n + L_{n-1}))$
$a_3 = (a_2 + (L_n + L_{n-1} + L_{n-2}))$
$a_n = (a_{n-1} + L_n + L_{n-1} + \ldots + L_1)$

e.g.

$$\sum^{3} a_i = (6L_3 + 3L_2 + L_1) = \frac{1}{2} \sum^{3} (i^2 L_i + iL_i)$$

But

$$V = \frac{\pi}{4} T_2 \sum^{n} ((i^2 L_1 + iL_i) + L_i/3)$$

$$V = \frac{\pi}{2} T^2 \left(\sum^{n} a_i + \sum^{n} L_i/6\right)$$

Similarly for paraboloids:

$$v = \frac{\pi}{2} T^2 \left(\sum^{n} a_i + \sum^{n} L_i/4\right)$$

and for neiloids:

$$V = \frac{\pi}{2} T^2 \left(\sum^{n} a_i + \sum^{n} L_i/8\right)$$

APPENDIX 4: The Derivation of Schneider's Formula for the Current Annual Increment Percent in Basal Area from an Increment Boring

d = current dbh ob, cm
Δd = average annual diameter increment in cm over outer 1 cm radius
m = number of rings in outer cm of radius
Δg = current annual increment in basal area

Averaging the basal area increment of the year past and the next year, gives

FIGURE 63 Volume by height accumulation

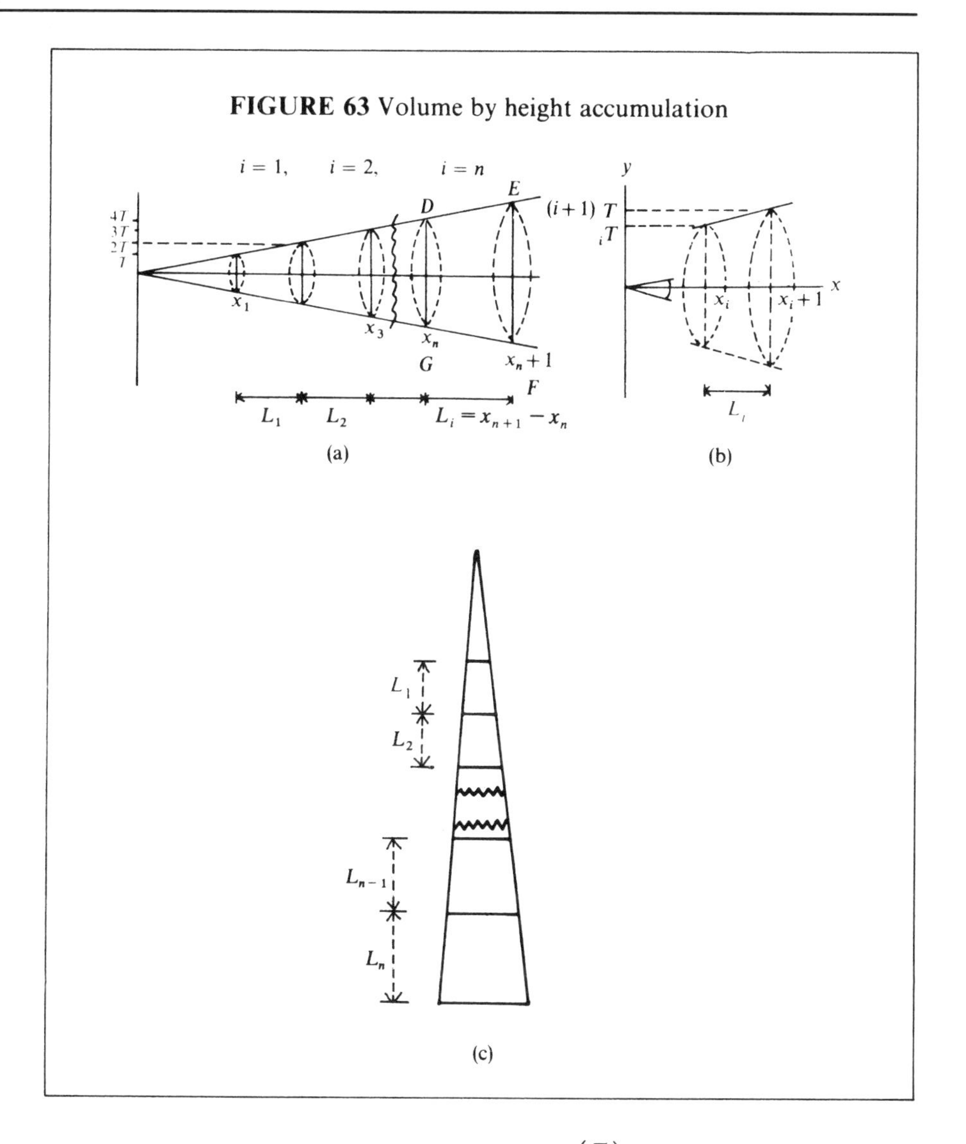

(a) (b) (c)

$$\Delta g = \tfrac{1}{2}((d + \Delta d)^2 - (d - \Delta d)^2)\left(\frac{\pi}{4}\right)$$

$$\Delta g\% = ((d + \Delta d)^2 - (d - \Delta d)^2)\,\frac{100}{2d^2}$$

$$= (d^2 + 2d\Delta d + \Delta d^2 - d^2 + 2d\Delta d - \Delta d^2)\,\frac{100}{2d^2}$$

$$= (4d\Delta d)\,\frac{100}{2d^2}$$

$$= 200\,\frac{\Delta d}{d} \quad \text{but} \quad \Delta d = \frac{2}{m}$$

$$\Delta g\% = \frac{400}{md}$$

FIGURE 64 Derivation of Schneider's formula

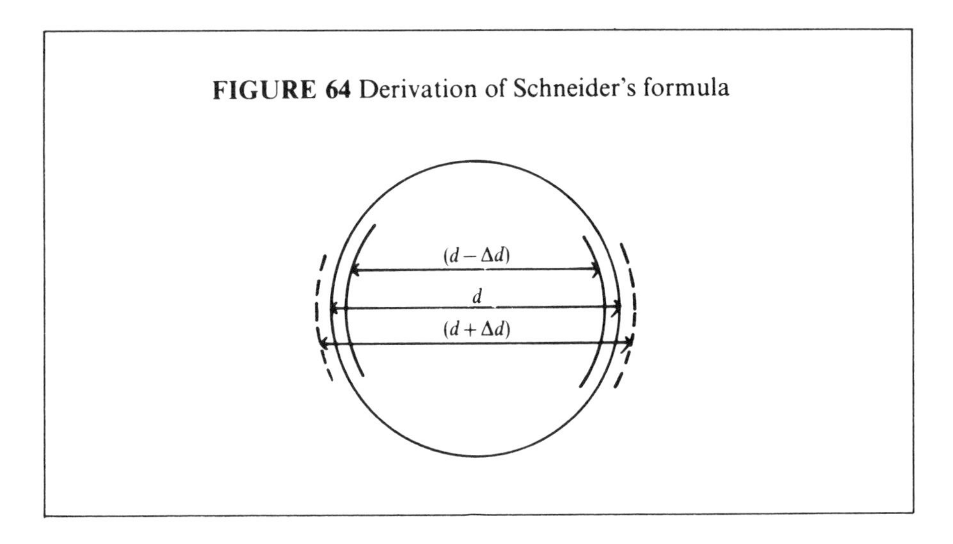

N.B. m = number of rings in outer cm of radius, *not* number of years between measurements which is always 2 in this calculation.

Also $\Delta g\% = \dfrac{200\Delta d}{d}$ may be used where Δd is the mean width of the annual rings over a set period of time.

APPENDIX 5: Derivation of Formulae Quoted in Section 5.1.1.7

The number of samples

(1) $$n = \frac{t^2 s_x^2}{E^2 \bar{x}^2}$$

If the confidence limits are defined as a proportion E of $\bar{x}$, then

$$t s_{\bar{x}} = E\bar{x}$$

$$\frac{t s_x}{\sqrt{n}} = E\bar{x}$$

$$n = \frac{t^2 s_x^2}{E^2 \bar{x}^2}$$

(2) $$n_i = \frac{n(P_i s_{xi})}{\sum^{m} P_i s_{xi}}$$

For two strata:

$$s_{\bar{\bar{x}}}^2 = \frac{P_1^2 s_{x1}^2}{n_1} + \frac{P_2^2 s_{x2}^2}{n_2}$$

$$ns_{\bar{\bar{x}}}^2 = \left[\frac{(P_1^2 s_{x1}^2)}{n_1} + \frac{(P_2^2 s_{x2}^2)}{n_2}\right](n_1 + n_2)$$

$$= P_1^2 s_{x1}^2 + P_2^2 s_{x2}^2 + \frac{n_2 P_1^2 s_{x1}^2}{n_1} + \frac{n_1 P_2^2 s_{x2}^2}{n_2}$$

but

$$n_2 = \frac{n_1 P_2^2 s_{x2}^2}{(n_1 s_{\bar{\bar{x}}}^2 - P_1^2 s_{x1}^2)} \quad \text{from the first equation}$$

Therefore

$$ns_{\bar{\bar{x}}}^2 = P_1^2 s_{x1}^2 + P_2^2 s_{x2}^2 + \frac{P_1^2 s_{x1}^2 P_2^2 s_{x2}^2}{n_1 s_{\bar{\bar{x}}}^2 - P_1^2 s_{x1}^2} + n_1 s_{\bar{\bar{x}}}^2 - P_1^2 s_1^2$$

$$= P_2^2 s_{x2}^2 + n_1 s_{\bar{\bar{x}}}^2 + \frac{P_1^2 s_{x1}^2 P_2^2 s_{x2}^2}{n_1 s_{\bar{\bar{x}}}^2 - P_1^2 s_{x1}^2}$$

Taking the first derivative in respect of n_1 and setting to 0

$$0 = s_{\bar{\bar{x}}}^2 - \frac{P_1^2 s_{x1}^2 P_2^2 s_{x2}^2 s_{\bar{\bar{x}}}^2}{(n_1 s_{\bar{\bar{x}}}^2 - P_1^2 s_{x1}^2)^2}$$

$$(n_1 s_{\bar{\bar{x}}}^2 - P_1^2 s_{x1}^2)^2 = P_1^2 s_{x1}^2 P_2^2 s_{x2}^2$$

$$n_1 s_{\bar{\bar{x}}}^2 - P_1^2 s_{x1}^2 = P_1 s_{x1} P_2 s_{x2}$$

$$n_1 = \frac{(P_1 s_{x1} P_2 s_{x2} + P_1^2 s_{x1}^2)}{s_{\bar{\bar{x}}}^2}$$

$$n_1 = \frac{P_1 s_{x1}(P_1 s_{x1} + P_2 s_{x2})}{s_{\bar{\bar{x}}}^2}$$

and generalizing from the particular case of 2 strata to M strata

$$n_i = \frac{\left(P_i s_{xi}\left(\sum^M P_i s_{xi}\right)\right)}{s_{\bar{\bar{x}}}^2} \quad \text{(see below)}$$

$$\frac{n_i}{\sum^M n_i} = \frac{\left(P_i s_{xi}\left(\sum^M P_i s_{xi}\right)\right)}{\sum^M \left[P_i s_{xi}\left(\sum^M P_i s_{xi}\right)\right]}$$

$$= \frac{P_i s_{xi}}{\sum^M P_i s_{xi}}$$

but $\sum^M n_i = n$, so

$$n_i = \frac{n(P_i s_{xi})}{\sum^M P_i s_{xi}}$$

N.B. $\sum^{M} n_i = \dfrac{\left(\sum^{M} P_i s_{xi}\right)\left(\sum^{M} P_i s_{xi}\right)}{s^2_{\bar{\bar{x}}}}$ see above

Therefore

$$s^2_{\bar{\bar{x}}} = \frac{\left(\sum^{M} P_i s_{xi}\right)^2}{\sum^{M} n_i}$$

and for stratified random sampling

$$n = \frac{t^2\left(\sum^{M} P_i s_{xi}\right)^2}{E^2 \bar{\bar{x}}^2}$$

Where:
t = Student t statistic
s^2_x = esimated variance of a random sample
E = confidence limits (expressed as a ratio of the mean)
$\bar{x}$ = estimated mean from a random sample
n = total number of sampling units
$\bar{\bar{x}}$ = estimated mean from a stratified random sampling design with M strata
n_i = number of sampling units in the ith stratum
s^2_{xi} = estimated variance of the ith stratum
P_i = proportion of stratum i of the whole population

N.B. These formulae ignore the finite population correction factor.

APPENDIX 6: The Derivation of Formulae for Ratio Sampling

$$R = \frac{\sum^{N} y_i}{\sum^{N} x_i} \text{ which is estimated by } r = \frac{\sum^{n} y_i}{\sum^{n} x_i}$$

The variance of this estimated ratio is estimated by

$$s^2_r = r^2 \left[\frac{s^2_y}{\bar{y}} + \frac{s^2_x}{\bar{x}} - \frac{2s_{xy}}{\bar{y}\bar{x}}\right]$$

$$= \frac{(r^2)}{(n-1)} \left[\frac{\left(\sum^{n} y_i^2 - n\bar{y}^2\right)}{\bar{y}^2} + \frac{\left(\sum^{n} x_i^2 - n\bar{x}^2\right)}{\bar{x}^2} - \frac{2\left(\sum^{n} x_i y_i - n\bar{x}\bar{y}\right)}{\bar{x}\bar{y}}\right]$$

$$= \frac{(r^2)}{n-1} \left[\frac{\left(\sum^{n} y_i^2\right)}{\bar{y}^2} - n - \frac{\left(\sum^{n} x_i^2\right)}{\bar{x}^2} - n - \frac{\left(2\sum^{n} x_i y_i\right)}{\bar{x}\bar{y}} + 2n\right]$$

$$= \frac{r^2}{(n-1)\bar{y}^2} \left[\left(\sum^{n} y_i^2\right) + r^2\left(\sum^{n} x_i^2\right) - 2r\sum^{n} x_i y_i\right]$$

$$= \left(\frac{r^2}{(n-1)\bar{y}^2}\right)\left[\sum^{n}(y_i - rx_i)^2\right] = \left(\frac{1}{(n-1)}\,\bar{x}^2\right)\left[\sum^{n}(y_i - rx_i)^2\right]$$

and

$$s_{\bar{r}}^2 = \frac{s_r^2(1-f)}{n}$$

whilst the estimated variance of the ratio estimate of total Y where Y is estimated by $Y = Nr\bar{x}$

then

$$s_{\bar{y}}^2 = N^2\bar{x}^2 s_{\bar{r}}^2 \frac{(N-n)}{N} = \frac{N(N-n)}{n(n-1)}\left(\sum^{n}(y_i - rx_i)^2\right)$$

N.B. $$s_r^2 = \frac{r^2}{(n-1)\bar{y}^2}\left[\left(\sum^{n} y_i^2\right) + r^2\left(\sum^{n} x_i^2\right) - 2r\sum^{n} x_i y_i\right]$$

$$= \frac{r^2}{(n-1)\bar{y}^2}\left[\sum^{n} y_i^2 - n\bar{y}^2 + \frac{\bar{y}^2}{\bar{x}^2}\left(\sum x_i^2 - n\bar{x}^2\right) - 2r\sum^{n} x_i y_i + n\bar{y}^2 + \frac{\bar{y}^2}{\bar{x}}\, n\bar{x}^2\right]$$

$$= \frac{r^2}{\bar{y}^2}\left[s_y^2 + r^2 s_x^2 - \left(2r\sum^{n} x_i y_i + 2n\bar{y}^2\right)/(\mathrm{n}-1)\right]$$

$$= \frac{r^2}{\bar{y}^2}\left[s_y^2 + r^2 s_x^2 - \left(2r\left(\sum^{n} x_i y_i - \frac{n\bar{y}^2}{r}\right)/(\mathrm{n}-1)\right)\right]$$

but

$$\frac{n\bar{y}^2}{r} = n\bar{x}\bar{y}$$

so $$s_r^2 = \frac{r^2}{\bar{y}^2}\left[s_y^2 + r^2 s_x^2 - 2r\,(\mathrm{cov}\ xy)\right] = \frac{r^2}{\bar{y}^2}\left[s_y^2 + r^2 s_x - 2r\rho s_y s_x\right]$$

Where:

s_y^2 = estimated variance of y
s_x^2 = estimated variance of x
s_{xy} = estimated covariance of x and y
r = estimated ratio of y to x
n = number of sampling units
f = sampling fraction = $\left(\frac{N-n}{N}\right)$
N = number of sampling units in the sampling frame
s_x = estimated standard deviation of x
s_y = estimated standard deviation of y
ρ = estimated correlation coefficient of x and y

APPENDIX 7: Formulae in Recurrent Forest Inventory with Partial Replacement of Sampling Units

When
$\bar{x}_p$ = means of permanent plots on the first measurement
$\bar{y}_p$ = means of permanent plots on the second measurement
$\bar{x}_t$ = mean of temporary plots on the first measurement
$\bar{y}_t$ = mean of temporary plots on the second measurement
n_p = number of permanent sample plots
n' = number of temporary sample plots at the first measurement
n'' = number of temporary sample plots at the second measurement
$n_1 = n_p + n'$
$n_2 = n_p + n''$

and

$$\text{Expected value of } \bar{x}_p = \text{expected value of } \bar{x}_t = \mu_x$$

and

$$\text{Expected value of } \bar{y}_p = \text{expected value of } \bar{y}_t = \mu_y$$

and

$$\bar{x}_b, \bar{y}_b \text{ are the best estimates of } \mu_x \text{ and } \mu_y$$

then

$$\bar{y}_b = a_y\bar{x}_t + b_y\bar{x}_p + c_y\bar{y}_p + d_y\bar{y}_t$$

$$\bar{x}_b = a_x\bar{y}_t + b_x\bar{y}_p + c_x\bar{x}_p + d_x\bar{x}_t$$

where

$$(c + d) = 1 \quad \text{therefore} \quad d = (1 - c)$$

and

$$(a + b) = 0 \quad \text{therefore} \quad b = -a$$

so

$$\bar{y}_b = a_y(\bar{x}_t - \bar{x}_p) + c_y\bar{y}_p + (1 - c_y)\bar{y}_t$$

$$\bar{x}_b = a_x(\bar{y}_t - \bar{y}_p) + c_x\bar{x}_p + (1 - c_x)\bar{x}_t$$

The most efficient estimate is that which minimizes the variance of the best estimate of μ_x and μ_y. Hence the problem is to set the parameters a and c to provide the minimum variance; these are estimated by:

$$a_y = \frac{\left(\frac{rs_y}{s_x}\right)(n_p n')}{(n_1 n_2 - n'n''r^2)}$$

$$a_x = \frac{\left(\frac{rs_x}{s_y}\right)(n_p n'')}{(n_1 n_2 - n'n''r^2)}$$

$$c_y = \frac{(n_p n_1)}{(n_1 n_2 - n'n''r^2)}$$

$$c_x = \frac{(n_p n_2)}{(n_1 n_2 - n'n''r^2)}$$

where r = estimated correlation coefficient of y_i and x_i

and

$$s^2_{\bar{y}_b} = a_y^2 s_x^2 \left(\frac{1}{n'} + \frac{1}{n_p}\right) + \frac{c_y^2 s_y^2}{n_p} + \frac{(1-c_y)^2 s_y^2}{n''} - \frac{2a_y c_y r s_y s_x}{n_p}$$

As a result the best estimate of the change $\bar{z}$ is

$$\begin{aligned}\bar{z}_b &= \bar{y}_b - \bar{x}_b \\ &= a_y(\bar{x}_t - \bar{x}_p) + c_y\bar{y}_p + (1-c_y)\bar{y}_t - (a_x(\bar{y}_t - \bar{y}_p) + c_x\bar{x}_p + (1-c_x)\bar{x}_t) \\ &= \bar{y}_t(1-(a_x+c_y)) - \bar{x}_t(1-(a_y+c_x)) + \bar{y}_p(a_x+c_y) - \bar{x}_p(a_y+c_x)\end{aligned}$$

or if

$$A = (a_x + c_y); \quad B = (a_y + c_x);$$

$$\bar{z}_b = \bar{y}_t(1-A) - \bar{x}_t(1-B) + \bar{y}_p A - \bar{x}_p B$$

but the regression coefficient β_{yx} of y on x is estimated by

$$b_{yx} = rs_y/s_z$$

and similarly $$b_{xy} = rs_x/s_y$$

and therefore

$$A = n_p(b_{xy}n'' + n_1)/(n_1n_2 - n'n''r^2)$$

and

$$B = n_p(b_{yx}n' + n_2)/(n_1n_2 - n'n''r^2)$$

therefore

$$s^2_{\bar{z}} = (A^2 s_y^2 + B^2 s_x^2 - 2ABrs_x s_y)/n_p + (1-A)^2 s_y^2/n'' + (1-B)^2 s_x^2/n'$$

Again, as previously, the closer that r approaches 1, the smaller is the standard error of $\bar{z}_b$.

Derivation of a_y, c_y, a_x, c_x to minimize variance of $s^2_{\bar{y}_b}$

$$s^2_{\bar{y}_b} = \frac{a_y^2 s_x^2}{n'} + \frac{a_y^2 s_x^2}{n_p} + \frac{c_y^2 s_y^2}{n_p} + \frac{s_y^2}{n''} + \frac{c^2 s_y^2}{n''} - \frac{2a_y c_y r s_x s_y}{n_p} - \frac{2cs_y^2}{n''}$$

Differentiating $s^2_{\bar{y}_b}$ with respect to a_y

$$0 = \frac{d(s^2_{\bar{y}_b})}{d(a_y)} = \frac{2a_y s_x^2}{n'} + \frac{2a_y s_x^2}{n_p} - \frac{2c_y r s_x s_y}{n_p}$$

$$0 = \frac{2s_x}{n'n_p}\,(a_y s_x n_p + a_y s_x n' - c_y r s_y n')$$

$$a_y = \frac{c_y r s_y n'}{s_x(n_p + n')} = c_y r\left(\frac{s_y}{s_x}\right)\left(\frac{n'}{n_1}\right)$$

$$0 = \frac{d(s^2_{\bar{y}_b})}{d(c_y)} = \frac{2c_y s_y^2}{n_p} + \frac{2c_y s_y^2}{n''} - \frac{2a_y r s_x s_y}{n_p} - \frac{2s_y^2}{n''}$$

$$0 = \frac{2s_y}{n''n_p}\,(c_y s_y n'' + c_y s_y n_p - a_y r s_x n'' - s_y n_p)$$

$$c_y = \frac{a_y r s_x n'' + s_y n_p}{s_y(n'' + n_p)} = \frac{a_y r s_x n''}{s_y n_2} + \frac{n_p}{n_2}$$

Substituting c_y in equation for a_y above

$$a_y = \left(\frac{a_y r s_x n''}{s_y n_2} + \frac{n_p}{n_2}\right)\left(r\left(\frac{s_y}{s_x}\right)\left(\frac{n'}{n_1}\right)\right)$$

$$= \frac{a_y r^2 n'n''}{n_1 n_2} + \frac{n_p r s_y n'}{n_2 n_1 s_x}$$

$$a_y\left(\frac{n_1 n_2 - r^2 n'n''}{n_1 n_2}\right) = \frac{r s_y n_p n'}{s_x n_1 n_2}$$

$$a_y = \frac{r s_y}{s_x}\left(\frac{n_p n'}{n_1 n_2 - r^2 n'n''}\right)$$

Similarly as

$$c_y = \frac{a_y r s_x n''}{s_y n_2} + \frac{n_p}{n_2} \quad \text{and} \quad a_y = c_y r\left(\frac{s_y}{s_x}\right)\left(\frac{n'}{n_1}\right)$$

$$c_y = \frac{c_y r^2 n'n''}{n_1 n_2} + \frac{n_p}{n_2}$$

$$\frac{c_y(n_1 n_2 - r^2 n'n'')}{n_1 n_2} = \frac{n_p}{n_2}$$

$$c_y = \frac{n_p n_1}{(n_1 n_2 - r^2 n'n'')}$$

and similarly

$$a_x = \frac{r s_x}{s_y}\left(\frac{n_p n''}{n_1 n_2 - n'n''r^2}\right)$$

$$c_x = \frac{n_p n_2}{(n_1 n_2 - r^2 n'n'')}$$

APPENDIX 8: Examples of Static and Dynamic Growth Models from Tanzania

The first example is taken from the *Pinus patula* plantations at Sao Hill using data from temporary sample plots only (Adegbhin & Philip, 1979).

The stages in construction were as follows.

1. Site index curves were derived from data of dominant height and age obtained by stem analyses using Tveite's method (Section 6.2.3.2.2). Three site classes were recognized. (Equation 1)

2. The relationship describing total volume production as a function of h_d was established for all three site classes combined. In thinned plots the removals were estimated from the stump diameters. (Equation 2)

3. Three separate relationships were then derived using Equation 1 to provide predictions of total volume production on age for each site. (Basal area production could have been treated similarly, but Adegbhin followed the procedure outlined below.)

4. In unthinned and under-thinned stands, Adegbhin marked and measured thinnings in accordance with the recommended local procedures. The following equations were derived from the data collected:

Number of stems thinned on dominant height (Equation 3)
Number of stems after thinning on dominant height (Equation 4)
h_L of thinnings on dominant height (Equation 5)
h_L after thinning on dominant height (Equation 6)
Cumulative volume thinned on dominant height (Equation 7)
(where h_L = Lorey's mean height)

Knowing the total volume thinned and numbers, Adegbhin calculated the mean volume thinned, m^3 per tree; he then used the standard volume tables for *Pinus patula* in Tanzania and the mean height derived from Equation 5 to give the basal area of the tree a mean volume and hence the basal area removed in the thinning. Main crop figures were derived by subtracting thinnings from the total production using Equation 2.

Consequently the columns of a table similar to that illustrated in Table 3 in Chapter 7 were compiled as follows:

Column 1 age from records and tabulated at suitable intervals
2 from original spacing and as adjusted by the the removals of numbers thinned – see Equation 3
3 h_d tabulated from Equation 1
4 G calculated from the volume per hectare of the main crop and numbers in col. 2 giving $\bar{v}$ per tree. Then, using $\bar{v}$ and h_1 from Equation 6, $\bar{g}$ was derived from the standard volume table. This figure multiplied by numbers of stems gave basal area per hectare
5 $d_{\bar{g}}$ corresponding to the $\bar{g}$ calculated above
6 calculated above (see 4)
7 calculated by (col. 6 × col. 2)
8 predicted through Equation 3
9 Equation 7 predicted total thinning to date, hence the current

volume thinned was calculated by subtracting the volume previously removed. Using the numbers thinned as calculated above, the mean volume of the thinnings was calculated; then the total basal area of the thinning was calculated in a manner similar to that used for column 4, emloying Equation 5 to predict h_L

10 Already calculated above

11 Already calculated above

12 by (col. 11 × col. 8)

13 (col. 4 + the sum of col. 9), but would be better derived from field measurements in a manner similar to Equation 2

14 from Equation 2

15 by calculation using successive entries in col. 13 divided by the period between entries

16 as for 15, but using the entries from col. 14.

17 col. 13 divided by the age in col. 1

18 col. 14 divided by the age in col. 1

The steps are given here to illustrate how, using a relatively few basic predictive equations, all the entries in a yield table can be derived. This particular method is not recommended for general adoption as it would be more reliable to base the total basal area production on a predictive equation from dominant height derived from original field measurements.

One of the weaknesses of a static growth model well illustrated by this example is that Equation 2 predicting total volume production has been derived from a set of stands managed under a uniform regime to give a relatively restricted range of stocking. Consequently extrapolation to spacing and thinning regimes outside this range is unwarranted.

In contrast Winston Mathu (Mathu & Philip, 1979) used a dynamic model for *Cupressus lusitanica* in Kenya which was based on data from permanent sample plots. He derived the following relationships:

1. A single tree volume table based on dbh and crop dominant height (Equation 1)

2. Dominant height on age for three site quality classes (Equation 2)

3. Basal area per hectare before the first thinning as a function of dominant height and stem numbers (Equation 3)

4. Annual basal area increment per hectare on dominant height (Equation 4)

5. Diameter of mean basal area tree thinned as a function of diameter of tree of mean basal area before thinning (Equation 5)

The model was run by incrementing basal area per hectare each year and superimposing a recommended thinning schedule defined by numbers thinned. The volume per hectare was calculated using the single tree volume table applied to the tree of mean basal area and the stand's dominant height. Following the layout given in Table 3, the entries were calculated as follows:

Column 1 age, tabulated at suitable intervals

2 provided by the thinning schedule

3 from Equation 2
4 the basal area before thinning as derived from Equation 3; the annual basal area increment was derived from Equation 4 coupled with Equation 2. The diameter of the tree of mean basal area before thinning was derived knowing the stem numbers, and the diameter of the tree of mean basal area thinned was predicted through Equation 5. The basal area thinned and the basal area of the main crop remaining were then calculated by multiplying the mean basal area by the number of stems thinned given in the thinning schedule. Subtraction of the basal area thinned from the accumulated basal area gave the basal area of the main crop after thinning
5 the diameter corresponding to $\bar{g}$ calculated from col. 4 divided by col. 2
6 from Equation 1 using the entries in col. 5 and col. 3 as predictor variables
7 from (col. 6 × col. 2)
8 from the thinning schedule
9 from (col. 8 × the $\bar{g}$ calculated through Equation 5, as shown above)
10 as shown above in the calculation of the entry in col. 4
11 from Equation 1 using the entries in col. 10 and col. 3 as predictor variables
12 from (col. 11 × col. 8)
13 from Equation 3 plus the sum of the annual increments in basal area since the first thinning
14 from (col. 7 plus accumulated sum of col. 12)
15 from Equation 4 coupled with Equation 2
16 by calculating the increase in total volume production over a two year period, say years 6–8, expressing this difference as an annual change corresponding to the intermediate year, say year 7
17 from (the sum of col. 4 and the accumulated sum of col. 9) divided by the age in col. 1
18 from (the sum of col. 7 and the accumulated sum of col. 12) divided by the age in col. 1

Such a dynamic model has similar faults to the static model outlined above, as the prediction of basal area increment is independent of stocking. The model would be more flexible and applicable to a wider range of conditions if Equation 4 included a term that adjusted the growth, possibly predicting lower increments with very low stocking. Similarly greater flexibility would be introduced if Equation 5 was amended to predict basal areas thinned and remaining for different types of thinning; for example if a non-selective thinning was done then the expected mean diameter before and after thinning would be identical, whereas with low thinnings the expected diameter of the thinnings would be less than that of the trees remaining and vice versa with a crown thinning.

BIBLIOGRAPHY

Adegbhin, J.O. and Philip, M.S. (1979) *Studies of Dominant Height Development and Yield of* Pinus patula *at Sao Hill Forest Project, Southern Tanzania.* Record no.6, Division of Forestry, University of Dar es Salaam, Morogoro, Tanzania.

Adlard, P.G. (1990) *Procedures for Monitoring Tree Growth and Site Change: A Field Manual.* Tropical Forestry Paper no. 23, Oxford Forestry Institute.

Adlard, P.G. (1992) Research strategy for monitoring tree growth and site change. (And also paper on: Observed growth patterns for eucalypts.) In: Calder, I.R., Hall, R.L. and Adlard, P.G. (eds), *Proceedings of the International Symposium held at Bangalore, India, Feb 4–7, 1991, through the joint action of the Oxford Forestry Institute, The Institute of Hydrology, Wallingford, UK, the Karnataka Forest Department and the Mysore Paper Mills Ltd.* John Wiley & Sons, Chichester, UK.

Adlard, P.G. and Rondeux, J. (1989) (eds) *Forest Growth Data: Capture, Retrieval and Dissemination.* Proceedings of the joint IUFRO S4.02.03/4.02.04 workshop, 3–5 April, 1989 at Gembloux, Belgium.

Adlard, P.G., Spilsbury, M.J. and Whitmore, T.C. (1988) Current thinking on modelling. Tropical Moist Forest Meeting of IUFRO. Sections S4.01.02 & 1.07. Growth and yield in tropical mixed/moist forests. Forest Research Institute of Malaysia.

Alder, D. (1977) A growth and management model for coniferous plantations in East Africa. DPhil. thesis, Oxford University, Oxford.

Alder, D. (1980) *Forest Volume Estimation and Yield Prediction,* vol. 2. FAO Forestry Paper 22/2, Rome.

Alder, D. (1992) Simple methods for calculating minimum diameter and sustainable yield in mixed tropical forest. Proceedings of the Oxford Conference on tropical forests *Wise Management of Tropical Forests.* Oxford Forestry Institute.

Alder, D. and Synott, T.C. (1992) *Permanent Sample Plot Techniques for Mixed Tropical Forest.* Tropical Forestry Paper No. 25, Oxford Forestry Institute.

Amateis, R.L., Burkhart, H.E. and Burk, T.E. (1986) A ratio approach to predicting merchantable yields of unthinned loblolly pine plantations. *Forest Science* 32(2), 287–296.

Applegate, G.B., Gilmore, D.A. and Mohns, B. (1988) The use of biomass estimations in the management of forests for fuelwood and fodder production. *Commonwealth Forestry Review* 67(2), 141–147.

Arney, J.D. (1985) A modelling strategy for the growth projection of managed stands. *Canadian Journal of Forest Research* 15(3), 511–518.

Avery, T.E. and Berlin, G.L. (1992) *Fundamentals of Remote Sensing and Airphoto Interpretation.* Macmillan, Basingstoke.

Avery, T.E. and Burkhart, H.E. (1983) *Forest Measurements.* McGraw-Hill, New York.

Ayhan, H.O. (1977) A crown radius instrument. *Scottish Forestry* 31(2), 67–70.

Bailey, G.R. (1970) A simplified method of sampling logging residues. *Forestry Chronicle* 46(4), 288–294.

Bailey, R.L. and Dell, T.R. (1973) Quantifying diameter distribution with the Weibull function. *Forest Science* 19(2), 97–104.

Banyard, S.G. (1987) Point sampling using constant tallies is biased: a tropical rain forest case study. *Commonwealth Forestry Review* 66(2), 161–163.

Beers, T.W. and Miller, C.I. (1964) *Point Sampling: Research Result, Theory and Application.* Research Bulletin 786, Purdue University, Agricultural Experimental Station, Lafayette, Indiana.

Bell, J.F. (1973) Choice of values for the concomitant variable in 3P sampling. In: Warren, W.G. (ed.), *Statistics in Forest Research.* IUFRO subject group S6.02. Vancouver.

Bella, I.E. (1971) A new competition model for individual trees. *Forest Science* 17(3), 364–372.

Besley, L. (1967) The importance of variation and the measurement of density and moisture. In: *Wood Measurement Conference Proceedings.* Faculty of Forestry, University of Toronto, Technical Report no. 7, pp. 112–142.

Biggs, P.H., Wood, S.B., Schreuder, H.T. and Brink, G.E. (1985) Comparison of point-model based and point-Poisson sampling for timber inventory in Jarrah forest. *Australian Forest Research* 15(4), 481–493.

Biging, G.S. (1985) Improved estimates of site index curves using a varying parameter model. *Forest Science* 31(1), 248–259.

Bitterlich, W. (1976) Volume sampling using imperfectly estimated critical heights. *Commonwealth Forestry Review* 55(4), 319–330.

Bitterlich, W. (1984) *The Relascope Idea.* CAB International, Wallingford.

Bitterlich, W. (Undated) *Spiegel-Relaskop + Telerelaskop.* FOB Postfach 12, A-5035 Salzburg, Austria.

Blyth, J.F. (1974) The importance of initial check and tree form in the estimation of yield class as a growth index for site assessment. *Scottish Forestry* 28(3), 198–210.

Borders, B.E. and Bailey, R.L. (1986) A compatible system of growth and yield equations for slash pine fitted with restricted three stage least squares. *Forest Science* 32(1), 185–201.

Bose, K.J.C. (1988) Potential productivity–growth model of tropical forest (thermodynamic approach). Proceedings of the IUFRO meeting on growth and yield of tropical moist/mixed forests. Forest Research Institute of Malaysia.

Botkin, D.B., Janak, J.F. and Wallis, J.R. (1972) Some ecological consequences of a computer model for forest growth. *Journal of Ecology* 60, 849–873.

Brandiani, A., Hartshorn, G.S. and Orians, G.H. (1988) Internal heterogeneity of gaps and species richness in Costa Rica tropical wet forest. *Journal of Tropical Ecology* 4(2), 99–119.

Bruce, D. and Schumacher, F.X. (1942) *Forest Mensuration.* McGraw-Hill, New York.

Bruce, D., Curtis, R.O. and van Coeverind, C. (1968) Development of a system of taper and volume tables for red alder. *Forest Science* 14(3), 339–350.

Bruner, H.D. and Moser, J.W. Jr. (1973) A Markov chain approach to the prediction of diameter growth in uneven aged forest stands. *Canadian Journal of Forest Research* 3, 409–417.

Buchner, E.R., Richeson, J.L., Rennie, J. and Boyd, W. (1977) Modifying the Christen hypsometer for use in dense conifer plantations. *Journal of Forestry* 75(3), 139–190.

Burkhart, H.E. and Knoebel, B.R. (1983) A growth and yield model for thinned stands of yellow poplar – *Liriodendron tulipifera.* In: *Forest Growth Modelling and Simulation.*

IUFRO, Mitteilungen der Forstlichen Bundesversuchsanstalt.

Burkhart, H.E. and Strub, M.R. (1971) A model for simulation of planted loblolly pine stand. In: Fries, J. (ed.) *Growth Models for Tree and Stand Simulation*. Proceedings of meetings in 1973 of IUFRO working party S4.01-4. Royal College of Forestry, Stockholm.

Busgen, M., Munch, E. and Thomson, T. (1929) *The Structure and Life of Forest Trees*. Chapman and Hall, London.

Byrne, J.C. and Reed, D.D. (1986) Complex compatible taper and volume estimation systems for red and loblolly pine. *Forest Science* 32(2), 423–443.

Cajander, A.K. (1926) The theory of forest. *Acta Forestalia Fennica* 29, 1–108.

Calder, I.R., Hall, R.L. and Adlard, P.G. (1992) (eds) *Proceedings of the International Symposium held at Bangalore, India, Feb 4–7, 1991, through the joint action of the Oxford Forestry Institute, the Institute of Hydrology, Wallingford, the Karnataka Forest Department and the Mysore Paper Mills Ltd*. John Wiley & Sons, Chichester, UK.

Campbell, J.S., Lieffers, V.J. and Pielou, E.C. (1985) Regression equations for estimating single tree biomass of trembling aspen: assessing their applicability to more than one population. *Forestry Ecology and Management* 11(4), 283–295.

Cao, Q.V., Burkhart, H.E. and Max, T.A. (1986) Segmented polynomial compatible taper equations for loblolly pine. *Forest Science* 26, 71–80.

Caplewski, R.L. and McClure, J.P. (1988) Conditioning a segmented stem profile model for two diameter measurements. *Forest Science* 34(2), 512–522.

Chang, S.J. (1984) A simple production function model for variable density growth & yield modelling. *Canadian Journal of Forest Research* 14(6), 783–788.

Chaturvedi, M.D. (1926) *Measurement of the Cubical Contents of Forest Crops*. Oxford Forestry Memoir no. 4.

Christie, J.M. (1970) The characterisation of the relationships between basic crop parameters in yield table construction. In: *Proceedings of the 3rd Conference of the Advisory Group of Forest Statisticians*. IUFRO, Jouy-en-Josas, France.

Cochran, W.G. (1977) *Sampling Techniques* (2nd edn). John Wiley & Sons, New York.

Cunia, T. (1964) Weight least squares method and construction of volume tables. *Forest Science* 10(2), 180–191.

Cunia, T. (1965) Continuous forestry inventory, partial replacement of samples and multiple regression. *Forest Science* 11(4), 480–502.

Cunia, T. (1973) Dummy variables and some of its uses in regression analyses. In: *Proceedings of IUFRO Subject group 4.02 – Forest Resource Inventories*. Nancy, France.

Cunia, T. and Briggs, R.D. (1984) Forcing additivity of some biomass tables; some empirical results. *Canadian Journal of Forest Research* 14(3), 376–384.

Cunia, T. and Briggs, R.D. (1985) Forcing additivity of biomass tables: use of the generalized least squares method. *Canadian Journal of Forest Research* 15(1), 23–28; 15(2), 331–340.

Cunia, T.and Michelakackis, J. (1983) On the error of tree biomass tables constructed by a two-phase sampling design. *Canadian Journal of Forest Research* 12(2), 303–313.

Curtin, R.A. (1964) Stand density and the relationship of crown width to diameter and height in *Eucalyptus obliqua*. *Australian Forestry* 28(2), 91–105.

Daniel, W.W. (1978) *Applied Non-Parametric Statistics*. Houghton Miffin Co., USA.

Davidov, M.V. (1978) Feature of the growth and site class of stands of fast growing wood species. In: *Forest Abstracts*. Lesnoi Zhurnal 5.3–8, Ukrain. Sel'skokhoz. Akad, USSR.

Dawkins, H.C. (1958) *The Management of Natural Tropical High Forest with Special Reference to Uganda*. Commonwealth Forestry Institute Paper no. 34, Oxford.

Dawkins, H.C. (1963) Crown diameters: their relation to bole diameter in tropical forest trees. *Commonwealth Forestry Review* 42(4) no. 114, 318–333.

Dawkins, H.C. (1975) *Statforms-'Pro-formas' for the Guidance of Statistical Calculations*. Edward Arnold, London.

Dawkins, H.C. (1978) *A Long Term Surveillance System for British Woodlands*. Occasional Paper no. 1. Commonwealth Forestry Institute, Oxford.

Dawkins, H.C. (1985) Sampling errors in extensive forest inventory. *Commonwealth Forestry Review* 64(1), 78–79.

Demaerschalk, J.P. (1972) Converting volume tables to compatible taper equation. *Forest Science* 18, 241–245.

Demaerschalk, J.P. (1973) Integrated systems for the estimation of tree taper and volume. *Canadian Journal of Forest Research* 3(1), 90–94.

Demaerschalk, J.P. and Kozak, A. (1974, 1975) Suggestions and criteria for more effective regression sampling, I and II. *Canadian Journal of Forest Research* 4, 341–348; *Canadian Journal of Forest Research* 5, 496–497.

Dick, J. (1963) Forest stocking determined by sequential stocked quadrat tally. *Journal of Forestry* 61(4).

Dilmy, A. (1971) The primary productivity of equatorial forests in Indonesia. In: *Proceedings of the Brussels Symposium on Productivity of Forest Ecosystems*. UNESCO, Paris.

Edwards, P.N. (1983) *Timber Measurement: A Field Guide*. Forestry Commission Booklet no. 49. HMSO, London.

Edwards, P.N. and Christie, J.M. (1981) *Yield Models for Forest Management*. Forestry Commission Booklet no. 48, Edinburgh.

Ek, A.R. and Monserud, R.A. (1973) Trials with program FOREST, growth and reproduction simulation of mixed species, even the uneven-aged forest. In: Fries, J. (ed.) *Growth Models for Tree and Stand Simulation*. Proceedings of meetings in 1973 of IUFRO working party S4.01-4. College of Forestry, Stockholm.

Eugene, C. (1989) The ultrasonic angular caliper – a promising new tool for computerized forest surveying. In: Adlard, P.G. and Rondeux, J. (eds), *Forest Growth Data: Capture, Retrieval and Dissemination*. Proceedings of the joint IUFRO S4.02.03/4.02.04 workshop. Gembloux, Belgium.

Evert, F. (1969) Use of form factors in tree volume estimation. *Journal of Forestry* 67, 126–128.

Evert, F. (1973) New form class equations improve volume estimates. *Canadian Journal of Forest Research* 3(3), 338–348.

Fairweather, S.E. (1985) Sequential sampling for assessment of stocking adequacy *Northern Journal of Applied Forestry* 2(1), 5–8.

FAO (1968) *Manual of Forest Inventory Operations Executed by FAO*. (B Husch.) Rome.

FAO (1973) *Manual of Forest Inventory with Special Reference to Mixed Tropical Forest*. (Lanly.) Rome.

FAO (1980) *Forest Volume Estimation and Yield Prediction*, vols 1 and 2. (Coordinator Joran Fries.) Rome.

Findlayson, W. (1969) *The Relascope – Bitterlich's Spiegel relaskop*. FOB, Postfach 12, A-5035 Salzburg, Austria.

Flewelling, J.W. (1981) Comparable estimates of basal area and basal area growth from the remeasurement of point samples. *Forest Science* 27, 191–203.

Frayer, W.E. (1974) The increasing complexity of forest inventories. *Journal of Forestry* 72(9), 578–579.

Freese, F. (1960) *Elementary Forest Sampling*. US Department of Agriculture Handbook 232.

Freese, F. (1967) *Elementary Statistical Methods for Foresters*. US Department of Agriculture Handbook 317.

Freese, F. (1984) *Statistics for Land Managers: an Introduction to Sampling Methods for*

Foresters, Farmers and Environmental Biologists. Paeony, Jedburgh.

Fries, J. (ed.) (1973) Proceedings of meetings in 1973 of IUFRO working party S4.01-4. Royal College of Forestry, Stockholm.

Furnival, M.G. (1961) An index for comparing equations used in constructing volume tables. *Forest Science* 9(4), 337–341.

Furnival, M.G., Valentine, H.C. and Gregoire, T.G. (1986) Estimation of log volume by importance sampling. *Forest Science* 32(4), 1073–1078.

Gambil, C.W. and Wiant, H.V. Jr. (1985) Optimum plot size and BAF. *Forest Science* 31(3), 587–594.

Goulding, C.J. and Lawrence, M.E. (1992) Inventory practice for managed forests. *Forest Research Institute Bulletin* no. 171, Rotorua, New Zealand.

Gray, H.R. (1956) *The Form Taper of Forest Tree Stems.* Imperial Forest Institute Paper no. 32, Oxford.

Greaves, A. (1978) *Site Index Curves for Gmelina Arborea Roxb.* Occasional Paper no. 3, Commonwealth Forestry Institute, Oxford.

Gregoire, T.G. (1987) Generalized error structure for forestry yield models. *Forest Science* 33(2), 583–590.

Gregoire, T.G., Valentine, H.T. and Furnival, G.M. (1986) Estimation of bole volume by importance sampling. *Canadian Journal of Forest Research* 16(3), 554–557.

Grey, D.C. (1989) Environmental factors and diameter distributions in *Pinus radiata* stands. *South African Forestry Journal* no. 149, 36–43.

Griffin, R.H. and Johnson, C.E. (1980) Polymorphic site index curves for spruce and balsam fir growing in even spaced stands in northern Maine. *Bulletin of the Maine Life Sciences and Agriculture Experimental Station,* no. 765, University of Maine. (In *Forest Abstracts.*)

Grosenbaugh, L.R. (1952) Potless timber estimates – new fast, easy. *Journal of Forestry* 50(1), 32–37.

Grosenbaugh, L.R. (1964) *STX-Fortran 4 Program for Estimates of Tree Populations from 3P Sample Tree Measurements.* US Department of Agriculture Forest Service, Pacific South-west Forest and Range Experimental Station, Research Paper PSW-13.

Grosenbaugh, L.R. (1966) Tree form, its definition, interpolation, extrapolation. *Forestry Chronicle* 42(4), 444–457.

Grosenbaugh, L.R. (1967) The gains from sample tree selection with unequal probabilities. *Journal of Forestry* 65(3), 203–206.

Grosenbaugh, L.R. (1976) Approximate sampling variance of adjusted 3P estimates. *Forest Science* 22(2), 173–176.

Grut, M. (1971) Estimating the yields from *Pinus radiata* stands. *Forestry in South Africa,* no. 12, 41–8.

Hafley, W.L. and Schreuder, H.T. (1977) Statistical distributions for fitting diameter and height data in even aged stands. *Canadian Journal of Forest Research* 7, 481–487.

Hall, J.B. (1977) Forest types in Nigeria: an analysis of pre-exploitation forest enumeration data. *Journal of Ecology* 65(1), 187–199.

Hamilton, G.J. (1971) *Timber Measurement for Standing Sales.* Forestry Commission Booklet no. 36. HMSO, London.

Hamilton, G.J. (1975) *Forest Mensuration Handbook.* Forestry Commission Booklet 39 (3rd impression, 1981). HMSO, London.

Hamilton, G.J. and Christie, J.M. (1973) Construction and application of stand yield models. *Forestry Commission Research and Development Paper 96,* Edinburgh.

Hamlin, D. and Leary, R. (1987) Method for using an integro-differential equation as a model of tree height growth. *Canadian Journal of Forest Research* 17(5), 353–356.

Hansen, M.H. (1985) Line intersect sampling of wooded strips. *Forest Science* 31(2), 281–288.

Hetherington, J.C. (1982) 3P sampling – the devil you don't know. *Scottish Forestry Journal* 36(1), 25–35.

Holdridge, L.R., Grenke, W.R., Hatheway, W.H., Liang, T. and Tosi, J.A. Jr. (1971) *Forest Environments in Tropical Life Zones: a Pilot Study.* Pergamon Press, New York.

Horn, H.S. (1971) *The Adaptive Geometry of Trees.* Monographs in population biology, no. 3. Princeton University Press, Princeton.

Howard, J.A. (1991) *Remote Sensing of Forest Resources: Theory and Application.* Chapman & Hall, London.

Hummel, F.C. (1955) *The Volume–Basal Area Line.* Forestry Commission Bulletin, no. 24. HMSO, London.

Husch, B. (1968) *Manual of Forest Inventory Operations Executed by FAO.* FAO, Rome.

Husch, B., Miller, C.I. and Beers, T.W. (1982) *Forest Mensuration.* Ronald Press, New York.

Iles, K. and Beers, T.W. (1983) Growth information from variable plot sampling. International Conference: *Renewable Natural Resource Inventories for Monitoring Changes and Trends.* Oregon State University, Corvallis, USA.

Iles, K. and Fall, M. (1988) Can an angle gauge really evaluate borderline trees accurately in variable plot sampling? *Canadian Journal of Forest Research* 18(6), 774–781.

Iles, K. and Wilson, W.H. (1988) Changing angle gauges in variable plot sampling: is there a bias under ordinary conditions? *Canadian Journal of Forest Research* 18(6), 768–773.

Jeyaratnam, S., Bowden, D.C. Graybill, F.A. and Frayer, W.E. (1984) Estimation of multi-phase designs for stratification. *Forest Science* 30(2), 484–491.

Jonsson, B. (1981) An electronic caliper with automatic data storage. *Forest Science* 27(4), 765–777.

Karani, P.K. (1978) Pruning and thinning in a *Pinus patula* stand at Lendu plantation, Uganda. *Commonwealth Forestry Review* 57(4), 269–278.

Kirkpatrick, D.J. (1978) Growth models for Sitka spruce in Northern Ireland. *Forestry* 51(1), 46–56.

Kirkpatrick, D.J. and Savill, R.S. (1981) Top height curves for Sitka spruce in Northern Ireland. *Forestry* 54(1), 31–40.

Kitamura, M. (1964) (unseen) A simple method of estimating the volume of trees in a stand by the method of the sum of critical heights. Translation by the Canadian Department of Forestry and Fish from *Journal of the Japanese Forestry Society* 50(11), 331–335.

Knuchel, H. (1953) *Planning and Control in the Managed Forest.* Translated by M. Anderson. Oliver and Boyd, Edinburgh.

Kofod, O.E. (1982) *Stand Table Projections for the Mixed Dipterocarp Forest of Sarawak.* FAO Report FO:MAL/76/008 Field Document no. 9. Forestry Development Project, Sarawak.

Kollman, F.F.P. and Cote, W.A. Jr. (1968) *Principles of Wood Science and Technology.* George Allen & Unwin, London.

Korsgaard, S. (1988) The standtable projection simulation model. In: *Growth and Yield in Mixed/Moist Tropical Forests.* Proceedings of the IUFRO meeting. Forest Research Institute of Malaysia.

Kozak, A. (1988) A variable exponent volume table. *Canadian Journal of Forest Research* 18, 1363–1368.

Kozak, A., Munro, D.D. and Smith, J.H.G. (1969) Taper functions and their application in forest inventory. *Forest Chronicle* 45(4), 278–283.

Lanly, J.P. (1973) *Manual of Forest Inventory with Special Reference to Mixed Tropical Forest.* FAO, Rome.

Larson, P.R. (1963) *Stem Form Development of Trees.* Forest Science Monograph, no. 5.

Lieberman, D. and Lieberman, M. (1985) Simulation of growth curves from periodic increment data. *Ecology* 66, 632–635.

Lieberman, D., Lieberman, M. and Hartshorn, G.S. (1985) Growth rates and agesize class relationships of tropical wet forest trees in Costa Rica. *Journal of Tropical Ecology* 1(2), 97–109.

Little, S.N. (1984) Weibull diameter distributions for mixed stands of different conifers. *Canadian Journal of Forest Research* 13(1), 100–107.

Loetsch, F. and Haller, K.E. (1964) *Forest Inventory,* vol. I. BLV Verlagsgesellschaft, Munich.

Loetsch, F., Zohrer, F. and Haller, K.E. (1973) *Forest Inventory,* vol. II. BLV Verlagsgesellschaft, Munich.

Lynch, T.B. (1990) Stand volume estimation from tree counts in the context of vertical line sampling. *Canadian Journal of Forest Research* 20(3), 274–279.

Lynch, T.B. and Moser, J.W. Jr. (1986) A growth model of mixed species stands. *Forest Science* 32(3), 697–706.

McClure, J.P. and Czalewski, R.L. (1986) Compatible taper equation for loblolly pine. *Canadian Journal of Forest Research* 16(6), 1272–1276.

McTague, J.P. and Bailey, R.L. (1985) Critical height sampling for stand volume estimation. *Forest Science* 31(4), 899–911.

McTague, J.P. and Bailey, R.L. (1987) Simultaneous total and merchantable volume equations and a compatible taper function for loblolly pine. *Canadian Journal of Forest Research* 17(1), 87–92.

Malla, K.G., Hetherington, J.C. and Kassab, J.Y. (1984) Sampling with partial replacement: a literature review. *Commonwealth Forestry Review* 63(3), 193–206.

Marshall, P.L. and Demaerschalk, J.P. (1986) A strategy for efficient sample selection in simple linear regression problems with unequal per unit sampling costs. *Forestry Chronicle* 62(1), 16–19.

Martin, G.L. (1983) The relative efficiency of some forest growth estimators. *Biometrics* 39(3), 639–649.

Mathu, W.L. and Philip, M.S. (1979) *Growth and Yield Studies of* Cupressus lusitanica *in Kenya.* Record no. 5, Division of Forestry, University of Dar es Salaam, Morogoro, Tanzania.

Meyer, H.A. (1953) *Forest Mensuration.* Penns Valley Publishers Inc., State College, DA, USA.

M'Hirit, O. and Postaire, J.G. (1985) A non-parametric technique for taper function estimation. *Canadian Journal of Forest Research* 15(5), 862–871.

Mitchell, K.J. (1975) *Dynamics and Simulated Yield of Douglas Fir.* Forest Science Monograph, no. 17.

Monserud, R.A. (1984) Height growth and site index curves for inland Douglas fir based on stem analysis data and forest habitat type. *Forest Science* 61(1), 19–22.

Moser, J.W. (1976) Specification of density for the inverse J-shaped diameter distribution. *Forest Science* 22(2), 177–780.

Munro, D.D. (1973) Forest growth models – a prognosis. In: Fries, J. (ed.) *Growth Models for Tree and Stand Simulation.* IUFRO meeting of S4.01-4. College of Forestry, Stockholm, Sweden.

Munro, D.D. and Demaerschalk, J.P. (1974) Taper based versus volume based compatible estimating systems. *Forestry Chronicle* 50(5), 197–199.

Munyuku, F.C.N. (1980) Compatible volume and taper estimating systems for *Pinus patula* at Sao Hill in Tanzania. MSc Thesis, University of Aberdeen, UK.

Murchison, H.G. and Ek, A.R. (1989) The efficiency of two-phase and two-stage sampling for tree heights in forest inventory. In: *State-of-the-Art of Methodology for Forest Inventory.* IUFRI/SAF Symposium ESF, SUNY, Syracuse, New York, USA.

Newberry, J.D. and Burkhart, H.E. (1986) Variable form stem profile models for loblolly pine. *Canadian Journal of Forest Research* 16(1), 109–114.

Newberry, J.D., Burkhart, H.E. and Amateis, R.L. (1989) Individual tree merchantable volume to total volume ratios based on geometric solids. *Canadian Journal of Forest Research* 19(5), 679–683.

Newnham, R.M. (1965) Stem form and the variation of taper with age thinning regime. *Forestry* 38(2), 218–224.

Newnham, R.M. (1988) A modification of the Ek–Payandeh nonlinear regression model for site index curves. *Canadian Journal of Forest Research* 18(1), 115–120.

Nokoe, Sagary (1978) Demonstrating the flexibility of the Gompertz function as a yield model using mature species data. *Commonwealth Forestry Review* 57(1), 35–42.

Nye, P.H. and Greenland, D.J. (1960) *The Soil Shifting Cultivation.* Technical Communication no. 51, Commonwealth Bureau of Soils, Commonwealth Agriculture Bureaux, Farnham Royal, England.

Opie, J.E. (1968) Predictability of individual tree growth using various definitions of competing basal area. *Forest Science* 14(3), 314–323.

Ormerod, D. (1986) The diameter point method for tree taper description. *Canadian Journal of Forest Science* 16(3), 484–490.

Overseas Development Administration (1972) *The Land Resources of Northern Nigeria.* Land Resources Division, Overseas Development Administration, London.

Payandeh, B. and Ek, A.R. (1986) Distance methods and density estimators. *Canadian Journal of Forestry Research* 16(5), 918–924.

Philip, M.S. (1976) The role of forest inventories in the development of tropical moist forest. *Commonwealth Forestry Review* 55(1), 57–64.

Philip, M.S., Chamshama, S.A.O., Enyola, M.K.L. and Zimba, S.C. (1979) Studies of volume estimation of *Pinus patula* in Tanzania. Record no. 11, Division of Forestry, University of Dar es Salaam, Morogoro, Tanzania.

Pickford, S.G. and Hazard, J.W. (1978) Simulation studies on line intersect sampling of forest residues. *Forest Science* 24(4), 469–483.

Pryag, V.R. and Gore, A.P. (1989) Cost efficient density estimation based on nearest individual distances in natural forest (i = 1,2,3. . .6). Method to optimise (maximum likelihood estimates of N) i to provide minimum least squares estimate for a given cost. *Biometrical Journal* 31(3), 331–337.

Rebner, C.A. and Ek, A.R. (1983) Variance estimation from systematic samples in Minnesota timber stands. *Canadian Journal of Forest Research* 13(6), 1255–1257.

Reed, D.D. and Green, E.J. (1984) Compatible stem taper and volume ratio equations. *Forest Science* 30(4), 977–990.

Reed, D.D. and Green, E.J. (1985) A method of forcing additivity of biomass tables when using linear models. *Canadian Journal of Forest Research* 15(6), 1184–1187.

Rennie, J.C. (1976) Point-3P Sampling: a useful timber inventory design. *Forestry Chronicle* 52(3), 145–146.

Rennolls, K. (1978) Top height: its definition and estimation. *Commonwealth Forestry Review* 57(3), 215–219.

Rennolls, K., Geary, D.N. and Rollison, T.J.D. (1985) Characterising diameter distributions by the use of the Weibull distribution. *Forestry* 58(1), 57–66.

Richards, F.J. (1959) A flexible growth function for empirical use. *Journal of Experimental Botany,* 10, 290–300.

Roeder, A. (1979) Sequential sampling for reducing sampling work – a case study. In: *Forest Resource Inventories*. Workshop Proceedings Vol. 1, pp. 279–280. Colorado State University, Fort Collins.

Rollet, B. (undated) *L'Architecture des forêts denses, humides, sempervirentes de plaine.* Centre Technique Forestier Tropical, Nogent-sur-Marne, France.

Rollison, T.J.D. (1986) Don't forget production class. *Scottish Forestry* 40(4), 250–258.

Romesberg, H.C. and Mohai, P. (1990) Improving the precision of tree height estimates. *Canadian Journal of Forest Research* 20(8), 1246–1250.

Schmid-Haas, P. and Werner, J. (1970) *Swiss Continuous Forest Inventory Instruction.* Swiss Forest Research Institute, Birmensdorf.

Schreuder, H.T. and Anderson, J. (1984) Variance estimation for volume when D2*H is the covariate in the regression. *Canadian Journal of Forest Research* 14(6), 818–821.

Schreuder, H.T. and Wood, G.B. (1986) The choice between design-dependent and model-dependent sampling. *Canadian Journal of Forest Research* 16(2), 260–265.

Schreuder, H.T., Brink, G.E. and Wilson, R.L. (1984) Alternative estimators for point-Poisson sampling. *Forest Science* 30(3), 803–812.

Schumacher, F.X. (1939) A new growth curve and its applications to timber yield studies. *Journal of Forestry* 37, 819–820.

Snedecor, W.G. and Cochran, G.W. (1967) *Statistical Methods* (6th edn). Iowa State University Press, Ames, Iowa.

Scott, C.T. (1984) A new look at sampling with partial replacement. *Forest Science* 30(1), 157–166.

Smith, V.G. (1983) Compatible basal area growth and yield models consistent with forest growth theory. *Forest Science* 29(2), 279–288.

Smith, S.H. and Bell, J.F. (1983) Using competitive stress index to estimate diameter growth for thinned Douglas fir stands. *Forest Science* 29(3), 491–499.

Snowdon, P. (1985) Alternative sampling strategies and regression models for estimating forest biomass. *Australian Forest Research* 15(3), 353–366.

Snowdon, P. (1986) Sampling strategies and methods of estimating the biomass of crown components in individual trees of *Pinus radiata* D. Don. *Australian Forest Research* 16(1), 63–72.

Space, J.C. (1974) *3P Forest Inventories. Design, Procedures, Data Processing.* USA Forest Service, State and private forestry-SE Area.

Spurr, S.H. (1960) *Photogrammetry and Photo-interpretation.* The Ronald Press Co., New York.

Strand, L. (1954) (Unseen) Relaskopisk hoyote- og kubikkmassebestemmelse (Determination of height and volume by relascope.) *Norsk Skogbruk* 3(2), 535–538.

Strand, L. (1958) (Unseen) *Sampling for Volume along a Line.* Meddelelser Norske Skogfersoksveresen No. 51.

Strickland, R.E. and Binns, W.O. (1975) Measuring tree heights in dense stands with sectional poles and a horizontal arm. *Commonwealth Forestry Review* 54(3&4), 257–265.

Sullivan, A.D. and Reynolds, M.R.J. (1976) Regression problems from repeated measurement. *Forest Science* 22(4), 382–385.

Synott, T.J. (1979) *A Manual of Permanent Plot Procedures for Tropical Rain Forests.* Tropical Forest Paper no. 14, Commonwealth Forestry Institute, Oxford.

Synott, T.J. (1980) *Tropical Rainforest Silviculture.* Occasional Paper no. 10, Commonwealth Forestry Institute, Oxford.

Temu, A. (1979) *Estimation of Millable Timber Volume in Miombo Woodland.* Record no. 7, Division of Forestry, University of Dar es Salaam, Morogoro, Tanzania.

Temu, A. (1980) *Fuelwood Scarcity and Other Problems Associated with Tobacco Production in Tabora Region*, Tanzania. Record no. 12, Division of Forestry, University of Dar es Salaam, Morogoro, Tanzania.

Thornthwaite, C.W. and Mather, J.R. (1963) Average climatic water balance data of the continents. *Climatology* XVI(1), Centerton, New Jersey.

Titus, S.J. and Morton, R.T. (1985) Forest stand growth models: what for? *Forestry Chronicle* 61(1), 19–22.

Trapnell, C.G. and Griffiths, J.E. (1960) The rainfall–altitude relation and its ecological significance in Kenya. *East African Agricultural Journal* 25(4), 207–213.

Tweite, B. (1969) *A Method for Construction of Site Index Curves*. Meddelelser fra Norsk Institut for Skogsforskning, 27.

Tweite, B. (1977) *Site Index Curves for Norway spruce (Pices abies (L) Korsk.)*. Meddelelser fra Norsk Institut for Skogsforskning, 33.

Usher, M.B. (1966) A matrix approach to the management of renewable resources. *Journal of Applied Ecology* 3, 355–367.

Valenti, M.A. and Cao, Q.V. (1986) Use of crown ratio to improve loblolly pine taper equations. *Canadian Journal of Forest Research* 16(5), 1141–1145.

Valentine, H.T., Tritton, L.M. and Furnival, G.M. (1984) Subsampling trees for biomass, volume or mineral content. *Forest Science* 30(3), 673–681.

Vales, D.J. and Bunnell, F.L. (1988) Comparison of methods for estimating forest overstorey cover. 1 Observer effects. *Canadian Journal of Forest Research* 18(5), 606–609.

Van Deusen, P.C. (1987) Combining taper functions and critical height sampling for unbiased stand volume estimation. *Canadian Journal of Forest Research* 17(11), 1416–1420.

Van Deusen, P.C. and Meerschaerrt, W.J. (1986) On critical height sampling. *Canadian Journal of Forest Research* 16(6), 1310–1313.

Van Deusen, P.C., Dell, T.R. and Thomas, C.E. (1986) Volume growth estimation from permanent horizontal points. *Forest Science* 32, 415–422.

Van Laar, A. (1978) The growth of *Pinus patula* in relation to spacing. *South African Forestry Journal*, 107, 3–11.

Van Wagner, C.E. (1968) The line intersect method in forest fuel sampling. *Forest Science* 14(1) 20–6.

Van Wagner, C.E. and Wilson, A.L. (1976) Diameter measurement – the line intersect technique. *Forest Science* 22(2), 230–232.

Vanclay, J. (1992) Before you begin your inventory. *International Tropical Timber Organisation* 2(4), 4–6.

Veldhoen, B.J. (1968) *Interim report on two thinning trials on* Pinus patula. Technical Note 111, Kenya Forest Department, Nairobi.

Veldhoen, B.J. (1970) *Interim on RE:105 – A Thinning Trial in* Cupressus lusitanica. Technical Note 105, Kenya Forest Department, Nairobi.

Vuokila, Y. (1966) Functions for variable density yield tables of pine based on temporary sample plots. *Communicationne Instituti Forestalis Fenniae* 60(4).

Wadsworth, F.H., Parresol, B.R. and Colon, J.C.F. (1988) Tree increment indicators in subtropical wet forest. In: *Growth and Yield in Tropical Mixed/Moist Forest*. Proceedings of the IUFRO, Forestry Research Institute of Malaysia.

Waichuri, P. and Wanene, A.G. (1975) *Variable Density Yield Tables Cypresses of the* Cupressus lusitanica *Group in Kenya*. Technical Note 174, Kenya Forest Department, Nairobi.

Wald, A. (1947) *Sequential Sampling*. John Wiley, New York.

Walter, H., Hornickell, E. and Mueller-Dombois, D. (1975) *Climate Diagram Maps*. Springer-Verlag, Berlin.

Wan Razali, M. (1988) Modelling the mortality in mixed tropical forest of peninsular

Malaysia. In: *Growth and Yield in Tropical Mixed/Moist Forests.* Proceedings of the IUFRO, Forestry Research Institute of Malaysia.

Ware, K.D. and Cunia, T. (1962) *Continuous Forest Inventory with Partial Replacement.* Forest Science Monograph, no. 3.

Waters, W.E. (1955) Sequential sampling in forest insect surveys. *Forest Science* 1(1), 68–77.

Webb, D.B., Wood, P.J. and Smith, J.P. (1980) *A Guide to Species Selection for Tropical and Sub-tropical Plantations.* Tropical Forestry Paper no. 15, Commonwealth Forestry Institute, Oxford.

Wetherill, G.B. (1975) *Sequential Methods in Statistics.* Chapman & Hall, London.

Wheeler, P.R. (1962) Penta prism caliper for upper stem diameter measurements. *Journal of Forestry* 60(12), 877–878.

Whyte, A.G. and Tennent, R.B. (1975) Improving estimates of stand basal area in working plan inventories. *New Zealand Journal of Forestry* 20(1), 134–147.

Wiant, H.V. (1976) *Elementary 3P Sampling.* Bulletin 650T. West Virginia University Agriculture and Forestry Experimental Station.

Wiant, J.V. Jr., Wood, G.B. and Miles, J.A. (1989) Estimating the volume of a radiata pine stand using importance sampling. *Australian Forestry* 52(4), 286–292.

Williamson, J.D.A. (1976) Decision making in British forest management for the private woodland owner by means of simulation and integer programming. PhD Thesis, University of Aberdeen, UK.

Wormald, J.T., Malla, Y.B. and Pradhan, P.R. (1983) *Estimating Tree Fodder Yields.* Forestry Technical Bulletin no. 9, 21–4, Nepal.

Worrell, R. (1987) *Predicting the Productivity of Sitka Spruce on Upland Sites in Northern Britain.* Forestry Commission Bulletin no. 72, HMSO, London.

Worrell, R. and Malcolm, D.C. (1990) Productivity of Sitka spruce in northern Britain. *Forestry* 63(2), 105–128.

Wright, T.M. and Will, G.M. (1958) The nutrient content of Scots and Corsican pine growing on sand dunes. *Forestry* 31(1), 13–25.

Yang, Y.C. and Chao, S.L. (1987) Comparisons of volume growth calculation methods for horizontal line sampling. *Forest Science* 33(4), 1062–1067.

Yates, F. (1954) *Sampling Methods for Censuses and Surveys.* C. Griffin and Co. Ltd., London.

Young, H.E. (1967) Complete tree mensuration. *Forestry Chronicle* 43(4), 360–364.

Young, H.E. (1968) Challenge of complete tree utilisation. *Forest Products Journal* 18(4), 83–86.

INDEX